HUMAN DEVELOPMENT
AN EMERGENT SCIENCE

McGRAW-HILL BOOK COMPANY

New York St. Louis San Francisco Auckland Düsseldorf
Johannesburg Kuala Lumpur London Mexico Montreal New Delhi
Panama Paris São Paulo Singapore Sydney Tokyo Toronto

HUMAN DEVELOPMENT

AN EMERGENT SCIENCE

JUSTIN PIKUNAS, PH. D.
professor and chairman, department of psychology
University of Detroit

with chapters revised by
ROBERT P. O'NEIL, PH.D
associate professor, department of psychology
University of Detroit

Third Edition

HUMAN DEVELOPMENT: AN EMERGENT SCIENCE

1 2 3 4 5 6 7 8 9 0 K P K P 7 9 8 7 6

Library of Congress Cataloging in Publication Data

Pikunas, Justin.
Human Development, an emergent science.

Published in 1961 under title: Psychology
of human development, and in 1969 under
title: Human development, a science of growth.
Bibliography: p.
Includes index.
1. Developmental psychology. I. Title
[DNLM: 1. Adult. 2. Child development.
3. Psychology. BF713 P639p]
BF713.P54 1976 155.2'5 75-25744
ISBN 0-07-050015-0

This book was set in Times Roman by Black Dot, Inc.
The editors were Richard R. Wright and Susan Gamer;
the cover was designed by Scott Chelius;
the production supervisor was Dennis J. Conroy.
New drawings were done by Eric G. Hieber Associates Inc.
Kingsport Press, Inc., was printer and binder.

To the college student who is
interested in a most fascinating subject:
the development of the human being

Contents

Part Three Before and After Birth

Part Seven Adulthood

Preface

Seven years have passed since the appearance of the second edition of this book, originally entitled *Psychology of Human Development.* During these years research in most areas of human growth and behavior has moved rapidly in terms of both theoretical models and refined empirical findings. A new society is emerging, and cultural change is accelerated. Some grim realities, such as corruption and delinquency, the energy crisis and inflation, and drug addiction and bitter racial conflicts are marring the start of the third century of our nation. Some traditional topics have lost their significance; many important new topics have evolved. Converging contributions from various fields—psychology, biology, medicine, education, sociology, and anthropology—are establishing a science of human development.

In this new edition, which has been completely rewritten, many current systems, hypotheses, and facts from numerous branches of science have been incorporated to revitalize the text and bring it up to date. Indeed, a serious attempt has been made to show the whole of human behavior in American society and culture. Developmental patterns found at each level of life from the prenatal period through senescence are discussed in order to facilitate understanding of the development of the person at every major phase of life.

This third edition, like the first and second editions, emphasizes the most important aspects of the psychological factors and processes. The continuous interaction of the individual within his social environment is viewed as the basis for both personality development and adjustment. The evolving self-concept and its effects on behavior control as well as interpersonal relationships are brought to the fore. By emphasizing the changing yet continuous sense of self of the individual at the major phases of life, I have tried to build a basic structure for the interpretation of changes in personality and behavior. By recognizing individual differences and various styles of life and the factors contributing to these differences, the discussion moves well beyond the typical ontogenetic pattern of human development.

Many texts on human development cover childhood and adolescence as if human development stopped at eighteen or twenty years of age. Others have only tail-like extentions into maturity and old age. This text gives extensive treatment to both adulthood and senescence. Also, it meets the need for a better understanding and genuine appreciation of the tasks and hazards of early adulthood—since most of the students for whom the book is intended are themselves young adults.

It is hoped that this revision will justify its existence by providing a fuller understanding of the typical changes undergone by individuals living in the current American society and culture—and of some deviant changes. Furthermore, its facts, illustrations, and models, drawn from many sources, will prove useful as guidelines in the prediction of behavior at various levels of human life. The text contains many well-supported hypotheses but also many generalizations into which there has been less research. Such generalizations are a necessary part of a comprehensive textbook at the current level of scientific achievement in the study of behavioral growth. Despite voluminous empirical literature, the body of experimentally verified knowledge is meager for most levels and aspects of human development. An overall model of behavioral growth may be emerging, but its sketchiness must be recognized.

Acknowledgments

Without the work of several past and present contributors this revision would not have been possible. Dr. Eugene J. Albrecht participated in the original planning of the first edition and wrote four chapters of it. Dr. Robert P. O'Neil wrote two chapters of the original edition and revised three chapters of the present edition. Dr. Mimi LaDriere and Dr. Martha Johnson read several chapters each and made many improvements. I am also grateful to my developmental psychology classes and to several of my former students, including John Bernardo and Crystal Noftz, who read parts of the draft and suggested practical modifications. Appreciation is expressed to Drs. Nancy Bayley, Luella Cole, Evelyn Duvall, Lawrence K. Frank, Robert J. Havighurst, Bradley M. Patton, Robert R. Sears, and Harold Shane, and to the following publishers and organizations, for permission to reprint copyrighted material: McGraw-Hill and its Blakiston Division, John Wiley & Sons, Basic Books, National Society for the Study of Education, American Academy of Arts and Sciences, Russell Sage Foundation, American Medical Association, Society for Research in Child Development, Psychological Corporation, and a number of professional journals.

JUSTIN PIKUNAS

To the Student

The changing pattern of human growth and behavior throughout life, and the multitude of influences affecting it, is a very difficult subject to master, even if some of the subject appears obvious to, or is already known by, the reader. Such subjects as human anatomy and physiology often exhaust the medical student's time and patience. Yet they seem very simple when compared with the perplexing varieties of human motivation and behavior and individual differences at various stages of life. Today, there are even more reasons for studying human growth and behavior than there were when Alexander Pope said: "The proper study of mankind is Man."

I feel it is advantageous to begin this study by a survey of major concepts, methods, and issues as well as major determinants and related influences before moving on to the various facets of the human life story—to follow this story as the human being contends against and copes with environmental forces and forces within himself. For a systematic course of study, it is suggested that the student read a chapter before it is presented in class. Facts, concepts, and models emphasized by the instructor can later be underlined, possibly in different colors, to signify different degrees of importance. Additional concepts, illustrations, and theories introduced by the instructor should be transferred into a notebook as fully and accurately as possible.

Taking notes is an essential part of the learning process. In class, it is efficient to make notations on the right-hand pages of the notebook and use the left-hand pages for notes from the textbook and related readings. This method makes it easy to tie up information on the subject. Questions at the end of each chapter of this text give you an opportunity to check your comprehension of the major ideas presented.

A mere knowledge of fundamentals is not sufficient at the college level; yet this comes first, because details without general organization do not produce meaningful knowledge. In your first reading of a chapter, try to get the major ideas. Then read the chapter a second time—that is where your study really begins—underlining the major ideas and picking out any points you missed the first time. Keep in mind that the chapter headings and subheadings are designed as aids for organizing the material: the individual grows and acts as a whole, and his psychological processes are not divided, despite the rubrics.

Additional reading of past and current research studies is equally important for progress in mastering the subject matter. It is advisable to do some reading from the selection of references following each chapter. The volumes of *Psychological Abstracts*, *Annual Review of Psychology*, *Child Development*

Abstracts and Bibliography, *Science*, and *Developmental Psychology* are other comprehensive sources for selecting articles and reports on developmental research. At the college level, there is really no substitute for reading original research reports.

To sum up, the student uses the lecture material, the textbook, his own observations, and readings: these four sources of learning supplement each other. Although the student may consult with his instructor, integration of these materials is largely his own task.

JUSTIN PIKUNAS

Part One

Basic Approach to the Study of Human Development

Since the emergence of scientific psychology about a century ago, human beings have vastly extended their fund of knowledge about themselves. More important than the specific facts acquired, however, were the invention and refinement of various methods and techniques for furthering explanatory knowledge. By the ingenious use of these methods, psychologists can anticipate continuing progress toward the threefold goal of human growth science: understanding, predicting, and controlling the course of growth and behavior. They can look forward hopefully to new insights into the intricacy of the human organism and personality and the consistencies of development and behavior, as well as the factors affecting and determining them.

Intelligent comprehension of human development must be built upon clarification of the nature of the emergent science of human growth and cognizance of the methods for investigating developmental continuities and transitions. Part One presents these subjects.

Chapter 1

Nature of Human Growth Science

HIGHLIGHTS

The science of human growth centers on the ontogenesis of the individual throughout life. It is an interdisciplinary study largely based on developmental psychology.*

Lambert Quetelet originated the life-span study of man. G. Stanley Hall's theory of behavioral recapitulation and his writings on the mind and behavior of the child, the adolescent, and the senescent produced a major impetus toward research on various developmental levels of ensuing issues.

By his clinical study of children, supported by ad hoc experimentation, Jean Piaget formulated the currently dominant theoretical system of cognitive development. Following some major Freudian tenets on psychosexual development, Erikson construed a psychosocial version of psychoanalysis emphasizing core conflicts and gains or losses in ego strength.

The "critical periods" hypothesis postulates phases of heightened sensitivity during infancy to selected emotional and social stimuli. Lack of contingent stimulation drastically curbs the developmental possibilities of the individual.

The developmental-level approach, marked by a somewhat arbitrary division of

*Many terms in human development are defined in the Glossary (pages 403–411).

the life-span into tripartite periods within gestation, infancy, childhood, adolescence, adulthood, and old age, is the way the life-span model of ontogenetic changes is charted here.

In the modern world, with its expanding culture and technology, racial and ideological conflicts, population explosion, urban decay, and increasingly complex social stratification, human beings face many difficulties in their quest for healthy growth and maturity, for adjustment and happiness. New tasks—decisions, challenges, and hazards—often overtake them; multiple change rather than stability is the order of the day. These external problems complicate the frustration each person faces as he or she grows and lives from the earliest stages of life to the period of old age.

As a person reviews his own life, it becomes clear that he is not the same today as he was two or ten years ago. He is still the same individual, but many of his motivational and behavioral characteristics are far different. Such common expressions as "Don't be a baby" and "Act your age" indicate that what is approved behavior at one age is unacceptable at another. The needs and drives, desires and aspirations of the individual undergo not only frequent modification but also several major revisions. Beliefs and attitudes, emotional responses, and intellectual abilities—indeed, all the dimensions of personality—change throughout the human life-span. The kinds of changes that take place are circumscribed by both genetic and enviromental factors, particularly the kinds of stimulation a person receives during the early formative years of life. Since the person faces nearly constant changes, problems, and decisions in going through life, it is crucial that he study his inner and outer "universe," to better know and understand himself. Now that the genetic code has been unraveled, it is time to tackle and penetrate the "brain code," to uncover its ties to the "behavior code." The brain structure of 10 billion cells, each of which may have 60,000 interacting junction points, has eluded our scientific efforts (Edison, 1970).

The armchair speculations, the old wives' tales, and the maxims of earlier times—many of which have endured until the present day—have been found sadly wanting. True, literature demonstrates some deeply penetrating analyses of human nature and behavior in the past, but it also reveals many accepted absurdities. It has therefore become the task of science to evaluate critically what has been believed previously, in order to deepen current self-knowledge.

As a major branch of science, psychology deals with human beings' understanding of themselves. It approaches human behavior and emotion from its own point of view and applies its unique designs of investigation. Using scientific methods, it studies cognitive processes, motivation, and behavior in order to predict and control them. *Developmental* psychology is dedicated to one aspect of this search for knowledge; it seeks understanding and control of the basic processes and dynamics underlying human behavior at the various stages of life. This field encompasses the growth and maturation of the individual organism and its cognitive and emotional components, as well as its personality structure. Factors that promote or retard any aspects of development are also considered. Within this broad range of concerns, the types of interaction between the human organism and environmental factors also require careful study. As a core for the science of human growth and behavior, developmental psychology deals with all the processes contributing to becoming an infant, a child, an adolescent, and a mature adult. It emphasizes the scientific explanation of changes in human growth and behavior throughout all levels of life. A dynamic yet precise representation of human needs and goals can be sketched within the developmental framework of the various levels and phases of life. By now the field of developmental psychology includes "practically all topics in general psychology and uses nearly every method available for the study of behavior" (Stevenson, 1967, p. 102). Beyond that, developmental studies generate questions and issues of their own.

As early as 1910, Wilhelm Stern argued that genetic psychography should not restrict itself to childhood and adolescence, since "adulthood is not a state of stagnation" but a period of many refined developments. In an approach consistent with Stern, Hall, Buhler, and many others, this book assumes that developmental psychology is an *ontogenetic* study of the human organism and behavior from conception to death. In other words, not the species, but the individual and his or her direction of growth in the environment and culture form its major concern. So that human beings may better know and understand themselves and those around them, developmental psychology seeks to discover the sequential changes in human behavior and the patterns in personality organization; it seeks to find order in what appears to be chance, hazard, and even chaos. With this knowledge of order and sequence, the individual should be able to study behavior patterns at various phases of self-development. A serious study should better equip people to face future probabilities and contingencies. This expectation should not be interpreted to mean that the mere knowledge of facts and models automatically establishes an individual as a well-adjusted person. Nor should the reader assume that developmental psychology is the study of how to live successfully and happily. It is rather an area of systematic, logical observation leading to scientific research and interpretation of growth and behavioral patterns throughout life. Hence the growth and behavior patterns at various levels of life form the subject matter of developmental psychology.

Developmental psychologists see the human organism as destined for a normal course of growth which ensures a high degree of realization of genetic potential unless detrimental factors damage, distort, or prevent this actualization. Young people are especially susceptible to deviation in behavior since they are almost completely dependent upon their immediate environment. "The extraordinary dependence of the human young upon adult care and caring provides both an unparalleled opportunity for mental and emotional development and a period of vulnerability to profound distortion by neglect" (Eisenberg, 1972).

The sequential changes that occur in human growth and behavior include the progressive unfolding of different dimensions and powers, as well as the eventual decline of the organism's functional abilities in old age. Thus intellectual development encompasses the emergence of its functions in the years of childhood, its revision during adolescence, its perfection during early adulthood, and its changing effectiveness in the later years of adulthood and senescence. Though behavioral scientists usually stress the beginning and early phases of life, the significance of later periods cannot be disregarded.

Developmental psychology forms a nucleus for a budding interdisciplinary science of human growth embedded in the significant findings of biological and behavioral research (Charles, 1970, pp. 24–25; Frank, 1963; Looft, 1972; McGraw, 1970). Contributions to the science of human growth have come primarily from psychology and biology, but sociology, education, pediatrics, psychiatry, and even anthropology and public health have also participated.

As can be seen in Figure 1-1, developmental psychology is closely allied to social and comparative psychology. *Social* psychology studies various relation-

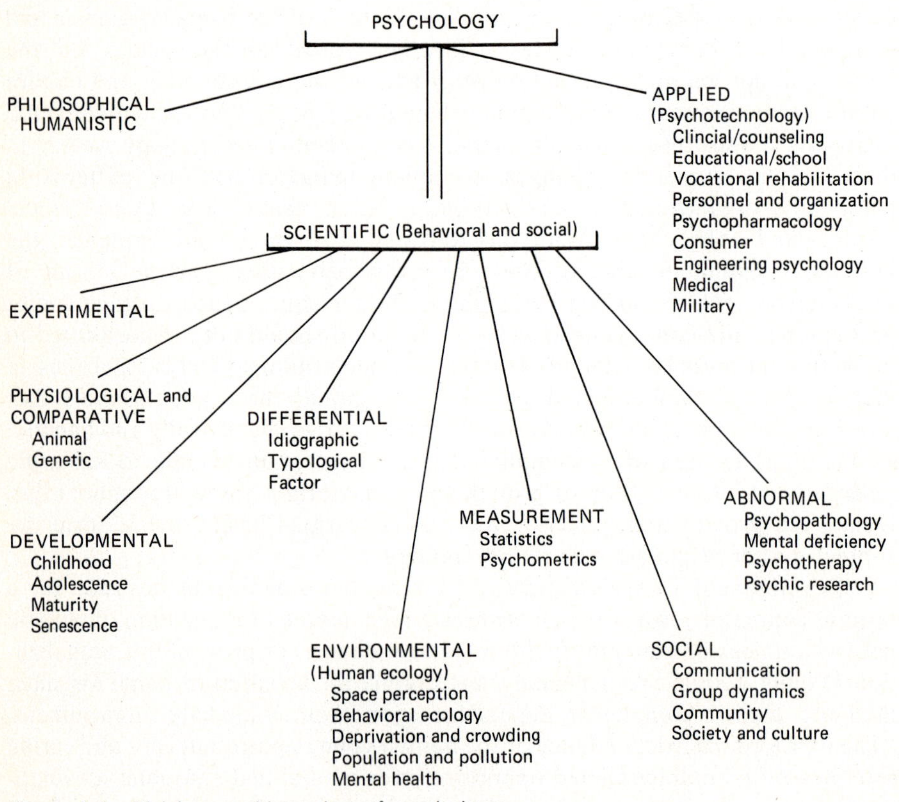

Figure 1-1 Divisions and branches of psychology.

ships among individuals and the influence other persons or groups exercise upon the individual. Since most individuals live in family groups and many are members of schools, churches, and other organizations, they cannot be studied or understood outside of their social milieu. In *comparative* psychology, the scientist explores various tendencies and forms of animal behavior in order to understand similar patterns of behavior as they appear in man. In an early study, the founder of the theory of evolution, Charles Darwin (1872), studied emotional reactions at human and animal levels and found considerable similarity in their expressions. He postulated that most human expressive actions are inherited from and so shared with a number of animal species. Wilhelm Preyer (1881/1888) also compared infants' reflexes, instincts, and emotional reactions with those of various animals, showing the similarities as well as the differences among them. In contrast to Darwin, Preyer found substantial differences in the growth of the human individual. In developmental psychology, comparisons are often made between infants and children and between adolescents and adults, as well as among various strata of the population.

THEORIES OF HUMAN DEVELOPMENT

Life-span psychology, as a distinct branch of the behavioral and social sciences, has relatively few historical foundations. However, several original theoretical and research orientations that were to influence the development of life-span psychology were formulated in the nineteenth century. Only those contributions from the past which were instrumental in raising critical issues of high interest at the present time are analyzed here.

An early foundation for the scientific study of life-span development was firmly laid by Lambert A. J. Quetelet (1796–1874) in his survey *Sur L'homme et le developpement de ses facultés* ("About man and the development of his faculties"), published in 1835 in Paris (Quetelet, 1835/1969). Quetelet studied physical, moral, intellectual, and emotional changes in the human organism as a function of age. Theoretical observations about patterns of growth and decline in human powers were accompanied by measurement and application of many statistics. As a result, Quetelet formulated the concept of the *average* man and assessed his powers at various ages. For Quetelet, the maximum of muscular strength, for example, coincided with the estimated maximum of delinquency at about twenty-five years of age (Groffmann, 1970). As a mathematician, Quetelet brought statistics into the assessment of developmental changes at various phases of life. He insisted on studying human beings through their actions by measuring them, and he estimated that all human traits were distributed according to the "law of accidental causes," or normal curve (Diamond, in Quetelet, 1969, Introduction). More than anyone else, Quetelet deserves the title of founder of life-span psychology. He was a worthy forerunner to G. Stanley Hall.

Behavioral Recapitulation

Hall's recapitulation model is an outgrowth of Darwin's well-known theory of biological evolution. For Darwin, the genetic pool of the highest species provides opportunities for some desirable mutations leading to the acquisition of new characteristics and behavior and ultimately to the emergence of new species. The last link in this apparently slow but unending process of evolution is the human species evolved from hominids, forebears of the human race. The civilized human being is just a recent descendant of *Homo sapiens*. Humanity is a species in dynamic evolution with biological and sociocultural assets toward a further evolutionary progress.

G. Stanley Hall expanded biological evolution into what was to become behavior genetics and formulated a psychological theory of behavioral recapitulation. The human individual passes through developmental stages and phases that correspond to the phylogenetic scale of ascendance from amoeba to civilized human being. Changes in the behavior of the individual occur as a result of stagelike maturation in a predetermined and universal pattern showing remarkably rapid progress as early individual (ontogenetic) development repeats the long and slow process of species (phylogenetic) development (Muuss, 1968, pp. 33–41). Hall emphasized that "ontogeny recapitulates phylogeny"—individual growth follows the lines of ascending animal evolution. Each stage and phase in the life of a child shows characteristics pertinent to major phylogenetic advances. Major animal stages of the phyletic scale occur during the prenatal growth and postnatal development of the child. A corollary to this is his assumption that throughout childhood, development is biologically determined (Riegel, 1972).

An example can be seen during the period of infancy, which Hall saw as lasting four years. When the infant crawls or creeps, he is reenacting the early mammal stage of phylogenesis. At this period socially unacceptable patterns of infant behavior must be accepted as normal and necessary steps toward later progress in socialization. Hall reassured parents and teachers that "nature is right" and that troublesome forms of behavior will disappear without disciplinary efforts as the child moves into higher levels of maturation. Hence little Johnny's "bad" behavior must not be curbed lest in the future some desirable quality should fail to unfold. The parable of the tadpole's tail, in which the hind legs fail to develop if the tail is amputated, illustrates Hall's concept of child management. According to Hall, the expression of "undesirable" traits in childhood would prevent their occurrence in adult years: children would be immunized against asocial behavior if toddlers were permitted to play as savages, children as barbarians, and adolescents as nomadic wanderers. Lenient permissiveness in child guidance is what Hall endorsed for the sake of complete phylogenetic synopsis, which forms the baseline for uniquely human development, occurring for the most part during the years of adolescence (Anandalakshmy & Grinder, 1970).

In the period of childhood from four to eight or nine years of age, the individual recapitulates the early epoch of prehistoric man. During the period of youth, the years from eight to twelve, the child passes through the early history of man—savage and cruel, yet moving toward higher forms of gregariousness and human aspirations. The oncoming pubertal changes introduce adolescence, which brings a "new birth, for higher and more completely human traits are now born" (Hall, 1904, p. xiii). According to Hall, the child invariably reenacts the experiential history of the human species to the rise of civilization and culture underlying the modern age of humanness. Before attaining adult maturity, the adolescent experiences within himself the blows of the modern era of advanced technology and culture. For Hall, the adolescent stage is the nascent period uplifting mankind to superanthropoid status, but it is also a period of great liability to reversion or retardation (Grinder, 1969).

Two years before his death, Hall's monumental volume on old age, *Senescence: The Last Half of Life* (1922), was published. Here Hall completes his broad foundation for life-span psychology. In analyzing the senescent pattern of life, Hall relied on questionnaire data and his own insights about the critical transition into old age, which differs widely from person to person "so that one man's norm would be another man's disaster" (p. 393). Claiming that "man remains essentially juvenile" (p. 411), he nevertheless comes to the affirmation that senescence calls "to construct a new self just as we had to do at adolescence, a self that both adds to and subtracts much from the old personality of our prime" (p. 403). Entrance into old age brings "no less tendencies to polymorphic perversity than before these were constellated into the normal sex life of maturity" (p. 394), but "it is the very nature of love to grow sublimated as years pass" (p. 397). Anxiety is the normal lot of man and "gives a tonic sense of responsibility which we all need" (p. 378). It is not easy to live an "Indian summer" in old age as moods swing and helplessness rises. Growing old is compared to

> walking over a bridge that becomes ever narrower so that there is progressively less range between the *licet* and the *non licet*, excess and defect. The bridge slowly tapers to a log, then a tight-rope, and finally to a thread. But we must go on till it breaks or we lose balance. Some keep a level head and go further than others but all will go down sooner or later (p. 437).

In myriad ways, old people describe their life as a downhill movement whose acceleration they cannot reverse. Hall's view of aging is quite negative and pessimistic and is not generally accepted today.

Hall's voluminous writings exerted a major forward thrust upon human beings' study of themselves throughout life. His recapitulation theory provides a basis for ordering the growth process in terms of ascending levels of developmental functioning. His influence was broadened by many students at Clark University, including John Dewey, James McKeen Cattell, Henry H. Goddard, Arnold Gesell, and Louis M. Terman. In spite of this influence, today Hall is largely recognized as a historical contributor to the field of life-span psychology. As the emphasis has moved toward environmental explanations of the origins of behavior, Hall's formulations, which emphasized phylogenetic and biological causes, have become less relevant.

Psychoanalytic Theories

A second major theory of human development that has had a great impact on life-span psychology is the psychoanalytic theory of Sigmund Freud (1856–1939). Psychoanalytic theory stemmed from Freud's experiences in treating frustrated and disturbed adults. Up to his death, Freud continued revising and remodeling psychoanalytic theory and its psychotherapeutic application. Fundamentally, his concept of human beings was thoroughly naturalistic, with much of behavior controlled by unconscious forces; its ideal type was the mature person free of infantilizing neuroticism, capable of constructive work and love, and finding his or her own way of living and relating to others by sublimating a large segment of sexual energy to service and to creative outlets. Freud's leading postulates on unconscious motivation, repression, anxiety, identification, psychosexual stages and Oedipus complex, aggression and projection, defined and related in different ways by Freud himself, provided a powerful impetus for related theoretical and clinical research. In diagnostic and therapeutic work, psychoanalytic concepts and tenets became a leading influence in the 1920s. Despite the fall of major schools and systems in the 1940s, psychoanalytic offshoots rooted in Freudian concepts continued exercising a powerful impact on many developmental exponents. In recent years Freudian analysis has lost more of its momentum, but its dominating influence still extends over many areas of developmental, abnormal, and clinical psychology (Strupp, 1972).

Reflecting Freud, Rosenblatt (1966) aptly says: "Every earlier stage of development continues alongside the later stage which has risen from it." Hence the lowest levels of sexuality and aggression "persist along with the higher ones without ever losing their potential for manifesting themselves again and again, in new and new situations, depending on circumstances, opportunities, trauma, regressions, and so on." Drawing examples from the behavior of primitive and savage people, Freud illustrated retarded ego growth and regressive forms of behavior (Anandalakshmy & Grinder, 1970).

In 1896, Freud introduced the concept of *psychoanalysis* to designate his method of scrutinizing the past to uncover traumatic and morbid experiences in order to discharge strong emotional complexes associated with them. Freud was satisfied with his free association and analysis of the past only when he came to understand the genesis of the symptoms of his patients in the early years of childhood. Developmental propositions and the distribution of psychic energy among id, ego, and superego constituted the core of Freudian psychoanalysis (Rosenblatt, 1966).

It is noteworthy that the concept of recapitulation is inherent in Freudian theory, since "each individual repeats in some abbreviated fashion during childhood the whole course of the development of the human race" (Freud, 1953, p. 209). Despite his emphasis on psychosexual stages in the individual's development, Freud did not overlook underlying continuity of life and postulated the notion that the human mind preserves the primitive, autistic, and archaic. Once formed in the mind, "everything survives in some way or other, and is capable under certain conditions of being brought to conscious light again" (Freud, 1930/1961, p. 15).

Basic Freudian Tenets Human sexuality is the basic source of energy to be discharged on selected (cathexed) objects, accompanied by pleasurable experiences. Libido *Trieb* (instinctual force) energy feeds Eros, the term used by Freud to denote sexual and erotic drives and desires directly or indirectly linked to sexual expression and gratification. In the early years of childhood the id generates impulses and desires linked to the cross-sexual parent. Oedipal strivings occur, and the resulting conflicts and problems play a leading role in creating normal or perverse patterns of lifelong adjustment. Fundamentally, a person continues to be what he or she became during the first five or six years of life. The past cannot be either disregarded or freely reshaped. Freud made it clear that adult disturbances are directly traceable to frightening, shocking, traumatic, and other unfortunate (and usually repressed) experiences in childhood, and that the early years of life represent a critical period for adult adjustment. The Freudian model of man is thus predictive, and this predictability has sustained its power to the present day.

Freud (1962, p. 97) maintains that "a disposition to perversions is an original and universal disposition of the human sexual instinct and that normal sexual behavior is developed out of it as a result of organic and psychical inhibitions occurring in the course of maturation." He postulates that the repression of sexual urges in childhood during the latency period and in adolescence is a necessity for sound psychosexual development. Since various amounts of libido are activated even when the libidinal object is missing, compensatory discharge routes are often found and used to gain substitute gratification. The plasticity of libidinal discharge is also emphasized by sublimation and symptom formation. Sublimation is an important process whereby base energies are redirected from their sexual outlets into other goal-directed and socially acceptable activities. Continued repression of sexuality leads to the formation of neurotic and psychotic symptoms. Art and athletic activities and the pleasures of resultant achievements are often embraced by children, adolescents, and adults alike. Such use of sublimation is a sign of normal growth, pointing toward a significant gain in maturity (Freud, 1910, 1962, p. 44).

The process of symptom formation is chiefly related to unconscious repression, the amount of which depends largely on the superego strength. Repression of sexual drives and desires, when continued for a long period, generates intense anxiety, the experience of which prompts additional repression. In such cases symptomatic discharge becomes a necessity. Its chronicity leads to a neurotic pattern of adjustment, often marked by phobic or compulsive reactions. Losses of ego strength thus incurred press the individual to withdraw from many challenging activities.

All human history is filled with events signifying expressions of hostility and aggression. So is the life of almost any individual from the early levels of childhood. In anger a six-year-old shouts to his mother, "I wish you were dead!" Similarly, an older adolescent shrieks, "How dare you tell me what I have to do!" At first Freud linked aggressive tendencies to libido as the source, since many sexual partners inflict pain, while others accept it as part of sexual stimulation. Since a complete explanation of all destructive tendencies and activities by the same source of libido was unacceptable to him, Freud later introduced the concept of Thanatos, the death instinct; but he never cared to explain precisely its source of energy, except that some libidinal energy was neutral and thus displaceable. Indeed, the death *Trieb* was never completely incorporated into the Freudian concept of operational personality structure consisting of id, ego, and superego. Nor is it possible to claim that Thanatos is analogous either to id or to libido. Eros and Thanatos act within the id but are not limited to it.

Freud (1930/1961) spoke of the primary hostility of one person toward another, which shows itself as the child grows and interacts with adults and peers. He dislikes or even hates many of them. Freud felt that society survives only by erecting powerful barriers against the aggressive human drives. Either the growing child must be subjected to control measures or the natural aggressive urges will gain in power and destroy those who become the objects of intense hostility. When a child's aggression—against a parent, for example—cannot be expressed directly, displacement of aggression onto innocent targets is the frequent outcome. A bully who takes advantage of smaller children is often displacing accumulated hostility against parents upon innocent victims.

Genetically, libidinal energy first operates on the id level, marked by a free expression of instinctual drives whenever such urges emerge. The id operates upon the *pleasure* principle. Since drive reduction is pleasurable, libido energy is released without any delay. The infant is not ready to wait or compensate but seeks instant gratification. It is the ego which then makes him differentiate his desires from the reality of the situation. As the child gains in ego strength, he or she can better adapt to the demands of reality. Reality demands order, and order implies some control of impulses and urges. Freud formulated the *reality* principle to account for the adjustment to diverse demands of the social and cultural order and to explain efforts toward the long-term goals and gratifications that older children and adults seek.

The force by which parental values and standards, wishes, and regulations become internalized and followed by the child is identified as the superego. It arises from parental, peer, and other social influences and helps the ego to enforce the control of base emotions and impulses in order to comply with acceptable social goals and expectations. The superego, comparable to an early form of conscience, is formed through the mechanism of defensive identification motivated by fear and castration anxiety. Anna Freud (1957) speaks of the son's identification with the aggressor. The oedipal conflict is resolved by the growth of the superego, as the child's search for approval expands to include the same-sex parent, other members of his family, and other adults as well as his peers. Guilt is experienced when the child falls short of parental standards as he or she sees them. An adolescent or adult is an id- or ego-dominated, or less frequently a superego-dominated, person. Conflicts occur within this tripartite division of the psyche but are usually settled unconsciously, possibly by compensation based on the strength of each segment.

Freud stressed the crucial significance of early psychosexual experiences in setting the pattern of adjustment for life. Frustration of oral urges in the first year of life distorts the process of maturation, since oral tendencies in such cases are not outgrown to a sufficient degree and the probability of fixation rises greatly. The same applies to anal tendencies during the second year of life and to phallic and oedipal trends during the later preschool years. More specific aspects of psychosexual development will be analyzed in the discussion of various levels of life.

Throughout the years of childhood, pleasure motivation retains much of its original strength. The child likes doing what is pleasant and avoids doing what is not. An adult is expected to be dominated by the reality principle, to be ready to delay gratification and work for long-range goals. The adolescent typically sways between pleasure demands and reality demands, often giving way to the first.

Freudian theory emphasizes early determination of behavior, which implies prediction of adult behavior if childhood motivation is carefully assessed. The structural personality aspects at five or six remain as main operational channels for life. Despite stages of sexual development, there is continuity of behavior throughout life resulting from an id-ego-superego balance of one kind or another. At present many psychologists accept this view and find it useful in their work as scientists or psychotechnologists.

Eriksonian System When Freud speaks of the libido and Thanatos, he recognizes them as animal-like instincts, the expressions of which, however, can and ought to be in large part transformed by societal regulations. One of the leading neopsychoanalysts, Erik Erikson (1963, pp. 95–96), embraces the concepts of libido and Thanatos but sees them in a different light. He refers to them as the "vague *instinctual* (sexual and aggressive) forces which energize instinctive patterns." These patterns in human beings, he recognizes, are "highly mobile and extraordinarily plastic," which means that they are subject to sociocultural influences because human beings possess minimal instinctive equipment. In the Erikson system libidinal energy is a force that directs the human system's epigenetic development and constitutes its all-inclusive motivating diversity of drives, urges, and desires. Libido generates "the drive to live, to gratify oneself, and to reach out beyond oneself." A polarity emerges through the functions of Thanatos, generating urges to regress and destroy, and the longing to return to the safety of the uterus. Since bisexuality is inherent in the organism, masculinity and femininity form another basis for polarization and conflict in each individual's life. With much behavior springing from polarities, Erikson emphasizes the recurrence of core crises and related conflicts as ever-present motives in life (Maier, 1969, pp. 22–23).

Erikson's theory (1963) of eight stages of man is primarily concerned with the psychosocial aspects of development. Maier (1969, p. 29) notes that these stages are "essentially a reformulation and expansion of Freud's psychosexual developmental stages." As a rule, Erikson begins where Freud ended but moves well beyond what is akin to Freud.

In *Childhood and Society* (1963), Erikson posits that (1) side by side with the stages of psychosexual growth as interpreted by Freud there are psychosocial stages of ego development; (2) personality growth continues throughout the whole life cycle; and (3) each stage has a positive as well as a negative component. In each stage a new dimension of social interaction becomes possible. Each stage has its own distinctive goal to be attained in that period. While working within a psychoanalytic frame, Erikson is chiefly concerned with ego development, for he sees attainment of ego identity as a key developmental task of puberty and adolescence and a sound foundation for adulthood. A healthy ego identity is acquired only as the child receives consistent and warm recognition of accomplishments that are meaningful in his culture. The underlying developmental conflict is universal, but the particular situation becomes culturally defined. As the core conflict of each stage is resolved positively, a new quality is incorporated into the ego, raising its adequacy and strength. If a core conflict is negatively resolved, specific damage to the ego results.

Erikson divides the total life-span into eight stages and assigns each stage a core conflict. Infancy and childhood span the first four core conflicts: (1) *trust* versus mistrust, (2) *autonomy* versus doubt and shame, (3) *initiative* versus guilt, and (4) *industry* versus inferiority. The adolescent period is marked by *identity* versus role diffusion. The core conflicts of adulthood include (1) *intimacy* versus isolation, (2) *generativity* versus self-absorption, and (3) *integrity* versus despair.

The foundation of all later developments and adjustments is centered on the first stage, extending through the first year of life. In this oral-sensory stage, the infant develops a sense of expectancy through a mixture of trust and mistrust. Physical comfort and frequent contact with the mother, affectionate and gratifying, ensure the dominance of trust. A growing sense of basic trust helps the infant to become more active and to accept new experiences willingly, especially when injuries and other painful events are minimal. A sense of basic mistrust becomes dominant when the infant's needs are poorly gratified and he is subjected to many frustrating and painful events. The management of the infant during the first year of life determines to a large degree whether he is to become a trusting and outgoing person or a mistrusting, suspicious, and demanding one. Extreme scarcity of maternal warmth and receptivity at this early age predisposes the infant to autistic isolation at a later age (Erikson, 1963, pp. 248–250).

During the anal-muscular stage, which occurs between the ages of about fifteen and thirty months, Erikson focuses on gains in autonomy and realization of will. The toddler is gaining much in refined perception of persons, objects, and situations and also in the ability to reach what attracts him and to move toward or away from the parents. The child's ability to control his or her body and to apply its parts to attain goals adds much to the sense of autonomy. When this sense outbalances feelings of shame and doubt, the child is prepared to be self-willed and autonomous in later phases of childhood. Encouragement by parental figures to seek and explore and the availability of something to explore are major factors in the self-gratification of the toddler. The child's success makes him feel a need to be himself; it adds to his ego strength. As Bettelheim (1967) and Ciaccio (1971) suggest, the establishment of autonomy is the focal crisis of the first five stages—it continues through the highly plastic phases of childhood well into adolescence as a significant "residue."

The child's initiative grows as he moves into the locomotor-genital stage of life several months prior to three years of age. At this age, the child becomes more aware of direction and purpose. For Erikson the social dimension that appears has *initiative* at one of its poles and guilt at the other. The child initiates actions by others and gets them involved in his own behavior. In some instances at least his trusting autonomy is frustrated to some degree and calls for feelings of guilt. Internalization of parental rules and regulations forms another source of guilt when the child trangresses them.

In the second part of this phase, the child becomes his own "parent" as he tries to supervise himself and manages to avoid parental disapproval (Maier, 1969, pp. 46–47). He anticipates and readies himself to merge into role behavior, a leading mark of the schoolchild.

At about six years of age the child enters the period of latency, in which his *industriousness* rises and his feelings of competence in doing what he does are deeply experienced. He learns how things are made, how they work, and what he can do with them. The child begins identifying his tasks and ways to attain them. What he learns he is eager to teach others at home and in his neighborhood. Kindergarten and school activities add a new dimension at this age, as he becomes a member of a large peer group and wants to be accepted.

With the onset of major pubertal changes, adolescents are faced with physiological and sexual revolution within themselves, and with tangible adult tasks ahead of themselves. At this stage they are "primarily concerned with what they appear to be in the eyes of others as compared with what they feel they are" (Erikson, 1963, p. 261). Young people often arrive at a definition of their *identity* by repeated projections of their diffused ego image on peers and peer groups. The resulting reflections have a clarifying effect and strengthen some aspects of their identity.

When the adolescent identity is defined and strengthened by peer support, the young adult is "ready for intimacy, that is, the capacity to commit himself to concrete affiliations and partnerships" marked by long-term companionships and love (Erikson, 1963, pp. 263–264). Productivity and even creativity follow during young and middle adulthood as the great majority of individuals realign themselves and become deeply concerned with maintenance of family life and rearing the next generation. In the final stage of later adulthood, the mature individual strengthens his ego integrity and feels gratified with what life has offered him, or despair gains the upper hand when it is too late to start another life with different objectives and goals. Erikson ties together infantile trust and adult integrity by emphasizing that "healthy children will not fear life if their elders have integrity enough not to fear death" (1963, pp. 268–269).

If solutions of the psychosocial conflicts at each period of life are marked by fundamental gains in trust, autonomy, initiative, and industry, the child gains much in *ego* strength and becomes ready for the adolescent struggle of identity formation. Each period then boosts the strength of his or her ego and its capacity to cope with adversities. According to Erikson, the child's adjustment will be adversely affected if the infant's experiences elicit and strengthen mistrust in the earliest months of his life, or if he is unable to gain in autonomy during the second and third years of life. When mistrust, doubt, and guilt continue to deflate the ego, this situation calls for some deviant forms of self-defense. In the later years of childhood, the individual must gain much in industriousness in order to raise his or her achievements in various areas of activity. Without achievement and success, feelings of inferiority arise and cut deeply into ego adequacy. Without significant gains in ego strength the child is simply not prepared for the adolescent task of identity formation and heterosexual adjustment.

If during the early years of life the child continues gaining in ego strength, the reality principle becomes a powerful modifier of behavior. As a result, reasonably efficient techniques for coping with the demands of reality will be established. If the adolescent does not succeed in forming a strong identity rooted in family tree, race or ethnicity, or creed or ideology, he will face many difficulties as an adult. The attainment of genuine intimacy will be impossible for him. When identity formation lags, even deep attachments and exuberant love are not likely to produce stable long-term relationships needed in marital adjustment and family life. The capacity to commit oneself to concrete partnerships also depends on the ethical strength to abide by such commitments. Erikson's themes of ego strength and weakness will be found in the sections on adolescence and later adulthood.

Piaget's Interactionistic Model

Jean Piaget constructed another major developmental theory based chiefly on ontogenetic changes in cognitive functioning from birth through early adolescence. Much of his theory derives from discoveries made through systematic observation and ad hoc experimentation with his own children (Laurent, Lucienne, and Jacqueline) and with subjects at the Rousseau Institute in Geneva. At the Institute his associates included Bärbel Inhelder and A. Szeminska, who made significant contributions to Piaget's studies. Piaget's studies with his associates are published in nearly thirty books and at least two hundred articles or reports. These studies describe, explain, illustrate, and empirically support the qualitative components constituting his theory of intelligence and several other aspects of development such as language, socialization, and morality.

Piaget's viewpoint represents a configuration of both nature and nurture, the biological givens and environmental influences; yet the stress remains on the former, since the capacity to learn is tied to an invariant sequence of neural growth and innervation and cognitive maturation. His model (marked by interaction between the genome and the physical milieu) is both epigenetic and predictive (of what will come next). In his empirical studies Piaget uncovered the constant order in which lower operational modes serve as roots for the higher ones. Neurological structures and the functional invariants, assimilation and accommodation, produce a unique mode of interaction between organism and environment. This interaction calls for a nearly continuous readjustment of a frequently disrupted structure-operation equilibrium (Piaget, 1973, pp. 145–152).

Piaget (1970, pp. 60–68) embraces a modified form of *structuralism* bounded into the constructionist hypothesis. There is a formation of initial and ever more complex structures regulated by equilibrium requirements. His structuralistic perspective is an all-inclusive and interdisciplinary one, in which the organism represents a complex "network" of organizational activity toward "a never completed whole," because there is no terminal or absolute form (p. 140). Piaget complicates his own theory because he sees structuralism as a method he accepts on the basis of its fruitfulness (pp. 142–143). Piaget's theoretical position can be understood only by a careful study of his book *Structuralism* (1970), or at least of the fourth chapter (pp. 52–73).

Development occurs in stages and phases that represent the regular spiral sequence in which the human being becomes able to pursue different kinds of operations and to apprehend objects of increasing complexity and abstractness. Piaget maintained that, while neurological and behavioral maturation occurs at differing rates for individuals, an invariant sequential order of qualitatively different stages is universal. Ontogenesis is simply transition from one structure to another of a higher complexity (Piaget, 1970, pp. 140–143). Learning efficiency is thus subordinated to maturational gains expressed in a sequence of stages and a hierarchy of superimposed structures (Piaget, 1973, pp. 10–11).

Piagetian theory is based on an *active* model of the organism which presupposes frequent changes in structures and functions or operations as it interacts with an equally complex environment. This transactive relationship is at the core of human development. The action of the individual on the environment produces increasingly more complex functional knowledge by creating new schemata. The essential element in behavior is the structure that makes a response possible and hence permits environmental elements to be stimuli (Boyle, 1969, p. 117). First of all, schemata refer to well defined sequences of behavior, e.g., a way of apprehending objects in the child's vicinity or a method of search for a lost ball or ring. As action and perception units become integrated into a purposeful sequence, the formation of the schemata is finalized.

Scanning renders perceptions more accurate and, combined with more motor activity, decreases systematic distortions and illusions that form part of the field effect. By forming schemata, the organism acquires new ways of relating to its environment. Through internalization of action by means of words, signals, and symbols, representational schemata emerge. Language replaces the necessity of many actions, and thinking grows through the continued internalization of activities and verbalization. From about seven years of age, the child is usually able to deal with his environment without overt action, since many actions have been internalized in the form of concrete intellectual operations. *Schemata* are the cognitive structures by which individuals intellectually adapt to and organize their environment; they may be viewed as categories of knowledge into which new information is classified. For Piaget, schemata include affective states, since emotions cannot be separated from cognitive processes. There is a fusion, with a predominance of either cognition or emotion.

Cognitive functioning at any age represents the performance of the total organism—its physical, emotional, and motivational capacities are heavily involved (Piaget, 1936/1952, p. 42; Piaget, 1973). Cognitive development consists in fitting (accommodating or assimilating) each successive experience into the functional cognitive structure of that moment. In grasping a nipple and sucking, the baby "simultaneously interrelates visual, tactile, and a host of motion-inducing physiological components with such psychological processes as motivational need, recognition-perception, memory, beginning learning, and awareness" (Matarazzo, 1972, pp. 60–62).

What does the child do when he is confronted with a new stimulus that cannot be placed into any existing schema? Then he must form a new schema into which this stimulus can be assimilated and comprehended. This structural change refers to accommodation—formation of new schemata or modification of one or more old ones. Assimilation is the end product that the active organism seeks—a mark of quantitative cognitive growth, while accommodation is a qualitative change (Piaget, 1973, pp. 69–72; Wadsworth, 1971, pp. 14–16). Figure 1-2 illustrates this qualitative change—the formation of a new schema.

Equilibration is an internal active state of balancing assimilation and accommodation. Since the organism does not exist in a vacuum and is often striving for new stimulation, the need for balance thus provides an intrinsic motivating force. Cognitive development then is the process of successive schematic (qualitative) changes, where a change in each schema is derived logically from the one which preceded it. Progress in cognitive maturation involves *decentering,* that is, consideration of many aspects of an object or situation. A young child centers his attention on one detail of a person or

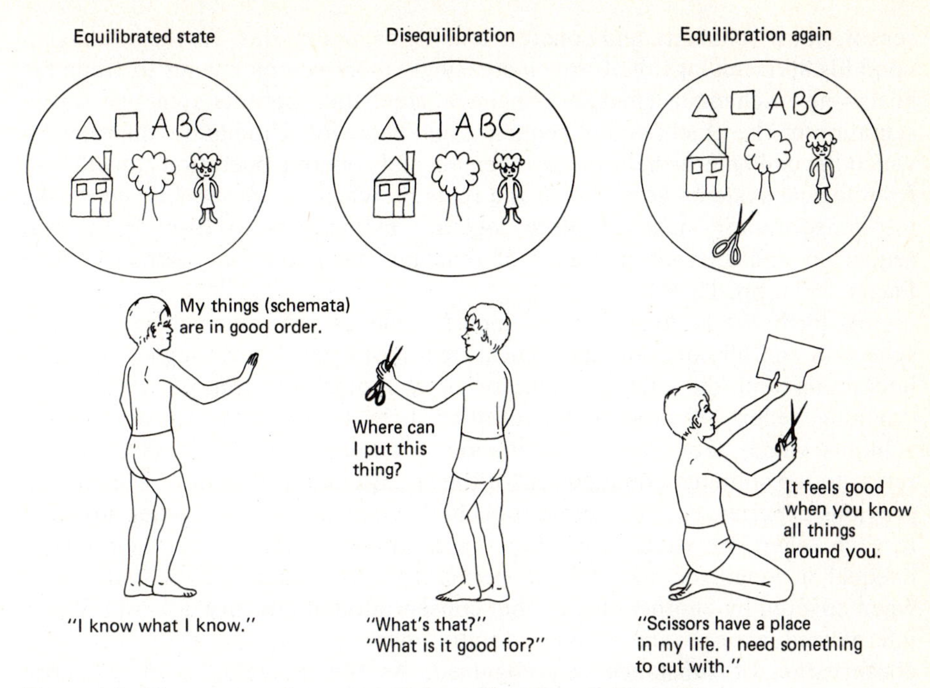

Figure 1-2 The process of accommodation—formation of schemata.

object. His apparent inability to decenter and consider all major aspects of an object or event leads to many distorted judgments. For young children there are no illusions or contradictions. An older child finds and tackles them. It takes time for him or the young adolescent to equilibrate his system of knowledge, as he often loses in combating the real world. Gains in reversibility represent steps toward equilibrium, since reversibility makes intelligence capable of detours and returns to the original points of departure (Looft, 1972; Piaget, 1967, 1973, pp. 60–61).

For Piaget, "to know an object is to act on it. To know is to modify, to transform the object and to understand the process of this transformation, and, as a consequence, to understand the way the object is constructed" (Piaget, 1964). From infancy through adulthood an object is known in an increasingly complex, organized, and interrelated way. It is known first, in a limited way, for its appearance and use, then for its permanence. Later it is known by its physical and related properties, and finally for its relationship to other objects in time and space. Only at the last stage of formal operations are all these attributes incorporated into a unified whole, so that a person relates to an object with all these properties and relationships explicitly (or implicitly) in mind.

The perception of reality thus changes from naïve realism to critical objectivity. An operation is an internalized action that modifies the object and increases knowledge. Intelligent behavior occurs as a result of applying sensorimotor schemata and concrete and logical operations. As the infant acts upon his surroundings, he forms increasingly more complex ways to attain his goals—new schemata emerge. Language acquisition permits symbolic representation of the object and of sequence and behavior. Ultimately, the rules by which the child pieces behavioral schemata and concrete operations together to form logical systems grow out of the rules by which he or she learned during the sensorimotor stage to piece together motor acts to form behavioral sequences and alternative means of reaching the same goal (Berlyne, 1966; Piaget, 1973, pp. 11–16).

At birth, the biological givens include reflexes and some schemata. Most schemata and all concrete and logical operations develop through a complex interaction and the ongoing activity of the organism (Piaget, 1973). The orienting reflex (Pavlov) and centration help to habituate (maximize the stimulus value). For Piaget, habituation is an early form of centration and refers to the stimulus-boundedness of the young child. He cannot rely much on previous experiences. For a preschool child two identical sticks, when adjacent to each other, are equal, but when one is moved farther away, they appear unequal in length or size. Object permanence is recognized—a stick exists when covered by another object—but conservation of quantity does not. When internalized actions are converted into operations at age seven or eight, conservation of substance is recognized. As the individual acquires more advanced schemata, he uses them extensively. At the same time he drops the primitive ones. By means of language, the preschool child internalizes objects and schemata and uses them to solve some of his problems and questions. Following Piaget, developmental maturity is the final equilibration (integration) of all growth trends in all aspects of human development. Fundamentally, Piaget's qualitative developmental theory distinguishes three major levels in the cognitive mode of operation: (1) sensorimotor, marked by simple behavioral schemas; (2) semiotic and conceptual representation and concrete operations; and (3) logically founded formal operations. Cognitive functioning will be more fully analyzed in Chapter 4 and in the presentation of the childhood and adolescent levels of life.

CRITICAL PERIODS OF LIFE

The concept of predetermined sequences and phases in human life was emphasized by Hall, Freud, Bühler, Werner (1957), Gesell (1952), and Piaget, among many others. G. Stanley Hall discussed predetermination in his analysis of recapitulation; that is, every human being passes through the key forms of early and vertebrate life in ascending the phyletic scale. When Arnold Gesell (1952) speaks of "intrinsic growth" and its normative aspects, he also expresses predetermination through rigid sequences in structural and behavioral differentiation, with individual differences springing from timing variations among individuals. Charlotte Bühler (1930) defined invariant maturational sequences and subsequently (1933) divided the life-span into functional levels and phases.

A penetrating study of the infant, child, adolescent, or senescent is possible only when it is recognized that he or she is a person with distinctive qualities and traits at each phase of life. Each level of early development involves the rise of new abilities and the reconstruction of the individual's motivation. Each level lays a sound or damaged foundation for the following one, and workable knowledge of early stages is crucial for an understanding of subsequent levels of life. More specifically, Murphy (1962, pp. 307, 315) assumes that if the necessary environmental stimuli do not appear at the time of emergence of a new function, the function may never fully develop.

The concept of the critical period was first used in animal studies, which demonstrated the critical nature of some early phases of life in setting a pattern for later life. Konrad Lorenz's concept (1935/1970) of *imprinting,* whereby young birds attach themselves to a moving object, living or nonliving, is pertinent here. Imprinting is thought to occur during a critical time in the physiological and social development of the bird. The duration of this critical period is about one-and-a-half days but varies somewhat for different species. Depending on resources at the critical time, goslings and ducklings have learned to follow people, hens, and a variety of decoy objects. No attachment will occur if there are no objects to follow during this period of heightened sensitivity.

The hypothesis of "critical periods" in early human development has been examined by McGraw (1935), Scott (1962, 1968), and Stendler (1952). In McGraw's largely experimental study of a pair of possibly identical twins (Johnny and Jimmy), in which special training was given to one of the twins and later differences in performance were assessed, it was concluded that there are critical periods for human learning that vary for different ontogenetic (but not phylogenetic) functions. James P. Scott (1968, p. 69) defines a critical period as "any period of life when rapid organization is taking place," and in 1962 placed the start of the critical period for major emotional disturbances at about seven months, after the end of primary socialization. Celia B. Stendler (1952) points out that during the first eight or nine months of life the infant is building up a set of expectations with regard to how his or her needs will be met. He or she is learning to depend. In the process of socialization there are two critical periods for forming overdependent behavior. The first such period starts when the child begins to test his mother to see if he can depend on her. For most children, Stendler says, this occurs toward the end of the first year of life, when they recognize the importance of the mother's proximity. The second critical period for the formation of dependency comes during the two-to-three-year-old period when parents demand that the child change his ways of acting in order to acquire the controls expected by society. Ordinarily the child's anxiety produces about the right amount of dependency. Excessive anxiety, when generated at this age, leads to overdependency.

After surveying imprinting studies on animals and humans, Philip H. Gray (1958) placed the critical period for social imprinting in human beings from the age of about six weeks to about six months (from the onset of conditionability to the fear reaction to strangers). At this time deprivation of perceptual and social stimuli can be seriously damaging. He provides some evidence suggesting that "wild children" and delinquent teenagers may have lacked imprinting, in addition to having stressful experiences, during the period of infantile helplessness. The main principle of the "critical periods" hypothesis is that certain dispositions and sensitivities appear in clusters rather than individually and grow when stimulated by proper influences. If relevant stimuli are absent or are present only at an earlier or a later period, contingent development fails to occur. When a fundamental growth does not take place, any advanced development of the same power or trait is impossible. If affection is not shown to the baby, he or she will fail to return it, and the capacity for genuine love in later life is thus curtailed. A lack of emotional and behavioral identification between the mother and her child handicaps the child for later identification with other individuals of the female sex in particular. Lack of identification with the father impedes identification with males.

Critical reviews of maternal deprivation studies (Bronfenbrenner, 1972; O'Connor, 1965) provide some substance for the notion of high sensibility to the power of the mother to stimulate growth and lowered sensibility in infants and young children after a long maternal absence. Any prolonged or very intense experience may cast the infant's personality into a mold that is difficult to reshape; just as if one branded the baby's body, the stamp endures throughout life (Orlansky, 1949).

In contrast to animal development, the critical periods for any human aspects of growth or learning are apparently elongated, ascending and descending over a comparatively long period of time. This pattern is compatible with the concept of stages and developmental tasks associated with them. Within certain limits, each major stage appears to give the individual a new opportunity to regain a major part of what was lost earlier (Kagan & Klein, 1973).

DEVELOPMENTAL PROCESSES AND LEVELS

Basically, development may be most clearly understood as a series of sequential changes in an organism leading to its maturity. Sequential changes unfold the organism's innate dispositions and permit their structural and functional realization. Development includes metabolic changes, structural increments, unfolding of functions, and increase in achievement as a result of experience. Development occurs as new kinds and levels of differentiation and integration, learning and maturation, emerge.

In general terms, development (D) is a function (f) of genetic (g) and environmental (e) stimulation (S). Occurring within a time continuum (T), this produces changes (C) in both organism (O) and behavior (B). Assuming that psychological attributes of development appear as responses (R) of the organism, the developmental paradigm may be expressed as

$$C_R = f(Sg, e, T)$$

Since not all past and present genetic and environmental elements are known, the experimenter with human subjects of any age cannot control satisfactorily even the basic determinants of C_R. Postulating that maturation processes are chiefly due to genetic factors and learning opportunities resulting from *Se* is helpful, but one must be aware that neither maturation nor learning represents any unitary process; both are conceived of as a series of more and less distinct mechanisms, like the unfolding of the helix in DNA, which emits genetically transforming information in the process of maturation, and transfer in the process of learning (Baltes & Goulet, 1970, pp. 9–12; Spiker, 1966; Watson, 1972).

Development is a broad term that refers to all the processes of change by which an individual's potentialities unfold and appear as new qualities, abilities, traits, and related characteristics. It includes the long-term and relatively irreversible gains from growth, maturation, learning, and achievement. Allied to development is the concept of *aging,* which starts at conception and ends with death. Aging increases with chronological age, but some organs and structures age at a faster rate than others. The same applies to the physical and functional powers of individuals. Aging implies involution or retrograde change and deterioration—various impairments of cells, organs, or systems, as well as a functional decline to the point of incapacity. Two persons, one at forty and the other at fifty, may have an equivalent biological age. Because of aging differences, for example, the life expectancy for mentally retarded persons is much lower than for intellectually bright groups (Moriya, 1971). Chapter 4 discusses major developmental processes in some detail.

For normative or comparative purposes, structural growth and functional achievement are often expressed quantitatively in standardized age limits. The quantifications of intelligence testing (IQ) and of achievement testing into mental age (MA) and educational age (EA) are too well known to need any explanation. The vital index (VI), less well known, is set by dividing the cubic centimeters of air exhaled after a maximum inhalation by the person's weight in kilograms. It is a good measure of athletic capacity. Carpal age (CA) is determined from the amount of calcium deposit in the wrist bones as seen in x-rays. Carpal age is often used as an index of ossification—a measure of physiological maturity. Prediction of height and of the age at which puberty occurs is also made on the basis of carpal age. Dentition or dental age (DA) is a measurement of dental growth usually based on the number of permanent teeth that have erupted. Height age (HA) and weight age (WA) are set by group standards and standard deviations for different ages which are set separately for males and females. As a total measure of physiological growth, *organismic* age (OA) refers to the mean of all the above measures for an individual (except the MA, IQ, and EA). The OA divided by chronological age yields the organismic quotient (OQ). There are a number of disagreements as to what the OA should include. For example, Olson and Hughes (1942) include in OA assessment some aptitude and achievement measures, such as mental and reading ages. The author's survey of the literature did not produce any recent studies on this ill-defined concept, which appears to have some usefulness in summarizing a person's physical status at a selected age.

ORGANIZATION OF THIS BOOK

In this text, many concepts of developmental psychology are defined or explained wherever they first appear. Terms and concepts that are not clarified either are self-explanatory in terms of general psychology or can be found in the Glossary starting on page 403. The student is urged to use the Glossary in order to get a fuller understanding of the terms used. Further information on developmental and related terms may also be found in the *Dictionary of Behavioral Science,* compiled and edited by Benjamin B. Wolman (1973), and in various encyclopedias.

In accordance with the leading theories of human development, this study of human growth and behavior examines comprehensively the human motivational and behavioral functions at each developmental level and phase. A developmental level may be defined as a period in a person's life at the onset of which certain powers and characteristics emerge which are not direct outgrowths of previous levels and which markedly change the person's behavior. The goal of this developmental-level organization is an understanding of the whole person with his or her various abilities and achievements, dynamics and motives, needs and problems as they appear in each phase of life. Rather than seeking to answer a question such as "What is the general course of intellectual growth and decline?" this approach attempts to clarify questions such as "What are the typical behavior patterns and dynamics of the infant, the preschool child, or the postpubertal adolescent?" The development of the person in American society at each level of life constitutes the content of all chapters of this book except the first four.

In dividing the human life-span into various levels or stages, the investigator encounters a number of problems. First, some stages are not sufficiently distinct from those which precede or follow them. Thus the infant does not suddenly enter childhood or the adolescent early adulthood; some differential characteristics typifying a new stage were acquired and shown less distinctly at preceding levels. Second, the criteria for judging what constitutes a level of development are difficult to discover. Should the length of infancy, for example, be defined in terms of chronological age, of bodily growth, or of speech ability? Moreover, what criteria should be used in considering bodily growth or the state of language development? Would an x-ray record of the calcification of the bones and success on a picture vocabulary test be sufficient indexes for these two aspects of growth? If not, how does one select the best measure for adequate appraisal of these factors?

Though some clarifying studies exist (Harris, 1971; Neugarten, 1969; Spiker, 1966) the issue of continuities and discontinuities will probably be debated for many years. Discontinuities permit the charting of segments of the age continuum into qualitatively different stages. Nonetheless, many outstanding exponents of human development from G. Stanley Hall up to the present have embraced the concept of stages. Moreover, Freud (1910), Erikson (1963), Piaget (1970, 1973), and Werner (1957) conceptualize fairly clear-cut stages in their studies of sexuality, socialization, cognition, and other major aspects of growth and maturation. A child who walks and talks, for example, has a different orientation to his environment from that of a child who merely creeps and babbles. The sexually mature adolescent has a different pattern of motivation from an older child whose sex glands are not as yet functioning (Skolnick, 1972).

At present there are two basic theoretical interpretations of human development. As Figure 1-3 illustrates, one can be represented by an ascending line, straight or somewhat curved, indicative of a gradually progressive pattern of growth; the second by elongated stairs, indicating a stagelike progression with short phases of transition toward higher levels of long-term functioning. In assuming continuity or dividing the life-span into various stages or levels of functioning, the investigator faces the dilemma of arbitrariness. With considerable justification many experts in human development see "striking leaps" forward moving the child into the next higher level of functioning, while others recognize only a gradual and rather even progression throughout the years of childhood and adulthood. Apparently the latter are nearly blind to any discontinuity, for any significant revision of behavior and personality structure.

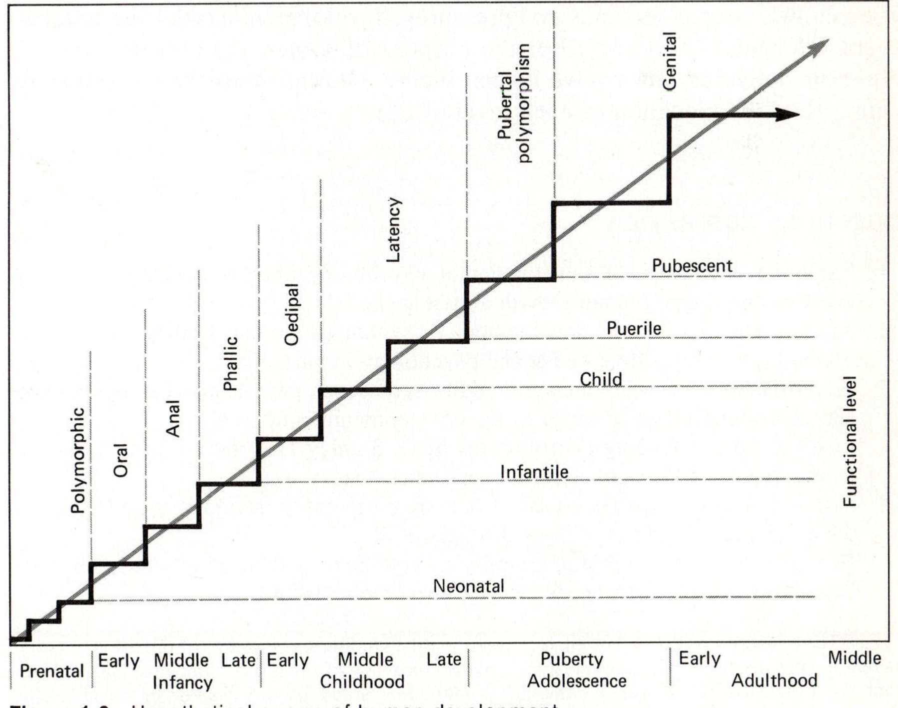

Figure 1-3 Hypothetical curves of human development.

Possibly the former disregard the continuity of accumulating experience and a relative sameness of the environment in which the child or adult functions. As the orthogenetic principle implies, human nature has an intrinsic tendency to develop in a definite direction and pattern primarily determined by the germ plasm and the genetic code. In actuality, development involves a balance of the two, continuous progression and periods of sudden growth and significant reorganization of motivation and behavior. There is a kind of fanning with regard to age, and less of a lockstep as a person grows older, but developmental order and predictability still remain.

In the study of the life-span to follow, a serious attempt has been made to define each developmental level in terms of behavior patterns and motivational variables distinguishing it from earlier and subsequent phases. In other words, the basis for dividing the life-span into developmental levels is demonstrated by the emergence of new abilities and motivational qualities, as well as significant changes in activities and achievements within the basic pattern of behavior at each given stage. Although general age limits for each of these stages are defined, one should keep in mind that such limits are only approximations of the standard: one person passes into the next developmental stage months or even years before another person of the same chronological age. Also, the passage may be partial, that is, in terms of only certain developmental aspects. It should also be recognized that considerable differences exist among various socioeconomic levels and racial groups within the same culture: social demands and pressures, developmental tasks and hazards, average longevity—all vary. For the purpose of systematic presentation, the life-span is divided into twelve levels; Figure 1-4 represents them graphically, with a brief identification of each level of life.

QUESTIONS FOR REVIEW*

1 Explain the concept of developmental psychology and its contribution to the evolving science of human growth and behavior.
2 Analyze the position of developmental psychology among leading divisions of psychology, comparative and social psychology in particular.
3 Who originated the scientific study of life-span development and what methodological contributions did he make to the developmental studies?
4 List several outstanding contributions of G. Stanley Hall and discuss the fundamental tenets of the behavioral recapitulation theory.
5 List and discuss Freudian concepts that are pertinent to developmental changes in the early years of life as well as throughout life.
6 How does the reality principle apply to childhood and to adulthood? Show the expected behavior differences at these two levels of life.

*The questions at the end of each chapter test the student's knowledge of the subject matter. If read before beginning the chapter, they provide a preview of the topics, concepts, and problems that are discussed. If answered either after individual study of the chapter or through class discussion, they summarize the principal points.

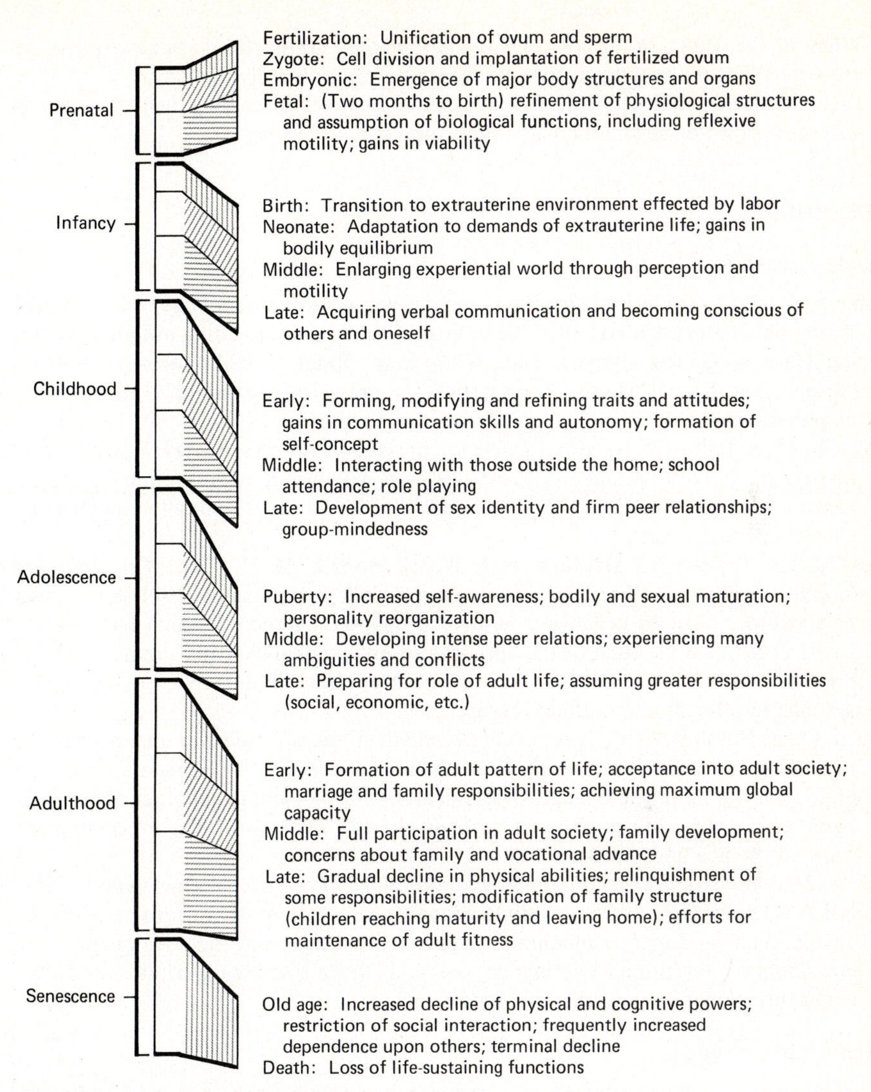

Figure 1-4 Developmental levels and phases.

7 Polarize *trust* versus *mistrust* and explain the significance of gaining trust during the first year of life.

8 List and clarify several fundamental tenets of Piaget's developmental psychology.

9 What practical guidelines might developmental psychology offer the student for his or her own development over and above those provided by a knowledge of general psychology?

10 Explain the concept of imprinting and discuss the "critical periods" hypothesis as it applies to the early stages of human development.

11 Evaluate Gray's survey of imprinting and consider the findings in the light of the "critical periods" hypothesis.

12 What is the value of discovering developmental norms for various aspects of growth? What is the danger in applying such norms to individuals?
13 List and discuss difficulties the investigator encounters in dividing the life-span into different stages or segments. Cite some supportive theories.

REFERENCES

Selected Reading

Dennis, W. (Ed.) *Historical readings in developmental psychology.* New York: Appleton-Century-Crofts, 1972. Selection of thirty-seven leading authors, including Darwin, Galton, Preyer, Hall, Thorndike, Binet, Freud, Terman, Watson, Gesell, Piaget, and Hebb. Each article is preceded by editor's introductory comments about the significance of the paper and often of the author himself.

Goulet, L. R., & Baltes, P. B. (Eds.) *Life-span developmental psychology: Research and theory.* New York: Academic Press, 1970. A volume of twenty contributors on history, theories, methods, and several aspects of development, emphasizing cognition and learning.

Jones, M. C., Bayley, N., Macfarlane, J. W., & Honzik, M. P. (Eds.) *The course of human development.* Waltham, Mass.: Xerox College Publishing, 1971. A selection of sixty-four articles organized in twelve chapters, three of which were newly written for this text. Most of the authors present original data and summarize their previous work. This volume reflects reasonably well what developmental psychology has to offer in longitudinal research.

Lugo, J. O., & Hershey, G. L. *Human development: A multidisciplinary approach to the psychology of individual growth.* New York: Macmillan, 1974. Presents historical, philosophical, anthropological, sociological, biological, and psychological perspectives, emphasizes life goals, and reviews the life cycle from the preconception stage to old age and death.

Nash, J. *Developmental psychology: A psychobiological approach.* Englewood Cliffs, N.J.: Prentice-Hall, 1970. A dimensional presentation of various aspects of development and of fundamental influences affecting them. Includes chapters on individuality, evolutionary influences, nervous system, critical periods, sex differences, identification, and the concept of self.

Specific References

Anandalakshmy, S., & Grinder, R. E. Conceptual emphasis in the history of developmental psychology: Evolutionary theory, teleology and the nature-nurture issue. *Child Developm.,* 1970, **41,** 1113–1123.

Baltes, P. B., & Goulet, L. R. Status and issues of a life-span developmental psychology. In L. R. Goulet & P. B. Baltes (Eds.), *Life-span developmental psychology: Research and theory,* pp. 3–21. New York: Academic Press, 1970.

Berlyne, D. E. Curiosity and exploration. *Science,* 1966, **153,** 25–33.

Bettleheim, B. *The empty fortress.* New York: Free Press, 1967.

Boyle, D. G. *A students' guide to Piaget.* Oxford: Pergamon Press, 1969.

Bronfenbrenner, U. Early deprivation in monkey and man. In U. Bronfenbrenner (Ed.), *Influences on human development,* pp. 256–301. Hinsdale, Ill.: Dryden, 1972.

Bühler, C. *The first year of life.* New York: John Day, 1930.

Bühler, C. *Der menschliche Lebenslauf als psychologisches Problem.* Leipzig: Hirzel, 1933.

Charles, D. C. Historical antecedents of life-span developmental psychology. In L. R. Goulet & P. B. Baltes (Eds.), *Life-span developmental psychology: Research and theory,* pp. 23–52. New York: Academic Press, 1970.

Ciaccio, N. V. A test of Erikson's theory of ego epigenesis. *Developm. Psychol.,* 1971, **4,** 306–311.

Darwin, C. *The expression of the emotions in man and animals.* London: J. Murray, 1872.

Edison, L. Science probes a last frontier. *Think,* 1970, **36** (6), 2–8.

Eisenberg, L. The human nature of human nature. *Science,* 1972, **176,** 123–128.

Erikson, E. H. *Childhood and society* (2d ed.). New York: Norton, 1963.

Frank, L. K. Human development: An emerging scientific discipline. In A. J. Solnit & S. A. Provence (Eds.), *Modern perspectives in child development,* pp. 10–36. New York: International Universities Press, 1963.

Freud, A. *The ego and mechanisms of defense.* London: Hogarth, 1957.

Freud S. *Three contributions to the sexual theory.* New York: The Journal of Nervous and Mental Disease, 1910.

Freud, S. *Civilization and its discontents* (standard edition, vol. 26). London: Hogarth Press, 1961. (Originally published, 1930).

Freud, S. *A general introduction to psychoanalysis.* New York: Permabooks, 1953.

Freud, S. *Three essays on the theory of sexuality* (J. Strachey, Ed. and trans.). New York: Basic Books, 1962.

Gesell, A. *Infancy and human growth.* New York: Macmillan, 1952.

Gray, P. H. Theory and evidence of imprinting in human infants. *J. Psychol.,* 1958, **46,** 155–166.

Grinder, R. E. The concept of adolescence in the genetic psychology of G. Stanley Hall. *Child Developm.,* 1969, **40,** 355–369.

Groffmann, K. J. Life-span developmental psychology in Europe: Past and present. In L. R. Goulet & P. B. Baltes (Eds.), *Life-span developmental psychology: Research and theory,* pp. 54–68. New York: Academic Press, 1970.

Hall, G. S. *Adolescence.* (2 vols.) New York: Appleton, 1904.

Hall, G. S. *Senescence: The last half of life.* New York: Appleton, 1922.

Harris, D. B. The development of human behavior: Theoretical implications for future research. In E. Tobach, L. R. Aronson, & E. Shaw (Eds.), *The biopsychology of development,* pp. 473–501. New York: Academic Press, 1971.

Kagan, J., & Klein, R. E. Cross-cultural perspectives on early development. *Amer. Psychologist,* 1973, **28,** 947–961.

Looft, W. R. The evolution of developmental psychology. *Hum. Developm.,* 1972, **15,** 187–201.

Lorenz, K. Z. Der Kumpan in der Umwelt des Vogels. Der Artgenosse als auslosendes Moment sociales Verhaltungsweisen. *J. Ornithology,* 1935, **83,** 137–213. (Trans. Robert Martin.) In *Studies in animal and human behavior,* Vol. 1, *Companions as factors in the bird's environment.* Cambridge: Harvard University Press, 1970. This is the first of K. Lorenz's three volumes.

McGraw, M. B. *Growth: A study of Johnny and Jimmy.* New York: Appleton-Century-Crofts, 1935.

McGraw, M. B. Major challenges for students of infancy and early childhood. *Amer. Psychologist,* 1970, **25,** 754–756.

Maier, H. W. *Three theories of child development* (2d ed.). New York: Harper & Row, 1969.

Matarazzo, J. D. *Wechsler's measurement and appraisal of adult intelligence* (5th ed.). Baltimore: Williams & Wilkins, 1972.

Moriya, K. A discussion of the biological model of developmental rotation. *Child Developm.* (Tokyo), 1971, **7,** 35–47.

Murphy, L. B. *The widening world of childhood: Paths toward mastery.* New York: Basic Books, 1962.

Muuss, R. E. *Theories of adolescence* (2d ed.). New York: Random House, 1968.

Neugarten. B. L. Continuities and discontinuities of psychological issues into adult life. *Hum. Developm.,* 1969, **12,** 121–130.

O'Connor, N. The evidence for the permanently disturbing effects of mother-child separation. *Acta Psychol.,* 1965, **12,** 174–191.

Olson, W. C., & Hughes, B. O. The concept of organismic age. *J. educ. Res.,* 1942, **35,** 525–527.

Orlansky, H. Infant care and personality. *Psychol. Bull.,* 1949, **46,** 1–48.

Piaget, J. *The origins of intelligence in children.* New York: International Universities Press, 1952. (Original French ed., 1936.)

Piaget, J. Development and learning. *J. Res. Sci. Teach.,* 1964, **2,** 176–186.

Piaget, J. *Six psychological studies* (David Elking, Ed.). New York: Random House, 1967.

Piaget, J. Quantification, conservation and nativism. *Science,* 1968, **162,** 976–979.

Piaget, J. *Structuralism* (Chaninah Maschler, Ed. and trans.). New York: Basic Books, 1970.

Piaget, J. *The child and reality.* New York: Grassman, 1973.

Preyer, W. *Die Seele des Kindes.* Leipzig: Grieber, 1881. [*The mind of the child* (J. W. Brown, Trans.). New York: Appleton, 1888.]

Quetelet, L. A. J. *Sur L'homme et le developpement de ses facultés.* Paris: Bachelier, 1835.

Quetelet, L. A. J. *A treatise on man and the development of his faculties.* Gainesville, Florida: Scholars' Facsimiles, 1969. (Reprint. Facsimile reproduction of Eng. trans. of 1842 with introduction by Solomon Diamond.)

Riegel, K. F. Influence of political and economic ideologies on the development of developmental psychology. *Psychol. Bull.,* 1972, **78,** 129–141.

Rosenblatt, B. Some contributions on the psychoanalytic concept of development to personality research. *Monogr. Soc. Res. Child Developm.,* 1966, **31** (5), 18–35.

Scott, J. P. Critical periods in behavioral development. *Science,* 1962, **138,** 944–958.

Scott, J. P. *Early experience and organization of behavior.* Belmont, Calif.: Brooks/Cole, 1968.

Skolnick, A. Historical perspectives on childhood: Some challenges to developmental psychology. *Proc. 80th Ann. Convention APA,* 1972, **7,** 89–90.

Spiker, C. C. The concept of development: Relevant and irrelevant issues. In H. W. Stevenson (Ed.), Concept of development. *Monogr. Soc. Res. Child Developm.,* 1966, **31** (5), 40–54.

Stendler, C. B. Critical periods in society. *Child Developm.,* 1952, **23,** 3–12.

Stern, W. Über Aufgabe and Anlage der Psychographie. *Z. angew. Psychol. psychol. Sammelforsch.,* 1910, **3,** 166–190.

Stevenson, H. W. Developmental psychology. *Ann. Rev. Psychol.,* 1967, **18,** 102–128.

Strupp, H. H. Freudian analysis today. *Psychol. Today,* July 1972, pp. 33–40.

Wadsworth, B. J. *Piaget's theory of cognitive development.* New York: David Day, 1971.

Watson, J. S. We assess lawfulness, but God knows what the law is. *Proc. 80th Ann. Convention APA,* 1972, **7**, 139–140.

Werner, H. The concept of development from a comparative and organismic point of view. In D. B. Harris (Ed.), *The concept of development,* pp. 125–148. Minneapolis: University of Minnesota Press, 1957.

Wolman, B. B. (Comp. and Ed.) *Dictionary of behavioral science.* New York: Van Nostrand Reinhold, 1973.

Chapter 2

Research Methods

HIGHLIGHTS

Methods open and build roads of inquiry. They include such procedures as "forming concepts and hypotheses, making observations and measurements, performing experiments, building models and theories, and providing explanations and making predictions" (Kaplan, 1964, pp. 23–24). The scientific method defines a procedure for all major research steps to gain valid results confirming or refuting hypotheses set for investigation.

Longitudinal and cross-sectional methods are two fundamental ways to observe and collect data about infants, children, adolescents, and adults. By correlational or factorial manipulation of such data, the investigator discovers what significant relationships, if any, exist among variables quantitatively assessed.

By experimental designs, the researcher aims to discover how one variable he manipulates (independent) affects a related variable (dependent) of interest to the experimenter. Any breakthrough or major discovery is usually at variance with accepted paradigm and past knowledge. It often calls for many additional studies to delineate its scope.

A research design combining several methods and measuring techniques is likely to cross-validate its findings to a high degree. When repetition of studies by competent investigators yields about the same results, the scientist accepts these findings as valid within the populations and parameters defined. Research in both natural settings and the laboratory is necessary to establish a scientific psychology of human growth and behavior.

The life-span science of human development utilizes biographical, correlational, experimental, and other research methods to help answer questions about behavioral changes that appear to be related to various levels of physical or functional maturity. The science of human growth seeks answers to hypotheses related to antecedent-consequent growth and behavior relationships. It explores trends in organismic, emotional, and cognitive functions at various ages. It seeks information about the ways in which present motivation or behavior modifies future activity and performance. Even partial answers to these questions contribute to the body of knowledge about the individual or group at the age studied. In many developmental studies, the investigator must show how a new reaction pattern has evolved from an earlier state and what contingencies are responsible for the change. These kinds of problems and questions have generated the need for appropriate paradigms and models to plan specific research designs for dealing with them. Compared with the early biographical, interview, and questionnaire methods of studying development, current research techniques represent a great advance in design complexity, objectivity, and verifiability of findings. Yet any reports of scientific inquiry of achievement are "invariably provisional, always subject to revision if and when better means of observation and measurement or other improvements in procedures of inquiry make possible more useful descriptions of what happens under specified circumstances" (Handy & Harwood, 1973, p. 14).

TOWARD CONTROLLED OBSERVATION

Until the relatively recent rise of controlled biochemical, psychological, and medical methodology, the techniques for studying human growth or behavior were for the most part limited to casual and subjective observation of children and acquaintances or to philosophically based essays on the nature of the growing child or aging adult. Although they were not subjected to the rigors of the scientific method and procedure, some of these early attempts led to the formulation of important questions and findings. Even today naturalistic field study, interviews, and questionnaires constitute a significant part of methodology in studying young and old, normal and deviant.

Psychological research needs both originality and diversity springing from various theoretical orientations. Each major theory assumes a particular viewpoint and provides its own research orientation about variables in human growth, behavior, or motivation. Acceptance of its tenets, usually supported by some research data, often involves the adoption of its paradigm (world hypothesis, a model that illustrates all the possible research design forms) as well. The term *paradigm* is defined by Kuhn (1970) as a conceptual framework and methodology providing direction and coherence for scientific research. A paradigm when applied shows the investigator what to expect in terms of the prevailing theory until new research produces a necessity for a paradigm shift by the progressing science. The paradigm usually involves an extension of the theoretical assumptions into many aspects of human activity (Anastasi, 1972). For example, a genuine acceptance of Skinner's operant-conditioning paradigm ultimately leads to the concept of engineered society, where the individual "loses" freedom and personal dignity by continually responding to environmental contingencies. Everyone in such a society becomes conditioned to react effectively in ways consistent with manifold reinforcement schedules regulated ultimately by a fully programmed central computer, which in turn is manipulated by those in power. Skinner (1971, p. 5) urges a technology of total behavior control for "preventing the catastrophe toward which the world seems to be inexorably moving."

Skinner (1969, 1971) sees human beings as organisms whose behavior depends more on environmental contingencies than on genetic endowments. These ever-present contingencies define the probabilities of human behavior. Simply stated, human behavior is seen as a function of its consequences: behavior that is reinforced (rewarded) is strengthened and behavior that is not reinforced or is punished is weakened. Contingencies of reinforcement specify the relationship between a behavior and its consequence. Behavior is thus learned, and learning is consistent with the operation of genetic factors. Human ethical behavior, compassion, and decency are the outcomes of sociocultural reinforcement: "Big Brother is still watching you, but from inside—a reminder of the Freudian censor." In psychoanalysis, censorship is a joint function of the individual ego and superego. Skinner assumes that the controlling self becomes the controlled self, since a person who introduces a change will also be affected by that change and so monitors the design to avoid errors. In order to change people's behavior, their contingencies must be changed. For Skinner, freedom appears to be limited to running away from noxious or harmful contacts. Support for Skinner's radical empiricism comes from efficiency of reinforcement contingencies in modifying behavior.

One of the strongest arguments against Skinner's basic paradigm $B = f(E_c)$ is that its reversal, expressed as $E_c = f(B)$, makes sense. According to Bandura (1974): "Human behavior to a significant degree partly creates the environment, and the environment influences the behavior in a reciprocal fashion . . . one and the same event can thus be a stimulus, response, or a reinforcer depending upon the place in the sequence at which the analysis begins." For example, aggressive children often create a hostile atmosphere, whereas friendly children usually further an amicable social milieu wherever they go.

At this stage of psychological knowledge, acceptance of Skinner's formulation or of any other single theory or paradigm is not justified. Every scheme of explanation leads to some question it cannot answer (Tyler, 1973). The current state of knowledge indicates that neither the deterministic nor the indeterministic hypothesis has demonstrated a satisfactory probability level (Myers, 1965; Tyler, 1973). Many psychologists would accept Burton L. White's notion (1969) that "we had better invest more of our resources in sharpening our observational tools and collecting twenty years of solid natural history first." For this reason probably the majority of psychologists today reject broad theories and paradigms, and rely on *pragmatic* models applied to limited domains, such as defining neonatal auditory acuity or assessing changes in visual perception over a limited period of time. Models facilitate research planning as well as communication among scientists. By restricting research to areas of empirical and experimental testing, the models can be expanded or delimited according to the accrued findings of those research designs which have been verified by replication or related studies.

Reliance on pragmatic models and narrow paradigms seems to fragment the study of the total person to findings of less and less significance. Reading of hundreds of laboratory research reports of the early 1970s would probably not make one more competent but rather more confused about dispersed bits of evidence. This information contributes little to the formulation of laws or theories (Bijou, 1972).

Carl Rogers (1964, 1973) has repeatedly raised the question of whether psychology will remain a narrow technological fragment of a past-oriented science, tied to an outdated theoretical conception of Newtonian framework, or will embrace the whole phenomenology of man. He provides an answer by urging psychology to become a "truly broad and creative science, rooted in subjective vision, open to all aspects of the human condition, worthy of the name of a mature science." Should psychology devise ingenious ways to test areas of "greatest mystery" or close its eyes to the possibility of another "reality" operating on rules very different from those discovered by empirical research in the past? Rogers' questions pose a problem to many psychologists. He feels that the psychologist ought to bc pullcd by his personal vision into a deeper and more significant view of the reality contained in the unified phenomenological world of man, responsive to *all* of himself and *all* of his environment (Rogers, 1973).

Burton L. White (1969) proposes that researchers first focus on a global analysis of experiential histories to "capture relevant causal factors in a rather large net at first" before they "sift the experiential clusters through finer and finer screens." Most present-day researchers construct minute experimental designs, keeping all but one variable constant or randomizing them while manipulating one or two, before exploring the total field and assessing most factors at work. White raises Piaget's approach of studying the ontogenesis of intelligence (Piaget, 1952) to a model of scientific inquiry, because in this study deductive and inductive elements are given their proper share. Piaget devised ingenious test situations to assess his preliminary hypotheses. In an approach consistent with White, Light and Smith (1971) offer a *cluster* procedure to combine the data of many related studies with similar and conflicting findings in order to extract conclusions supported by several or most studies analyzed. The flexibility of the cluster approach probably would lead to the accumulation of significant evidence.

In relying on reported research findings, caution is necessary. There is evidence that *E's* (the experimenter's) observations and measurements are influenced by his or her preconceptions, beliefs, and expectations. Since human subjects respond to very subtle cues, they often pick up the experimenter's biases and preferences. These uncontrolled variables sway their responses into the expected patterns (McGraw, 1970; Rosenthal, 1966). In psychological testing, for example, the permissible rewording of some questions occasionally becomes suggestive of scorable answers, and the standard rule of "no help" is sometimes disregarded by examiners as they test young children or attractive members of the opposite sex. Removal of *E* bias is thus a necessity to attain more objective results. *E* bias, of course, is only one problem, albeit a crucial one in many research designs. Other methodological problems include small sample size, subject bias, and uncontrolled variables. Successful design of exact techniques and procedures for systematic and critical investigation of specific developmental problems is rare, although found more frequently in recent decades of the twentieth century. Coan (1966, p. 732) assumes that practical realities in a normal social setting make the necessary experimental control unattainable. A brief examination of the scientific method will aid in understanding the specific models and designs used in modern developmental research.

SCIENTIFIC METHOD

In the various fields of empirical science, the scientific method offers a structure and procedure for acquiring valid knowledge. In its basic form this method may be summarized as follows:

1 *Problem.* Some definite problem is chosen for investigation. Broad generalizations and unobservable phenomena cannot be directly investigated. For example, the researcher might be interested in discovering whether children raised in an institutional setting differ in their language development from children raised at home. But what is meant by these terms? "Raised at home" may sound simple and clear, but is it? Is being raised at home with both parents the same as being raised with only one parent? Is it the same to have the mother at home during the day as to have her working outside? Any accurate statement of the problem must include an operational definition of "raised at home." Or again, what is "language development"? Has language developed when the child uses a simple word meaningfully, when he uses all parts of speech, or when he uses sentences of various types? Obviously the solution to the original problem will depend on the way in which its various elements are operationally specified.

2 *Hypotheses.* Considering related studies and the facts already known, the investigator makes an inductive inference or a calculated guess as a tentative explanation of the problem. The formulation of a hypothesis is helpful in carrying out the remaining steps of the scientific method because it aids the investigator in designing the total research project and in selecting methods for collection and analysis of data. The scientist may devise alternative hypotheses to test each method.

3 *Collection of data.* Testing of a particular hypothesis depends partly on the skillful selection of a method for systematic observation or measure-

ment. In developmental psychology, for example, operant conditioning, laboratory experimentation, and psychometric testing and retesting are often utilized. By applying the selected methods, the investigator collects data for later statistical analysis and discussion. Each method is utilized for the same purpose: accumulation of precise, reliable, and verifiable data.

4 *Analysis and interpretation of data.* The data accumulated by the application of the selected measurement device is subsequently subjected to a thorough analysis. Typically this involves application of certain statistical procedures. The investigator asks to what extent the set of data rejects or fails to refute the hypothesis originally stated. A definite relationship is often established, with stated probable errors and confidence level. It should be noted that the analysis of data usually neither proves nor disproves the hypothesis; it is always conceivable that the observed results might have been the product of some unknown factor. This is why the repetition of various studies, including experimental ones, is useful. Not all of them verify the original findings; some of them yield contrary results. This is why a discussion of results, in which findings of the study are compared with results of other pertinent studies and attempts are made to explain the differences, is necessary.

5 *Conclusions.* If feasible, conclusions based on the findings of the investigation and related sources of information are formulated. In view of propositions offered by other studies, new hypotheses may be formulated. Actually, each time some new relationship or lack of relationship is discovered, new problems arise requiring further investigation.

Generally, the basic reasoning process involved in scientific inquiry is *inductive*—going from specific observations or measurements to general hypotheses or propositions, or generalizations. There is a basic "uniformity of nature," and the patterns so far experienced will persist in what is yet to come (Kaplan, 1964, p. 20). Yet *deductive* reasoning also is an integral part of science. In this approach, the scientist infers specific relationships or explanations from a theory or a set of principles. For example, if we know that all live newborn infants cry, it is then logical to deduce that this is what will happen when an expectant mother gives birth to her offspring. Many research projects begin with deductive considerations, such as formulation of the problem and tentative answers or hypotheses for investigation. They end with interpretation of data and suggestions for new research, usually representing a fusion of inductive and deductive components. For Cattell (1966, pp. 710–718), scientific research in psychology is "an *inductive-hypothetico-deductive* spiral," in which a good array of alternative hypotheses are in constant development. In a concise article, Bronfenbrenner (1972) discusses various forms of hypotheses and ways to verify them, the measurement of variables studied, and the resultant systematic and variable errors. Moreover, White and Duker (1973) suggest certain standard characteristics for child samples studied so that readers can better understand any genuine discrepancies in their findings. After examination of experimental designs, a return will be made to methods of lesser control.

EXPERIMENTAL DESIGNS

The experimental method offers an excellent means of discovering how one factor modifies behavior of an individual or group. In an experimental study the variable that the researcher introduces is termed the *independent* variable. The resulting response being observed by the scientist is called the *dependent* variable. It is the aim of the investigator to determine whether the introduction and manipulation of the independent variable are accompanied by changes in the dependent variable. Naturally, any other factors or conditions that might influence the dependent variable must be taken into consideration. Whenever possible, such factors are held constant, or at least carefully observed, so that their influence will not distort the relationship between the variables. The resultant behavior, if recorded with a high degree of reliability, shows the effect of the single independent variable systematically manipulated by *E.*

If cause-and-effect conclusions are to be reached, the need for overall control of stimulation to the subject calls for a laboratory situation and its optimal conditions for experimentation, even when its findings may not be generalizable beyond a small population of subjects and beyond a narrow range of motivations within this population (Weber & Cook, 1972). Although the laboratory is the regular locus for experiments, other places, including nurseries and schools, may be adapted for a variety of experimental designs.

Though there is no limit to the complexity of experimental designs, some experiments are simple. By way of illustration, let us consider an experiment in reaction time. A child is instructed by the experimenter to press a certain button as quickly as he can after seeing a light signal or hearing a certain sound. A timing device records exactly how long it takes the subject to respond to either stimulus. By repeated measurements *E* discovers the relationship between the type of stimulus and the speed of response. This simple experiment, if repeated on a large number of subjects of different ages, would yield useful information about individual and age-group variations.

Complex experiments involve the simultaneous manipulation of a number of factors so that the patterns of interaction can be ascertained. The control-group technique, for example, is one of a number of experimental procedures that help to regulate various extraneous conditions. Two, three, or more groups of subjects are tested and equated to each other in mathematics, reading, or any other activity. The experimental groups are then exposed to the independent variable—noise, for instance—to see what effect it has on the performance of the activity selected, whereas the control group is not exposed to the independent variable.

As early as 1890, J. McKeen Cattell investigated individual differences in children and used statistics in analyzing the results. In psychological studies the experimental method spread slowly at first and then snowballed in the late 1950s. Since 1955, when J. Greenspoon reported on the reinforcing effect of two spoken sounds, hundreds of related experimental procedures have been constructed using Skinner's paradigm of operant conditioning with verbal and nonverbal forms of reinforcement. New strategies in energizing and modifying the child's behavior were discovered. The 1960s were marked by many sophisticated techniques in the empirical study of perceptual and learning changes from birth to adulthood. The early 1970s have seen the mushrooming of behavior-modification experiments. The first major textbook in experimental child psychology (Reese & Lipsitt, 1970) surveys and evaluates many of the significant studies of experimental design.

In specialized laboratory settings, several observers have the same opportunity to record the pattern of behavior of a child or adult. The use of the one-way vision screen provides such an opportunity. Children generally do not realize that the technique is being applied, but many adolescents and adults suspect this probability. When provided with stimulating objects or tempting attractions, children will tend to behave almost naturally. The accuracy of observation may be maximized by the use of hidden microphones and motion-picture cameras.

Various isolation and deprivation techniques represent a practical expansion of the experimental design. In isolation studies, for example, one group of children might be offered the opportunity to experience certain learning practices. A second group is not afforded the same opportunity. After a period of time the two groups are assessed and compared to discover the influence of the learning experience on the first group. In some studies this procedure is carried one step further: the control subjects are offered the learning experience later in order to discover if, and how quickly, they will match subjects of the first group. As a rule the first group, after its learning period, demonstates a marked advantage over the control group. If the time period is long, the control group usually requires far less training to reach the same degree of achievement. Apparently another factor is involved in attaining proficiency—maturation, which works throughout childhood and adolescence.

In some cultures and socioeconomic levels certain conditions are prevalent that are largely or totally absent in other cultures or levels. The latter, therefore, represent an isolated or control group. When other important variables are nearly equal, the differences in behavioral abilities or personality traits may be attributed to the variables or elements lacking. The tremendous complexity of any cultural group prevents attainment of the ideal of "all other things being equal," however; and the number of influences that can be isolated diminishes rapidly as the child grows older and finds exceptional opportunities to further the potentialities studied (Coan, 1966; Kagan & Klein, 1973). Human beings "cannot be contained uncontaminated under true experimental control for any long period of time" (Harris, 1971).

The advantages of the experimental method include its power to gather data at a fast rate and to verify its findings through repeated experiments. But despite the fact that an experiment contains precise procedures and optimum control, it is not without its limitations. Experimental procedures tend to introduce artificial situations far removed from real life, thus making human responses less natural. Ethical and humanitarian reasons restrict the application of harmful variables. The use of powerful fear- and conflict-producing stimuli, extended motor restriction, the use of new psychoactive drugs, and severance of verbal communication are examples of possible damage to the child or adolescent. The effects of long-term stimulus deprivation, for example, are of considerable interest, but moral considerations exclude the use of infants or children for any such experiments (APA, 1973; Barber, Lally, Makarushka, & Sullivan, 1973; Wolfensberger, 1967).

Co-twin control method Often employed in conjunction with the isolation technique, the co-twin method clearly illustrates the investigator's attempt to control as many significant variables as possible in determining the relationship between hereditary endowment and behavioral processes. This method seeks to control the influence of heredity to the greatest possible extent. Identical twins, persons who have developed from the same fertilized germ cell and therefore possess virtually the same hereditary makeup, are used as preferred subjects. In an experimental isolation study, one twin *(T)* is exposed to the experimental variable and the other twin *(C)* is not. Hence, both environmental and hereditary factors are maximally controlled.

The leading studies of co-twin control were conducted by Gesell and Thompson (1941) and by McGraw (1935). Both offer many practical insights into the nature of development, especially the effects of specific training on performance. Maximum rise in performance results from nurturing the desired response in a natural state of growth. McGraw observed that the behavior patterns do not grow all at once; each has its own period of rapid development. She made deliberate attempts to modify performance by giving one of a pair of possibly identical twins special training. In some activities, such as walking, performance was about the same despite early practice by one twin. In other activities, such as roller skating, the trained twin became adept soon after he began walking. On the basis of her varied experiments, McGraw concluded that there are critical periods for human learning, which vary for distinct ontogenetic activities; for each kind of behavioral pattern there is an optimum period for rapid and efficient learning.

This method has been used by some researchers supposedly to determine the relative importance of heredity and environment in the development of certain behavioral traits and abilities. However, its application is somewhat limited. The co-twin control method has the particular drawback of sampling limitation; too few sets of identical twins of the same age or developmental level are obtainable by the investigator. In addition, when older twins are used, another problem presents itself: the possible influence of different experiential backgrounds and personality factors. After several dramatic breakthroughs in genetics, this method is becoming more valuable for research in human behavior genetics (Vandenberg, 1966).

Still another use of the co-twin method is available to the researcher. Identical twins are employed as subjects, but with attempts to diversify their environment. The twins are reared in distinctive environmental settings—say in separate foster homes—or are exposed as teenagers or adults to decidedly different influences. They are then compared with respect to some specific trait or ability. If the twins manifest significantly greater similarity than persons who have comparable environments but lack this hereditary resemblance, then the greater similarity of the twins is probably due to some genetic variable. It can be noted in passing that research along this line has clearly demonstrated the important role of heredity in determining intellectual capacity and emotionality. Intensive research is currently being conducted to determine what hereditary basis, if any, exists for susceptibility to various illnesses, mental disorders, and types of personality.

LONGITUDINAL AND CROSS-SECTIONAL METHODS

In research on age-related variables and relationships, longitudinal and cross-sectional studies constitute the two basic methods of data collection. Both use many standard techniques with advantages and disadvantages of their own. As shown in Figure 2-1, a number of specific methods and tools can be used in either approach.

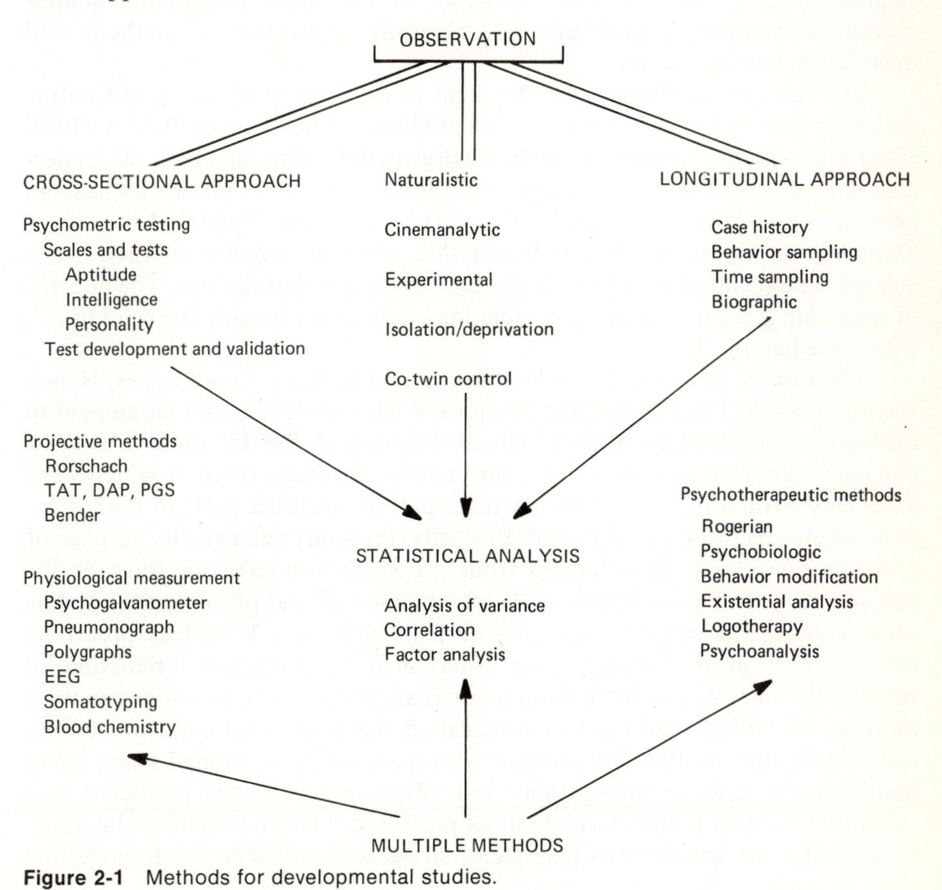

Figure 2-1 Methods for developmental studies.

Longitudinal Method

As the etymology indicates, the longitudinal method implies study of the same persons or samples of individuals over a long period of time—e.g., the same group of children studied from birth to the age of three years. The investigator observes and assesses the increase in intellectual or other growth and the environmental changes that occur during that period for each individual. Various growth aspects and patterns are charted, in an effort to learn what leads to what. Modern longitudinal investigations involve periodic assessments of comparatively large samples of children, adolescents, or adults. Usually some aspects of growth are extensively examined and tested through the use of a number of devices, while other aspects may be hardly considered; the approach depends on the objectives set by the investigators. Alan A. Stone and Gloria C. Onqué (1959) abstracted 297 longitudinal studies in a book: most of them are treatments of the social, emotional, and personality development of young children and schoolchildren. Jerome Kagan (1964) summarized ten major longitudinal research projects in the United States in terms of the samples used, methods applied, and goals set. For future longitudinal studies clearer delineation of problems and rigorous application of methods and techniques were suggested.

The Institute of Human Development at Berkeley (University of California) gave rise to several significant longitudinal studies. Nancy Bayley (1956, 1966) assessed the pattern of intellectual growth, adding much to the understanding of functional cognitive power. Jack Block (1971) related adolescent personality descriptions to adult life-styles by applying a modified version of Stephenson's Q technique. He found that adult personality characteristics echoed significantly but not uniformly the adolescent descriptions. The degrees of matching were indicated by correlation coefficients ranging from .78 to .17, with a median of .38.

The Course of Human Development, edited by Mary Cover Jones, Nancy Bayley, Jean W. Macfarlane, and Marjorie P. Honzik (1971), is a monument to the success and failure of longitudinal research at the Institute. The book contains sixty-four articles divided into twelve chapters, three of which were especially written for the volume. Articles published earlier (1932 to 1967) were extensively rewritten and abridged. Basically, the study achieved its purpose of predicting developmental changes from one stage to a later one for over 500 subjects. Physical growth such as height and sexual and physiological maturation could be predicted with considerable accuracy. With less precision, performance on intelligence tests could also be predicted. Prediction of personality variables suffered from many inaccuracies. The investigators were more often wrong than right in forecasting the general adjustment of their individual subjects. Possibly pleasant surprises were occasioned when about half turned out to be more stable and effective adults than predicted, and "slightly less than a third turned out as predicted." Often transitional pathognomic behavior was seen as persistent, stress was mistaken for trauma, and the potential for learning from experience was not taken into consideration (Jones et al., 1971, pp. 408–415; Rigler, 1973).

When case-history data are used to illustrate various points, the longitudinal method yields individualized analysis of the growth pattern. As in biographical or case-study techniques, the individual or group is not clocked extensively by norms or standards. As samples become large, however, individuals get "lost" in the findings or conclusions.

Case-study Technique This is a longitudinal survey concerned with the developmental aspects and influences of a single individual. As a rule, a detailed case-study outline is used to collect the same information about each subject. The taking of medical histories—surveys of the illnesses and treatments of the past—gave rise to the social and psychological case study of today. When the case study pertains to a child, it includes the family background, the developmental course in terms of psychomotor, language, social and cognitive growth, the present situation, and a survey of results of medical, school, and other examinations. Other aspects of child life included in this technique are unusual traits and attitudes, relationships to parents and siblings, emotions—especially fears and phobic reactions—and trends in play activities. Estimates are made of the child's level of maturity, assets, liabilities, and urgent needs, and possibilties for correction or improvement. The case study usually ends with a summary of the findings, including a recommendation based on the overall clinical picture of the child.

In covering the past, interviews with the child, the parents or guardians, or anyone who knows him or her yield information about significant events and relationships. The past must be carefully documented, since it is an inextricable factor of the condition and behavior of the subject today. In terms of present status, clinically oriented observations of appearance and behavior are made, often supplemented by a battery of psychodiagnostic tests. By synthesizing the data, the caseworkers attempt to draw a picture of the whole child in his specific conditions of life. As diagnosticians, they want to recognize the damage suffered by the child and provide corrective measures.

It is noteworthy that for the most part case studies represent children or adults who are deviant or disturbed in some way. A child who has failed to adjust satisfactorily to the given conditions of his life often represents in certain ways an exaggerated picture of a basically normal person. Hence case studies of such children or adolescents frequently advance understanding of what the individual needs for normal adjustment. Occasionally some damaging influences are clearly isolated, and recognition of them helps to protect other children from similar damage.

Cross-sectional Method

This widely used method is marked by the assessment of a large number of persons and selected variables as they are found at a specified age. Accurate height and weight tables of 1976 may be established by measuring representative samples of persons at various ages. Intellectual growth during the fifth year of life can be shown by comparing the performance on intelligence tests of representative samples of four- and five-year-olds. The differences between the two samples, if statistically significant, can be attributed to the age difference between the two samples.

If representative population samples of the United States were employed in standardizing intelligence and other tests, such tests would yield age norms for the total population of the United States. Subsequently any individual could be compared with the standard and its deviation. At present there are valid tests for the assessment of most structures and functions of child growth and maturation. Aptitude and achievement tests are widely used for cross-sectional testing of normal and deviant individuals and groups. The *Mental Measurement Yearbooks,* edited by Oscar K. Buros (1965, 1972), provide periodic evaluative analyses of most psychological and related tests. The leading tests are usually reviewed by two or more surveyors. The texts by Anne Anastasi (1968) and by Leo Cronbach (1970) are good guides for fundamental information about many psychological tests. The predictive power of a psychological test is usually recognized as a key criterion of its adequacy and validity.

The practical usefulness of the cross-sectional method consists chiefly in the speed with which even large studies of various ages may be begun and completed, as well as the ready availability of testing devices. A basic shortcoming of this method is the loss of the individuality and wholeness of the person. As individual curves fuse into a single one representative of the group, the resulting curve is not likely to match a single individual of that group; it masks the subtle shifts that occur from time to time in each subject. When diverse ages are compared, some outstanding traits and features of intermediate phases are often missed. The one-year-old, for example, is accepting and cooperative with those around him, the two-year-old is stubborn and negativistic, and the six-year-old is conflict-ridden and emotional. Were only the one-, four-, and seven-year-olds sampled, no basic changes would be indicated in acceptance and cooperation; only a gradual increase would be evident. Also, some four- or seven-year olds are likely to be significantly behind or ahead of their chronological age.

Several investigators (e.g., Baltes, 1968; Schaie, 1972; and Schaie & LaBouvie-Vief, 1974) noted explicitly that the cross-sectional method fails to account for generation effects (historical development) and therefore shows considerable losses of IQ with age. For example, if the intelligence of ten-, twenty-, thirty-, and forty-year-olds is tested in 1976, the differences found will reflect not only age changes but also cultural differentials, especially differences in the conditions under which the four age groups were reared. The cross-sectional study simply confounds age and cohort variables.

Schaie (1965) constructed a general developmental model that relates the three major components: response is a function of age *(A)*, cohort *(C)*, and time of measurement *(T)* differences. His trifactorial formula $R=f(A, C, T)$ is a basis for the distinction of cohort and generation effects, since *A, C,* and *T* are recognized as separate entities that represent three distinct sources of developmental change in the sense of different antecedent conditions. A distinct analysis of the confounded (interdependent) effects is obtained by algebraic substitution and by alternate empirical checks of assumptions. In this process, three sequential strategies are applied: (1) the cohort-sequential technique, (2) the time-sequential technique, and (3) the cross-sequential technique. These sequential techniques separate and unconfound the original sources of developmental change. Any conventional cross-sectional or longitudinal study represents only an incomplete special case of the sequential models. For specific information about the application of the developmental model and related corrective techniques the student is referred to Schaie (1965); Baltes (1968); Labouvie, Bartsch, Nesselroade, and Baltes (1974): and Nesselroade and Baltes (1974). Although it raises new issues, Schaie's work is a significant methodological advance in the refinement of developmental research, particularly where the growth or behavior under study is substantially influenced by environmental conditions.

Tools and Procedures

Naturalistic Observation Observation in natural settings—hospitals, nurseries, homes, schools, and recreational facilities—is an important source of information. The undetected investigator observes activities and relationships among infants, children, adolescents, and their parents in order to discover some decisive variables at work. Little if any effort is made toward control. At times the factual observation is accompanied by brief on-the-spot interviews or ad hoc experiments. Naturalistic observation often suggests hypotheses for further research through the use of experimental and testing methods.

Systematic observation of neonates and children in hospitals and nursery schools was applied extensively in the studies of Arnold Gesell, Charlotte Bühler, William E. Blatz, Jean Piaget, John B. Watson, and many others. The specialized time-sampling technique was first applied by W. C. Olson and Florence L. Goodenough. In the time-sampling technique a selected aspect of behavior is observed, and a representative sample of reliable data is secured for statistical analysis over a predetermined period of time. Many cross-cultural studies are chiefly based on naturalistic observation of children or adults, often supported by some data accumulated from the use of culture-fair tests (tests that are little affected by cultures and cultural differences among individuals and groups).

Interview and Questionnaire The interview (essentially a conversation during which information is elicited) has many diverse applications. Case-history data are collected in part by interviewing the subject and his or her parents and associates. Psychiatrists, clinical psychologists, and psychoanalysts use interviews as a key device for diagnostic and especially psychotherapeutic goals. Many clinical sessions are devoted to scrutiny of early experiences and relations with parents and other significant persons. Clinical interviewing often unearths the original events that have evoked maladaptive self-defenses and attitudes. It is well known that frustrating and traumatic occurrences in the early years may be causes of later symptoms and their patterns (Bowlby, 1953; Harlow & Suomi, 1970; Chess et al., 1965; Yarrow, 1964). Psychometric and projective tests applied by the clinical psychologist for diagnostic purposes greatly reduce the number of interviews needed for a well-supported diagnosis. A plan of professional treatment for the child or adult can then be set up without undue delay. Psychological testing represents a form of standardized interviewing, approaching experimental procedure.

Administration of questionnaires is another form of standardized interviewing. Oral questioning in an interview situation gives way to printed questions pertaining to various growth and adjustment areas. Selected sets of questions are a means of collecting desired information of any kind. They may pertain to past experiences and events and to present conditions and attitudes, to matters for which records are available to check reliability and to those for which no records can be found. In some instances mothers or fathers fill out questionnaires to complement data obtained directly from the child. In others, the child is asked the questions orally and his answers are recorded by the teacher or interviewer.

The advantages of questionnaires include easy administration to groups and the practically unlimited possibility of constructing questionnaires suited to the goals and hypotheses of the investigator. When approached with interest and concern, children like to be questioned, and their answers are usually less guarded than those of adolescents or adults. The questionnaire technique is well-suited to studying older children in their classrooms and club or chapter meetings. Some questionnaires call for simply classified answers that can be readily converted into statistical terms for assessment. Beginning with G. Stanley Hall (1891), who first extensively applied this "pencil-and-paper technique" in the United States to test children's knowledge on entering school, the questionnaire has continued to be used by many researchers in psychology and sociology.

Statistics Most methods and techniques of developmental study, when applied to representative age samples or large experimental groups, yield vast amounts of quantitative data. Analysis and interpretation of such data are facilitated by the application of pertinent statistics, which include the study of frequencies and distributions of scores and the computation of measures of variance or of central tendency (mean or median), all of which show where an individual or group stands with reference to the normal distribution or standard score. Correlational techniques and factor analysis add much to the understanding of relationships and the resultant structures. Tests of significance determine the acceptance or rejection of the hypotheses formulated at the outset of the research.

Each statistic adds something to the precision of findings. Assessing errors occurring in any measurement and generalizing from samples studied to pertinent populations are examples of the usefulness of statistics. From the refinement of experimental findings to the validation of psychological tests for children, adolescents, and adults, statistics make a significant part of most research designs.

Multiple Methods Increasingly, modern research studies apply a number of selected methods and techniques for the study of complex relationships and problems. When several methods are properly combined into a research design, each one contributes and offsets limitations of the others. Inclusion of interviews, questionnaires, and psychological tests is rather frequent when a large sample is used and depth analysis is part of the design.

In all research designs it is of paramount importance that the prescribed procedures and steps be executed by experts and that major findings be verified by repetition of the original design and procedure. Systematic and random errors occur in the application of all research methods; therefore, these errors should be estimated and discussed. It is also imperative to preserve any extensive data collected for further and later studies by the same or other methods.

In the designing of current developmental research projects, original Freudian postulates and Piaget's hypotheses are two of the most frequently used sources (Dasen, 1972; Wohlwill, 1963). Such theorists as Kurt Lewin, Alfred Bandura, Erik H. Erikson, B. F. Skinner, Robert R. Sears, Harry E. Harlow, Heinz Werner, Nancy Bayley, Charlotte Bühler, Arnold Gesell, Jerome Kagan, and Robert J. Havighurst also offer new concepts and constructs for further research. Practically every significant research study raises new questions to be answered by further studies. Thus data on the development of various aspects of human behavior have accumulated, but much more research is necessary to evaluate fully their significance for a comprehensive theory of human development.

QUESTIONS FOR REVIEW

1. What questions are usually raised and often answered in basic developmental research designs?
2. Explain the difference between the paradigm and pragmatic model as theoretical bases for research designs.
3. Give reasons why Burton White considers Piaget's approach to the study of intelligence a model of scientific inquiry.
4. List and explain the basic steps in the application of the scientific method.
5. What are the characteristics distinguishing the inductive from the deductive approach in scientific study?
6. What are the control advantages in the study of co-twins and identical twins?
7. Describe the isolation technique and explain the advantages of long-term isolation of individuals and groups in developmental studies.
8. Identify and explain the basic differences between the longitudinal and cross-sectional approaches.

9 Discuss and illustrate the application of the longitudinal method and consider the usefulness of its findings.
10 There are numerous cross-sectional studies but many fewer longitudinal ones. Why is there a comparative dearth of longitudinal investigations?
11 Explain how experimental method differs from correlational study.
12 Critically assess the usefulness of naturalistic observation for scientific research projects.
13 Explain the role clinical interviewing plays in collection of diagnostic data.
14 What are the advantages and limitations of the questionnaire technique?
15 Explain the necessity for statistics in developmental studies.
16 What are the advantages of applying several methods of research rather than just one in studying human development?

REFERENCES

Selected Reading

Hoffman, M. L., & Hoffman, L. W. (Eds.) *Review of child development research* (3 vols.). New York: Russell Sage Foundation, 1964, 1966, 1973. Volume 1 consists of twelve articles covering various research areas by thirteen authors, from genetics and the effects of infant care practices to the development of moral character and the effects of the mass media. Volume 2 contains eleven articles by seventeen authors dealing with family structure, socialization, development of language, attitudes, motives and roles, and deviant patterns of early growth. Volume 3, edited by B. M. Caldwell and H. N. Ricciuti and published by the University of Chicago, deals with infant-mother attachment, behavior modification, and disadvantaged families. It includes studies on aggression, adoption, and laws pertaining to children.

Kaplan, A. *The conduct of inquiry: Methodology for behavioral science.* San Francisco: Chandler, 1964. A sensitive yet critical interpretation of methods, laws, measurements, statistics, values, models, and theories as these apply to psychology.

Mussen, P. H. (Ed.) *Handbook of research methods in child development.* New York: Wiley, 1960. An extensive compilation of various methods and techniques of study by thirty experts, including research designs and applications of the techniques presented.

Sidowski, J. E. (Ed.) *Experimental methods and instumentation in psychology.* New York: McGraw-Hill, 1966. Over thirty contributors deal with a range of experimental techniques from conditioning to research on higher mental processes, offering suggestions for the use of many instruments, polygraphs, and apparatuses.

Specific References

Ad hoc Committee on Ethical Standards in Psychological Research. *Ethical principles in the conduct of research with human participants.* Washington, D. C.: American Psychological Association, 1973.

Anastasi, A. *Psychological testing* (3d ed.). New York: Macmillan, 1968.

Anastasi, A. The cultivation of diversity. *Amer. Psychologist,* 1972, **27,** 1091–1099.

Baltes, P. B. Longitudinal and cross-sectional sequences in the study of age and generation effects. *Human Developm.,* 1968, **11,** 145–171.

Bandura, A. Behavior theory and the models of man. *Amer. Psychologist,* 1974, **29,** 859–869.

Barber, B., Lally, J. J., Makarushka, J. L., & Sullivan, D. *Research with human subjects.* New York: Russell Sage Foundation, 1973.

Bayley, N. Individual patterns of development. *Child Developm.,* 1956, **27,** 45–74.

Bayley, N. Learning in adulthood: The role of intelligence. In H. J. Kalusmeier & C. W. Harris (Eds.), *Analysis of conceptual learning,* pp. 117–138. New York: Academic Press, 1966.

Bijou, S. W. The critical need for methodological consistency in field and laboratory studies. *Determinants Behav. Developm.,* 1972, 89–113.

Block, J. *Lives through time.* Berkeley, Calif: Bancroft, 1971.

Bowlby, John. Critical phases in the development of social responses in man. *New Biology* (Vol. 14). London: Penguin, 1953.

Bronfenbrenner, U. The structure and verification of hypothesis. In U. Bronfenbrenner (Ed.), *Influences of human development,* pp. 2–30. Hinsdale, Ill.: Dryden, 1972.

Buros, O. K. *The sixth mental measurement yearbook; The seventh mental measurement yearbook.* Highland Park, N.J.: Gryphon Press, 1965, 1972.

Cattell, J. McK. *Mental tests and measurements.* London and Edinburgh, 1890.

Cattell, R. B. (Ed.) *Handbook of multivariate experimental psychology.* Chicago: Rand McNally, 1966.

Chess. S., Thomas, A., & Birch, H. G. *Your child is a person.* New York: Viking, 1965.

Coan, R. W. Child personality and developmental psychology. In R. B. Cattell (Ed.), *Handbook of multivariate experimental psychology,* pp. 732–752. Chicago: Rand McNally, 1966.

Cronbach, L. J. *Essentials of psychological testing* (3d ed.). New York: Harper and Row, 1970.

Dasen, P. R. Cross-cultural Piagetian research: A summary. *J. Cross-cult. Psychol.,* 1972, **3,** 23–40.

Gesell, A. L., & Thompson, H. Twins T and C from infancy to adolescence: A biogenic study of individual differences by the method of co-twin control. *Genet. Psychol. Monogr.,* 1941, **24,** 3–121.

Greenspoon, J. The reinforcing effect of two spoken sounds in the frequency of two responses. *Amer. J. Psychol.,* 1965, **68,** 409–416.

Hall, G. S. The contents of children's minds on entering school. *Pedag. Sem.* 1891, **1,** 139–173.

Handy, R., & Harwood, E. C. *Useful procedures of inquiry.* Great Barrington, Mass: Behavioral Research Council, 1973.

Harlow, H. F., & Suomi, S. J. The nature of love—simplified. *Amer. Psychologist,* 1970, **25,** 161–168.

Harris, D. B. The development of human behavior: Theoretical considerations for future research. In E. Tobach, L. R. Aronson, & E. Shaw (Eds.), *The biopsychology of development.* New York: Academic Press, 1971, pp. 473–501.

Jones, M. C., Bayley, N., Macfarlane, J. W., & Honzik, M. P. (Eds.) *The course of human development.* Hinsdale, Ill.: Dryden, 1971.

Kagan, J., & Klein, R. E. Cross-cultural perspectives on early development. *Amer. Psychologist,* 1973, **28,** 947–961.

Kaplan, A. *The conduct of inquiry: Methodology for behavioral science.* San Francisco: Chandler, 1964.

Kuhn, T. S. *The structure of scientific revolutions* (2d ed.). Chicago: University of Chicago Press, 1970.

Labouvie, E. W., Bartsch, T. W., Nesselroade, J. R., & Baltes, P. B. On the internal and external validity of longitudinal designs: Dropout and retest effects. *Child Developm.*, 1974, **45,** 282–290.

Light, R. J., & Smith, P. V. Accumulating evidence: Procedures for resolving contradictions among different research studies. *Harvard educ. Rev.,* 1971, **41,** 429–471.

McGraw, M. B. *Growth: A study of Johnny and Jimmy.* New York: Appleton-Century Crofts, 1935.

Myers, C. T. Immergluck's freedom determinism. *Amer. Psychologist,* 1965, **20,** 93.

Piaget, J. *The origins of intelligence in children.* New York: International Universities Press, 1952.

Reese, H. W., & Lipsitt, L. P. *Experimental child psychology.* New York: Academic Press, 1970.

Rigler, D. A monument to longitudinal research. *Contemp. Psychol.*, 1973, **18**(7), 316–317.

Rogers, C. R. Toward a science of the person. In T. W. Wann (Ed.), *Behaviorism and phenomenology: Contrasting bases for modern psychology.* Chicago: University of Chicago Press, 1964.

Rogers, C. R. Some new challenges. *American Psychologist*, 1973, **28,** 379–387.

Rosenthal, R. *Experimenter effects in behavioral research.* New York: Appleton-Century-Crofts, 1966.

Schaie, K. W. A general model for the study of developmental problems. *Psychol. Bull.,* 1965, **64,** 92–107.

Schaie, K. W. Limitations on the generalizability of growth curves of intelligence: A reanalysis of some data from the Harvard Growth Study. *Human Developm.,* 1972, **15,** 141–152.

Schaie, K. W., & LaBouvie-Vief, G. Generational versus ontogenetic components of change in adult cognitive behavior: A fourteen-year cross-sectional study. *Developm. Psychol.* 1974, **10,** 305–320.

Skinner, B. F. *Contingencies of reinforcement.* New York: Appleton-Century-Crofts, 1969.

Skinner, B. F. *Beyond freedom and dignity.* New York: Knopf, 1971.

Stone, A. A., & Onqué, G. C. *Longitudinal studies of child personality.* Cambridge: Harvard University Press (for the Commonwealth Fund), 1959.

Tyler, L. E. Design for a hopeful psychology. *Amer. Psychologist,* 1973, **28,** 1021–1029.

Vandenberg, S. G. Contributions of twin research to psychology. *Psychol. Bull.*, 1966, **66,** 327–352.

Weber, S. J., & Cook, T. D. Subject effects in laboratory research: An examination of subject roles, demand characteristics, and valid inference. *Psychol. Bull.*, 1972, **77,** 273–295.

White, B. L. Child development research: An edifice without a foundation. *Merrill-Palmer Quart.,* 1969, **15,** 47–78.

White, M. A., & Duker, J. Suggested standards for children's samples. *Amer. Psychologist,* 1973, **28,** 700–704.

Wohlwill, J. F. Piaget's system as a source of empirical research. *Merrill-Palmer Quart.,* 1963, **9,** 253–262.

Wolfensberger, W. Ethical issues in research with human subjects. *Science,* 1967, **155,** 47–51.

Yarrow, L. J. Separation from parents during early childhood. In M. L. Hoffman & L. W. Hoffman (Eds.), *Review of child development research,* pp. 89–136. New York: Russell Sage Foundation, 1964.

Part Two

Fundamental Influences

Human growth, behavior, and personality are usually considered to be determined by heredity and environment. What the individual inherits organizes his or her genotype. Exposure to environmental factors such as family members, peers, school, community, and culture engenders modifications of the genotype into what we observe as a person—a phenotype.

Developmental occurrences per se and the resulting patterns at each phase of life set the stage not only for actual behavior but for later growth and adjustment. The self-concept, goals, and aspirations which the individual acquires, highly affected by relationships to parents and peers, gain in motivational power as the years of childhood advance. Most adolescents make significant efforts to free themselves from parental ties and influences, and later from peer pressures, in order to gain autonomy and to increase self-reliance. Many adolescents and young adults join ideological and political movements aimed at restructuring or even destroying the society and culture identified with their parents and the past. This section will deal with the functions and stimulation value of each of these major factors and show their influence on the total growth and behavior pattern of the individual in American society. It also surveys major dimensions of human growth, thereby providing a foundation for a more detailed analysis of the prenatal and postnatal periods of life.

Chapter 3

Factors Determining Growth and Behavior

HIGHLIGHTS

With the breakthrough discovery of the genetic code, the genesis of many hereditary influences is coming to light. The potential for genetic control of human development is thus increased greatly by the scientist's ability to manipulate DNA and RNA molecules, proteins, and amino acids.

Leading environmental influences continue to change rapidly. The dangers of spreading pollution and the energy crisis call for new and extensive controls. Dissolution of nutrients in food manufacturing, harmful additions to foods and drinks, and multitudes of artificial flavors and colors create a need for new curbs. Natural food rich in protein but low in calories is what most people need for their organismic welfare.

Children and adults need strongly bound families for their security and opportunities for personal growth. Maternal and paternal models are both needed for growth of socialization and sexual typing of the young. Violence, not only on the streets at night but also in the mass media, must be substantially reduced in order to ameliorate the social climate.

To a great extent, school is becoming a necessity throughout the life of the individual. A higher diversity of schools is needed to further what parents consider fundamental for the development of their children. Acceleration of scientific and technological advance presses for reeducational opportunities for many adults in professional and managerial vocations.

Social stratification, diversity of life-styles within each stratum, and increase of leisure time produce high group and subgroup complexities, difficult to adjust to for many persons. Is it any wonder that many young people experience conflict and alienation before they accept a definite pattern of adult life and find their niche in this highly urbanized yet impersonal society?

Being a conglomeration of many cultures and subcultures, American civilization includes many polarized traditions and expectations. Growing into this culture is a very complex task for most adolescents and young adults.

Assessment of the causes of and major influences on human growth and behavior in this chapter will facilitate understanding of a variety of patterns of development and adjustment. Recent findings in genetic research as well as leading studies of school, society, and culture will shed new light on these life-determining factors. The next chapter surveys major dimensions of growth and lays the foundation for more detailed examination of the prenatal and postnatal periods of life.

Recent advances in biogenetic research add much to the recognition of heredity as a key factor of human development. The great significance of many environmental factors is obvious but difficult to prove. Most of the purely human powers depend on stimulation from people—from the mother in particular. Growth in emotional, social, and moral sensitivity is determined mainly by others in the early years of life. Development of intellectual abilities is chiefly responsible for specifically human behavior, such as speaking, thinking, and creative self-expression. Among the behavior-organizing forces, emotional experience probably ranks the highest. It is followed rather closely by sex. These two factors determine largely the nature of interpersonal relationships and the satisfactions derived from them.

THE PERSON'S HERITAGE—A KEY TO DEVELOPMENT

Heredity comprises all influences biologically transmitted from parents to the sex cells, which fuse to form the offspring. What people inherit continues to predispose and stimulate them throughout life, because the hereditary code works for life. Heredity is a process, and genetic traits emerge in the course of development. Similar genes have different effects in diverse environments, as do dissimilar genes in similar environments. Genetic determination of the variation in intelligence, for example, does not necessarily mean that people's intelligence is irremediably fixed by their genes (Dobzhansky, 1973, pp. 8–9).

When describing the totality of biological inheritance, the term *genotype* is used, since the organism is a derivative of genes. The genotype, or genetic makeup, provides the basic matrix within which all human behavior is bounded (Thompson, 1972). The changing characteristics of the organism as it develops throughout life constitute the successive *phenotype*. Development is an orderly sequence of phenotypes, but these, in turn, are dependent on the succession of genotypes that result from the functioning of the genetic code. The phenotype represents the visible characteristics of an individual. A child or adult as we see him or her is a phenotype.

How heredity operates within the individual is not yet fully known, although a breakthrough discovery was made more than a hundred years ago. In 1869, Friedrich Miescher discovered the nucleic acids, first from the nuclei of lymphocytes and later from the spermatoza of the Rhine salmon. He isolated what is now designated as DNA, a deoxyribonucleic acid molecule consisting of a chain of nucleotides linked to one another by bridges of phosphate, a polynucleotide that contains a blueprint of the genetic code (Chargaff, 1971).

Progress in discovering the genetic code began with the studies on the chemical nature of the living substance by Oswald T. Avery and his associates, C. M. MacLeod and M. McCarty, in 1944. At the Rockefeller Institute for Medical Research they isolated deoxyribonucleic acid molecules (DNA) located in the nucleus of cells and assessed some of their transforming activities, indicative of genetic information in coded form. Nucleic acid molecules consist of polynucleotide chains of thousands, sometimes millions, of nucleotides in continuous chemical linkage. DNA is the gene structure, which "harbors its genetic information in the form of a precise sequence of the four bases along its polynucleotide chain" (Stent, 1970).

A fundamental genetic unit or gene is a linear array of DNA nucleotides that determines a linear array of protein acids. Genes are polymorphic, since, because of mutations, each gene is capable of functioning in different alleles (forms) with markedly different activity. Most human features and traits, whether physiological, temperamental, or intellectual, are highly polygenic—influenced by many genes in their potential values. On the basis of empirical findings, many indexes of heritability in human beings (as coefficients) have been estimated. For example, weight has a heritability index of 0.78; IQ has a heritability index of 0.81; and the cephalic index is 0.75 (Dobzhansky, 1973, pp. 16–18; Thompson, 1967, pp. 344–345). The heritability index, it may be added, is a population statistic, bound to a given set of environmental conditions at a given point in time (Scarr-Salapatek, 1971). Current scientific evidence does not permit a clear determination of heritability coefficients, and estimates of IQ heritability are subject to a variety of systematic errors (Layzer, 1974).

In 1953 the Crick-Watson model of the DNA molecules as double helixes of polynucleotide chains was formulated. It consists of two interwined DNA strands held together by hydrogen bonds. The deoxyribonucleic acid conveys genetic information from generation to generation. When the double helix separates into two single threads, each of these forms a copy of the original double structure by the proper coupling of its bases (adenine-thymine, guanine-cytosine). As a common source of life, DNA has the same double helical structure and the same four nucleotides but in a varying sequence.

Experimenting on the assumption that the primary genetic information is stored in the DNA molecules, Gall (1958) unraveled part of the genetic mechanism by eliciting certain enzyme actions with tissues of Drosophila (fruit fly) larvae. He found that segments of the chromosomal DNA unwind at specific intervals, transmitting bits of genetic decoding to a messenger strand of ribonucleic acid (RNA), which then carries them through the cell. The RNA molecule consists of a chain of half steps, which are responsible for the formation of enzymes from amino acids. Enzymes "act as catalysts and are vital to the operation of the organism. The reactions that are the very basis of behavior and indeed, of life itself, could not take place in their absence" (McClearn, 1968). DNA sequence specificity and the resultant RNA responses, "right" or "wrong," accelerated or delayed, probably add much to the biochemical individuality of each person. Various forms of behavior are phenotypical features of the maturing organism, largely dependent on central nervous excitatory potential, which, in turn, is limited by the changing amount of catecholamines in corresponding neural centers and mechanisms subjected to specific stimulation (Eibl-Eibesfeldt, 1970).

A major part of the genetic code came to light in 1963, when Francis H. C. Crick published his deductions from biochemical experimentation. In this revised genetic model (1963, p. 416), "the base sequence of the DNA—probably of only one of its chains—is copied onto RNA, and this special RNA then acts as the genetic messenger and directs the actual process of joining up the amino acids into polypeptide chains. . . . One cistron one polypeptide chain." In concluding, Crick implied that it is reasonable to hope that the entire genetic code will be unearthed as additional research designs are completed. Nirenberg's experimental work (1964), confirmed by other experiments (Kornberg, 1968), permits one to draw a model of the genetic code, the arrangement of which, as originally conceived by Crick, is given in Figure 3-1. This breakthrough in the field of genetics is comparable to Planck's formulation of the quantum theory and Einstein's of relativity in physics.

Nearly two decades of intense biochemical research has brought to light many aspects and some details of protein synthesis. An orderly interaction of three classes of RNA—ribosomal, soluble, and messenger—controls the assembly of amino acids into proteins, the prime substances of life. In July, 1968, Sol Spiegelman (University of Illinois) reported the achievement of "test-tube" evolution. As early as 1965 he had succeeded in synthesizing viral RNA. In December, 1967, Arthur Kornberg (Stanford University) reported a cell-free (test-tube) synthesis of the double helix by the stepwise addition of activated nucleotides to the polymer chain—a biologically active phi X 174 virus was produced. This opened the way for synthesis of the DNAs of viruses causing illness in humans and probably of the DNAs of multicellular organisms. Kornberg (1968) reported improved prospects for biochemists to assemble in the test tube not only complex viruses but also some major components of the cell, including components of the human chromosome. He also assumed the possibility of using nonpathogenic viruses to carry into man particles of DNA capable of replacing defective genes. In January, 1969, there were reports of the formation of an enzyme ribonuclease at Merck Sharp & Dohme and at

First \ Second	U	C	A	G	Third
U	PHE	SER	TYR	CYS	U
	PHE	SER	TYR	CYS	C
	LEU	SER			A
	LEU	SER		TRP	G
C	LEU	PRO	HIS	ARG	U
	LEU	PRO	HIS	ARG	C
	LEU	PRO	GLUN	ARG	A
	LEU	PRO	GLUN	ARG	G
A	ILEU	THR	ASPN	SER	U
	ILEU	THR	ASPN	SER	C
	ILEU	THR	LYS	ARG	A
	MET	THR	LYS	ARG	G
G	VAL	ALA	ASP	GLY	U
	VAL	ALA	ASP	GLY	C
	VAL	ALA	GLU	GLY	A
	VAL	ALA	GLU	GLY	G

Figure 3-1 The genetic code (Crick-Watson model). In this table the letters U, C, A, and G represent the four kinds of nucleotides, containing the bases uracil, cytosine, adenine, and guanine. The three- and four-letter abbreviations represent the twenty kinds of protein amino acids. The codon corresponding to any given position on this table can be read off according to the following rules: The base of the first nucleotide of the codon is given by the capital letter on the left, which defines a horizontal row containing four lines. The base of the second nucleotide is given by the capital letter on the top, which defines a vertical column containing sixteen codons. The intersection of rows and columns defines one line within any given horizontal row. The code is very nearly universal—it remained unchanged over a long period of organic and phyletic evolution. [*Source:* G. S. Stent. DNA. *Daedalus*, 1970, **99**, 909–937 (fig. 6, p. 931). By permission.]

Rockefeller University. Before the end of 1969, Harvard researchers announced success in isolating a single gene. The use of amniocentesis, electrophoresis, and ultracentrifugation increases the ability to detect changes and abnormalities in protein structure.

Manifold genetic control of human development is thus open, as scientists can now manipulate DNA, RNA, proteins, and amino acids (McClearn, 1968). In the long run, mutations of a desired kind controlled by medical practitioners will probably stamp out many if not most of about 1,600 diseases of genetic origin (Harris, 1971). Most genetic diseases are rare, but more frequently occurring genetic diseases include cystofibrosis (a Caucasian sex-linked syndrome with low life expectancy), sickle-cell anemia (a Negro disease), hemophilia (internal and external bleeding), phenylketonuria (amino acid accumulation), Down's syndrome (mongoloid retardation), Huntington's chorea (appearing in the thirties), galactosemia (a metabolic disturbance causing increased galactose in the blood), and heart defects.

The flow of genetic information apparently is not unidirectional (from

DNA to RNA to protein), as propounded by Watson and Crick. In 1970 Howard M. Temin reported reverse transcription of genetic information. It requires an enzyme called RNA-dependent DNA polymerase, found in the virion or core of the Rous sarcoma virus. The discovery of reverse transcription provides a new angle for studying the transfer of information from cell to cell and adds plausibility to the theory that viruses may cause human cancer (Culliton, 1971). Apparently a viral DNA adds new genetic material to the afflicted cells, which accounts for their transformation to the malignant state as well as to hyperplasia—uncontrolled proliferation of such cells. In 1974 National Cancer Institute scientists isolated a virus from a woman dying of acute myelogenous leukemia (a rare form of cancer of the white blood cells). The isolation of such viruses opens the possibility of producing cancer-preventive vaccines for at least some forms of cancer (Gallagher & Gallo, 1975).

But Chargaff (1971) suggests caution. Despite a hundred years of accelerating nucleic acid research, not even the actual grammar of the living cell is satisfactorily defined as yet, and several significant breakthroughs are necessary to permit definite control of the processes of life, illness, and death. Chapter 5 of this book deals with how genetic factors operate in the growth of the human organism from conception to birth, and Chapter 13 indicates how the genetic code acts in eliciting pubertal changes.

LEADING ENVIRONMENTAL INFLUENCES

Environmental influences consist chiefly of the intrauterine and physical environment, family, peers, school, community, and culture. The specific nature of the uterine surroundings is discussed in Chapter 5. The other environmental influences affect the growing child, who develops traits and features largely congruous with parental traits but modified in terms of the total family constellation and the external influences to which he or she is also exposed. The milieu in which a person lives is both physical and social. Various physical factors—e.g., climate, location in an open suburb or a congested inner city, size and convenience of the home—continually affect the child in myriad ways. The degree of urban congestion and environmental pollution, for example, affects his or her general welfare, health, and safety.

Environmental Ecology

The earth is both finite and vulnerable. Its rivers, lakes, and oceans as well as its atmosphere are subject to regional and total contamination in ever-rising degree. Current population growth and diminution of many natural resources compound the problem of pollution. Degradation, ugliness, and deadliness ensue when pollution gains the upper hand. Without global reconstructive efforts in the 1980s, the world ecological situation may become desperate, even catastrophic in some parts of the world. As a result, various forms of deprivation will spread, the detrimental effects of which are difficult to estimate. The computer projections of the Club of Rome, based on inputs, outputs, and accumulation or decumulation of people and products, show major catastrophes soon to come. The projections allow for hope, but only if early corrective measures are massively applied—immediate population control and global sharing of goods (Mesarovic and Pestel, 1974). Alvin Toffler (1975) speaks of the "eco-spasm"—the beginning of a major transformation of the total civilization whose dawning symptoms as yet cannot be explained.

The environmental situation is not yet critical, of course, and the world is not running out of energy or oxygen reserves; yet the energy crisis that started with the Mideast war of 1973 engulfed much of the industrial world in energy shortages and inflation spirals. This situation will probably continue until energy from nuclear fusion and geothermal and solar energy come to be produced in large quantities (Landsberg, 1974).

There are many sources of pollution, especially heavy industrial concentrations and the large quantities of dangerous gases emitted by various means of transportation, such as cars, snowmobiles, buses, trucks, boats, and airplanes. The number of pollutants is startling. A multitude of them are noxious and irritating, many are toxic and poisonous, some are mutogenic, and others are carcinogenic. In large amounts, all constitute a debasing hazard with many repercussions on human life. Decrease of the pollution potential of the major sources of pollution is one stratagem to meet this staggering problem of highly industrialized parts of the world. The rising density of population and the industrialization of nonindustrial nations will continue adding its share to the overall pollution level of our planet. Preservation of health may thus become a more serious problem.

Food and Chemicals

Increasing population in most parts of the world raises the question of nutritional sufficiency and the quality of food for human consumption. The basal metabolic rate is high in populations that have a diet of comparatively good quality; it is low in most poorly nourished populations. But some populations apparently thrive on small fractions of the regular minimum requirements in the United States. Adaptive mechanisms to meet nutrient shortages have been acquired in countries with frequent food insufficiencies and famines (Lasker, 1969; Newman, 1970). Recent estimates indicate that well over half the world's population is insufficiently nourished. In technologically undeveloped countries, protein-calorie malnutrition in infancy and early childhood is widespread. In these countries "there are about ten to twenty million young children with severe syndromes of kwashiorkor or marasmus at any one time—most of whom will die without treatment" (Jelliffe and Jelliffe, 1975). The situation is likely to worsen in the future because the population explosion in most undeveloped countries continues unabated (Poleman, 1975).

"Nearly all young children in developing countries experience some malnutrition during the preschool years" (Scrimshaw, 1969). As babies they usually have satisfactory breast feeding for the first four to six months, but breast milk does not provide adequate protein thereafter. Growth is slowed and susceptibility to infection increases. Children experiencing early malnutrition "perform less well on tests of learning and behavior in direct relation to the degree of retardation in height and weight for age" (Scrimshaw, 1969).

The majority of American infants ingest a hypercaloric diet. However, a recent national survey (HANES, 1974) shows that about 95 percent of all preschool children and women of childbearing age have an iron intake well below the standards set by the Food and Nutrition Board of the National Academy of Sciences, although only 10 percent of preschoolers score low on iron. For many persons the hypercaloric diet produces a major problem of excessive weight gain as age increases (Bogert, Briggs, & Calloway, 1973, pp. 473–474).

In the United States the food-manufacturing process unnecessarily reduces greatly the nutritional level of many foods. At the same time many harmful additions to bread, meat, and other food items constitute protracted food pollution. Some additives in the food supply, such as phosphoric or fumaric acid, sodium nitrite, monosodium glutamate, and dehydrated solids, when used in large amounts, have teratogenic effects. Depending on the amount or combination, they may form potent cancer-inciting compounds. Sugar is also extensively added to many foods, even though the human organism does not need it at all. Is there a realistic need for over 1,400 different artificial flavors and colors and many additional nonnutritive sweeteners, some of which are drugs often untested for their possible harmfulness, especially when combined with many other chemicals? Popular aerosal sprays, which contain fluorocarbons, are polluting the homes of most Americans. The increasing concentration of solid fluorocarbons may significantly reduce the ozone layer that shields the earth from excessive ultraviolet rays—another major hazard to plants, animals, and human beings may thus be created.

Fresh air and natural food with nutrient and healthful additives (if any) are what young and old people need to further their organismic welfare—a foundation for total development and adjustment. If optimal environments for each child, adolescent, and adult cannot be fashioned—present knowledge is insufficient for that—at least there is sufficient reason to provide those general requirements which are shared by all people (Eisenberg, 1972).

INTRAFAMILY INFLUENCES

The family is an enduring social group based on marriage and blood relationship, exercising hereditary and environmental influences of prime dimensions on the offspring. As a rule, two parents and their dependent children constitute the nuclear family, most common in the United States. The family's orientation and atmosphere grow out of commitments of husband and wife to each other and their commitments to their children. As a primary group, the family with children is bound together by kinship and intimate relations marked by care, affection, and support, as well as mutual sharing in various activities and concerns. Husband and wife, or father and mother, are crucial members of the family, and the family is considered incomplete or "broken" when either is absent.

The leading functions of the young family encompass (1) providing affection, support, and companionship, (2) bearing and raising children, (3) teaching and transmitting culture, religion, economics, and morals to the young, (4) developing personalities, and (5) dividing and discharging labor within the family and outside. The rapidly changing culture and values of society raise many problems for parents in setting realistic developmental goals for themselves and even for their children.

Parental Roles

The roles a mother assumes include nurturing and care, affection and protection, stimulation and tutoring. The child sees her from a deeply subjective point of view as a gratifier of needs. The physical separation at birth does not entail emotional or social separation. Infancy is a period of nearly total dependence, and childhood of substantial dependence, on the mother. Emotionally and socially the child remains dependent during the preschool and early school years, and to a lesser degree throughout adolescence.

For short intervals the mother may be substituted for by the father and also by nursery personnel and babysitters. However, such substitution for long intervals readily produces maternal deprivation, with many detrimental effects on the child's adjustment, including developmental retardation and regression (Bowlby, Ainsworth, & Rosenbluth, 1956; Caldwell, 1970; Yarrow, 1961). Statistics of 1970 indicate that about 43 percent of all mothers are working outside the home. With slight changes from year to year, this leaves approximately one-third of the children under six years of age (about 5.6 million) without mothers for long periods of time during the day. The working mother provides a different role model from the nonworking mother; the role of the working mother implies multiple mothering, in which the continuity of attachment that the infant needs often breaks, sometimes beyond repair (see Chapter 9, page 177). The infant needs a one-to-one relationship with an adult; without it, he or she suffers affective and cognitive loss (Bronfenbrenner, 1973; Caldwell, 1970; Hoffman, 1974). Probably worse than the mother's absence during the day is the situation of the approximately 15 percent of all children in the United States who are living in single-parent families—usually with their mothers, a third of whom are working full time (*Profiles,* 1970; U.S. Bureau of the Census, 1975).

The psychosocial development and emotional security of the child are based on the mother's reasonably consistent, patient, and tender care. Each year of full-time mothering tends to pay large dividends to both mother and children. The effectiveness of the mother's role is based on her affection and capacity for nurturing, stimulation, and affiliation, all of which bind the family together and enhance its development. "There is no other object in the child's universe that can attract and hold his attention the way a mothering adult can.

If attention has not been trained by interest in such a person, other objects have little chance" (Arnold, 1960, p. 213).

The basic developmental tasks of motherhood include (1) learning how to care for her child with competence, (2) providing opportunities for the child's development, (3) making satisfactory adjustments to the practical realities and pressures of family life, and (4) reconciling conflicting conceptions of the roles family members assume (Rollins and Feldman, 1970). Each task represents a challenge to many women entering motherhood.

The *father's* role begins with providing satisfactory companionship to his wife by meeting her emotional, social, and sexual needs. In most young families this includes the desire to have or at least to accept children. In addition to the traditional functions of protection and breadwinning, which are often shared with the wife, the father's role extends into various aspects of child management and education. The father usually shoulders a major part of discipline and arbitration as children grow. By his authority and discipline he stimulates their reality orientation. A mother, by her tendency to meet and gratify the child's desires and wishes, usually furthers the pleasure motivation of her children. Studies of paternal deprivation suggest a critical period when the father is very important to the child, lasting from the time of weaning until school entry. Identification with the father in later childhood is a major factor in boy's successful psychosexual typing as a male (Lynn, 1969; Nash, 1965).

A participating and active father adds much to the child's reality orientation by imposing tasks and chores on each member of the family. He teaches children to deal with life realistically. By unjust treatment a dominant but cruel father presses his children toward hostility, even delinquency, while a weak and passive paternal figure is often revealed in the case histories of homosexuals and schizophrenics. An inadequate father often sees children as intruders or competitors for the mother's attention and love. An adequate father contributes greatly to parental discipline, which offers children the experience necessary for internalizing the balance between expressing and controlling needs, emotions, and desires (Hoffman, 1975).

The majority of American families are probably too individualistic to be either patriarchal or matriarchal. In American society the nuclear family is largely egalitarian and uninhibited in the husband-wife relationship (Yorburg, 1973, pp. 124–127). In a minority of families, however, the mother becomes recognized as the decisive and leading person. Often it is the husband's passivity or insecurity that facilitates dominance by the wife and mother. The family situation becomes unsound if she takes advantage of her dominance.

In an increasing number of families, a nearly egalitarian leadership pattern becomes established, often with zones of dominant influence by either husband or wife and a common agreement on matters of significance to the total family. The general flexibility of young parents and the changing ages of their children produce many subtle shifts in the father-mother pattern of dominance and submission (Hobart, 1963).

Child Management

It is desirable for the father to share with the mother a common philosophy of child management. This promotes consistency and facilitates the child's adjustment to the parental approach. During the last few decades the dominant voice in child management has been that of psychoanalysts and progressive educators, who emphasized gratification of children's affectional needs and stressed facilitation of social adjustment and freedom from constraints and punishment. Benjamin Spock (1945/1968) became the chief spokesman for heightened permissiveness to prevent any possible frustration of the child. Instant gratification of children's desires became prevalent and paramount, while preparation for flexibly meeting a greatly changing society and culture, with its challenges and crises, was neglected. This is primarily a trend among middle-class parents; but as Becker's reinterpretation (1964) suggests, a great variety of parental management techniques continue to occur in most socioeconomic segments in the United States. Indeed, "parent-child relationships as they exist at any point in time are a function of so many factors that they almost defy explanation" (Walters & Stinnet, 1972).

In categorizing parental discipline by factorial analysis, Becker (1964) identified three major dimensions: restrictiveness versus permissiveness, warmth versus hostility, and calm detachment versus anxious emotional involvement. Different types of parents can be classified by combinations of these traits and their levels. For example, the anxious-neurotic parent is expected to rank high in hostility and in permissiveness, while the organized-effective parent will rank high in warmth and restrictiveness. Figure 3-2 shows Becker's model of parental disciplinary behavior.

The effects of parents on children are paramount. Much of what children inherit comes from their parents. The prenatal condition reflects the state and experiential currents of the mother. Parental personality traits and attitudes mold children's behavioral tendencies. For children feedback is an essential provider of what is acceptable to parents.

The self-concept is influenced greatly by parental models. Poor parent-child relationships are related to the child's aggressiveness and disciplinary problems in school. Affection, acceptance, and warmth are related to good adjustment and high academic achievement, as well as to creativity and leadership (Walters & Stinnett, 1972).

The child's exposure to parental influences declines slowly. The decline becomes accelerated only as preadolescent years pass and the child is emancipated socially and emotionally from the parents by subjection to peer and mass media influences to a much higher degree. In contrast to this notion of high child dependency on parents, Leon Eisenberg (1972) assumes that the human being is his own chief product from the early stages of life:

> The infant who discovers that he can control the movements of his own fingers transforms himself from observer into actor. The child who masters reading

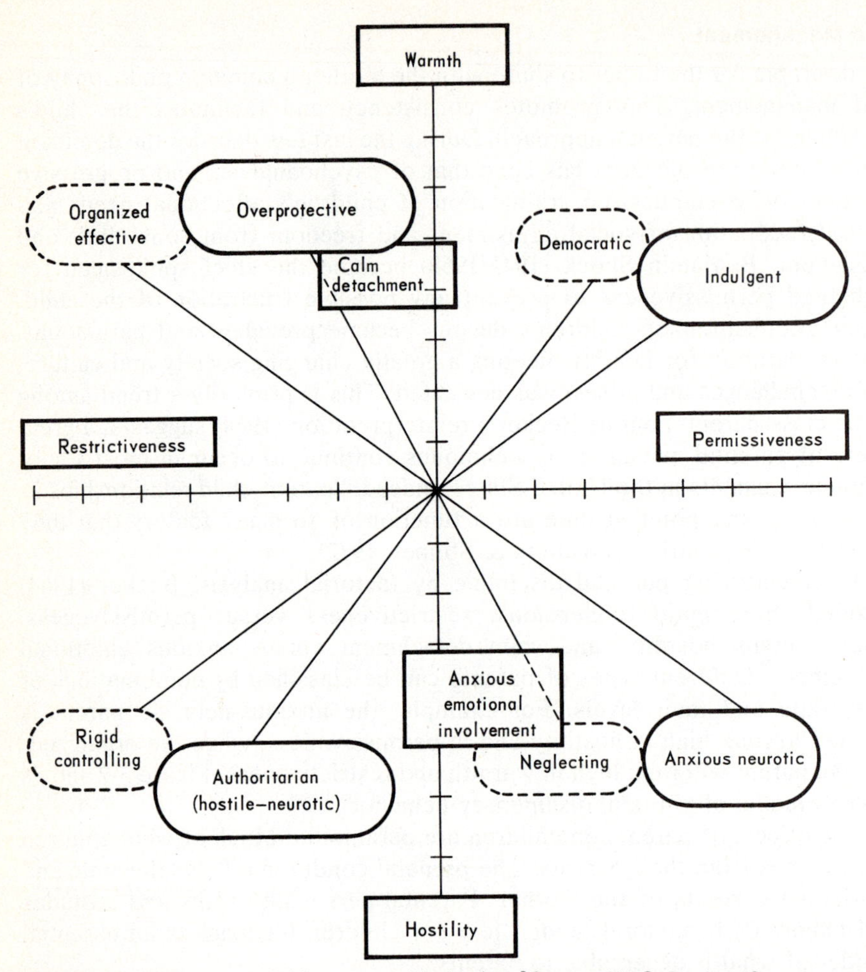

Figure 3-2 Becker's model for parental behavior. [*Source:* W. C. Becker. Consequences of different kinds of parental discipline. In M. L. Hoffman & L. W. Hoffman (Eds.), *Review of child development research* (Vol. 1). New York: Russell Sage Foundation, 1964, pp. 169–298. By permission.]

unlocks the treasury of the world's heritage. The adolescent who insists upon a critical reexamination of conventional wisdom is making himself into an adult. And the adult whose concerns extend beyond nation to mankind has become fully human. (Copyright 1972 by the American Association for the Advancement of Science.)

Television

American society and culture provide the general matrix and atmosphere in which children and parents live. Means of mass communication penetrate deeply into each family and reduce greatly the privacy of the home. Television is one of the most effective means of mass communication. It provides educational stimulation, but it also shows children many examples of violence and crime in realistic and compelling presentation. Consciously or not, the ideas communicated are transmuted into attitudes and desires to become operative in behavior (Jester, 1969; Stein, 1972). Another major effect of television is overstimulation, which produces "emotional blunting" and polymorphous desensitization, resulting in feelings of boredom so that lesser stimulation is viewed as meaningless (Cline, Croft, & Courrier, 1972). For many young children televised aggression reduces inhibition against similar aggressive acts. Surprisingly, children remember many details of depicted violence, and violent acts by humans elicit very powerful emotions of fear (Osborn & Endsley, 1971).

The effectiveness of film messages is well attested by findings of many experimental studies. A study by Alfred Bandura and Dorothea and Sheila Ross (1963) of 96 preschool boys and girls showed that the observation of models portraying aggression on film substantially increased the probability of aggressive reactions to subsequent frustrations: "*Ss* who viewed the aggressive human and cartoon models on film exhibited nearly twice as much aggression [as did] *Ss* in the control group." Similar experiments conducted by R. H. Walters and D. C. Willows (1968) confirmed these findings. Children's preference for violent television programs at ages eight and nine influenced their current aggressiveness and their aggressiveness ten years later. Eron et al. (1972) provide some evidence to suggest that regular viewing and liking of violent television leads to a more aggressive life-style for boys but not for girls.

Daily observation of blood and gore desensitizes children so that their empathetic reactions to the pain and suffering of others are blunted. Reduction of emotional responsiveness is a serious loss to anyone; it furthers dehumanization (Landsman, 1974). In contrast, the skillful combination of artistic creativity and knowledge about human development elicit intellectual curiosity and positive human relations. It would seem that this potential has been realized in several television programs, such as "Sesame Street," "Misterogers' Neighborhood," and "The Friendly Giant" (Stein, 1972).

Certain commercial television programs and many educational ones contribute much to lifelong human learning. By word and illustration, they expand the viewer's reservoir of ideas and provide powerful incentives for new interests and activities. "Sesame Street" shows the penetrating power of educational television. It was intended primarily for preschool disadvantaged children at home who were without the benefit of Head Start or similar educational experiences. Ball and Bogatz (1972) report that frequent viewers of "Sesame Street" made relatively large gains at ages three to five.

Before the typical child begins schooling, he or she has spent an estimated 3,000 hours watching television (McGovern, 1970). By the time the average American student leaves high school, he has watched 15,000 hours of television. With this, he has seen 500 movies in theaters. In comparison, he has spent 10,800 hours in school (Jester, 1969). Viewing long evening television shows contributes to listlessness in school and in the long run produces the "tired-child syndrome," also marked by high irritability, poor appetite, and other less obvious detrimental effects. How can the child grow in sensitivity and empathy toward others when daily television makes cruelty and violence exciting and attractive as means to solve problems? Is not a large portion of the child's viewing time a school for sadism and delinquency? The need for radical changes in television programming would appear to be a necessity, given the evidence suggesting that television viewing can have at least some effect on development.

Peers as Models

Most persons enjoy associating with those who are similar in age, maturity, and status. Peer interaction usually gives much gratification and need satisfaction from middle childhood to the years of late adulthood. From about the age of four, there is a growing need to meet and play with others of approximately the same age. Once the child learns to identify with those of his or her own age, parental identifications lose some of their earlier strength. Peers modify the child's thoughts, feelings, and aspirations as he learns to give and take, to wait for his turn, and to win and lose gracefully (Hartup, Glaser, & Charlesworth, 1967).

Peers, next to parents, are the most significant individuals in a person's life. In them one finds many qualities and traits for self-identification, because, as Harry S. Sullivan (1953, p. 22) puts it, "the peculiarity exists that one can find in others only that which is in the self."

EDUCATION FOR LIFE

The home is probably the most powerful educational institution, even when parents are not good teachers. Parents control most of the prime time during the critically formative years of childhood (Goodlad, 1973). During the preschool years, the child develops rapidly in all aspects of behavior, forming many enduring traits and attitudes and setting directions for his future behavior. Most major patterns of behavior are already functional by the time a child enters kindergarten or first grade. In many respects preschool children seem to follow Bruner's dictum: "The foundations for any subject may be taught to anybody at any age in some form." More recently, Bruner (1973, pp. 125–131) speaks about the learning that must first be acquired as a prerequisite, owing to the complexity of certain objects; and also about the constraints to early learning inherent in the nervous system, which is severely limited in the amount of information it can process at any time.

There is some evidence to suggest that many homes do not provide sufficient educational stimulation even during toddlerhood, not to speak of the ages from four to six. When children fail to attain the proper foundations for further education, they are likely to fail in school. Consequently there is a growing conviction that all children should be educated in areas of profound importance for later education and adjustment (Ainsworth, 1973; Bronfenbrenner, 1972; White, 1972). Well planned early education is expected to have a positive domino effect resulting in desirable changes at all subsequent educational levels (Anderson & Shane, 1972).

Kindergarten helps many parents to increase their children's readiness for school. It also furthers the child's socialization with peers—a form of assistance many children from lower socioeconomic levels need. Figure 3-3 shows the percentages of children enrolled in 1965 and 1970 according to family income; the lowest enrollment is in the lowest-income group. Many children in the lower income brackets are socially and culturally disadvantaged and greatly in need of regular preschool training to fill gaps in their cognitive and social experience so that they can respond properly to the demands of school.

At present, school *readiness* means possession of the qualities, traits, and skills by which the child adjusts to the regular requirements of the first grade. He is undisturbed at being away from home, relates well with other children, and shows no difficulties when the three R's are introduced; indeed, he readily accepts them as part of school activity. Many six-year-olds are ready for the first grade; many others are not ready but are required to attend because compulsory education laws arbitrarily set this age for school entrance. School learning is more than a mere transfer or extension of early home learning. Literacy training alone would be a serious burden for most parents.

The school's basic commitment centers on the cultivation of intellectual power. Socialization, group activities, and psychosexual adjustment are other major concerns of the school system. Provisions for play and sports, for creativity and self-expression form a major segment of the American commit-

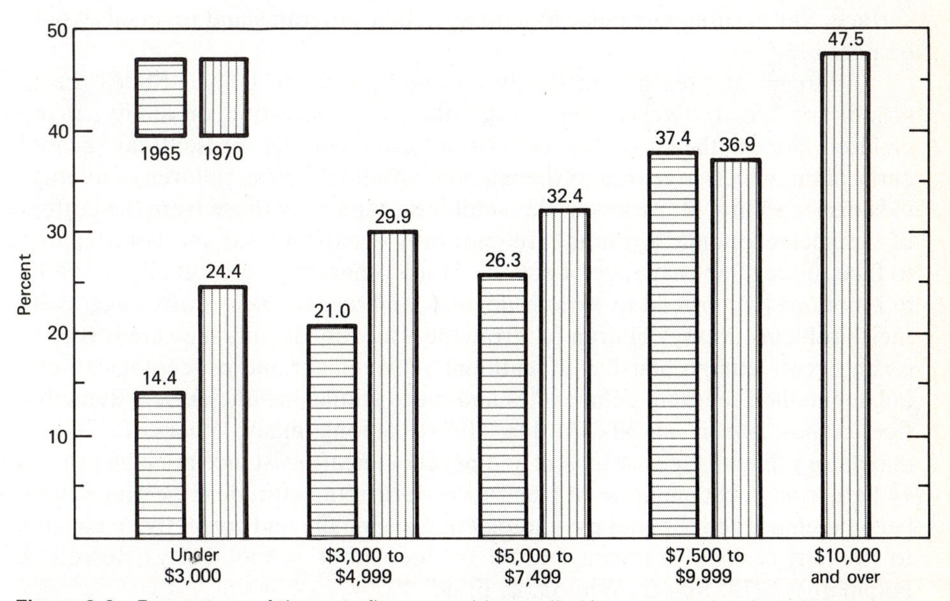

Figure 3-3 Percentage of three- to five-year-olds enrolled in nursery schools and kindergartens, by family income, 1965 and 1970. [*Source:* U.S. Office of Education. *Preprimary enrollment.* 1971 (annual).]

ment to children. Ungraded open schools utilizing the resources of the community are coming to the fore. At present many school systems are moving from ability subgroups to individualized instruction in order to help each child reach the educational objectives of which he or she is capable. A major step beyond individualization of instruction is suggested by Gerald R. Smith (1970), who feels that "the teacher must be engaged in turning out originals." Smith urges recreating most schools to serve uniquely human purposes, rather than the homogenization of the schoolchild that presses him to become a mass child. Fostering uniqueness is an emergent school function—a great challenge to the present school.

Shane (1973, pp. 386–394) suggests a model of an uninterrupted lifelong educational continuum, because "learning begins at birth and continues throughout life." Figure 3-4 presents Shane's model of an "infrastructure that seems consistent with what we know of learning and development and which educators should be able to implement with the resources and the talent to be found in the present educational community." The child's direct contacts with school would begin shortly before or after his or her second birthday. From then on the school would begin to collect and to analyze data from cognitive and physical examinations, background information, and the like, thus beginning to maintain computerized cumulative records for each child. Schooling with emphasis on developmental experiences probably should begin no later than age three and should be based on a *personalized* program of each learner's needs and potentialities to avoid any form of academic Procrustean bed. Corrective and preventive measures would be applied when potential problems surface. Shifts from curricular to paracurricular programs and reversals would be possible.

Although at present practically all children in the United States are in school for ten to twelve years, regardless of motivation or ability, many children are neither capable of nor adjusted to the present-day school curriculum, which is geared to the teaching of middle-class children of average to superior ability. Many lower-class children, especially those from the bottom of the socioeconomic pyramid, are culturally disadvantaged and not prepared to face the regular challenge of school. Many others are perceptually, socially, or emotionally deprived to various degrees, and regular classes often aggravate their problems. Many children fall from the educational cliffs they are forced to reach before comprehensive educational achievement and psychological testing define their learning deficiencies and motivational insufficiencies. Remediation comes late if at all. Cardon (1972), among many others, feels that mandatory diagnostic assessment and psychological assistance must be provided before any accumulation of frustration occurs and attitudes rejecting school become embedded. School psychology has diagnostic and corrective measures to prevent serious problems before or soon after school entry (Bessell & Palomares, 1970; Singer, Whiton, & Fried, 1970).

Any comprehensive educational program must involve parents, since most mothers and fathers need assistance in learning how to be teachers and

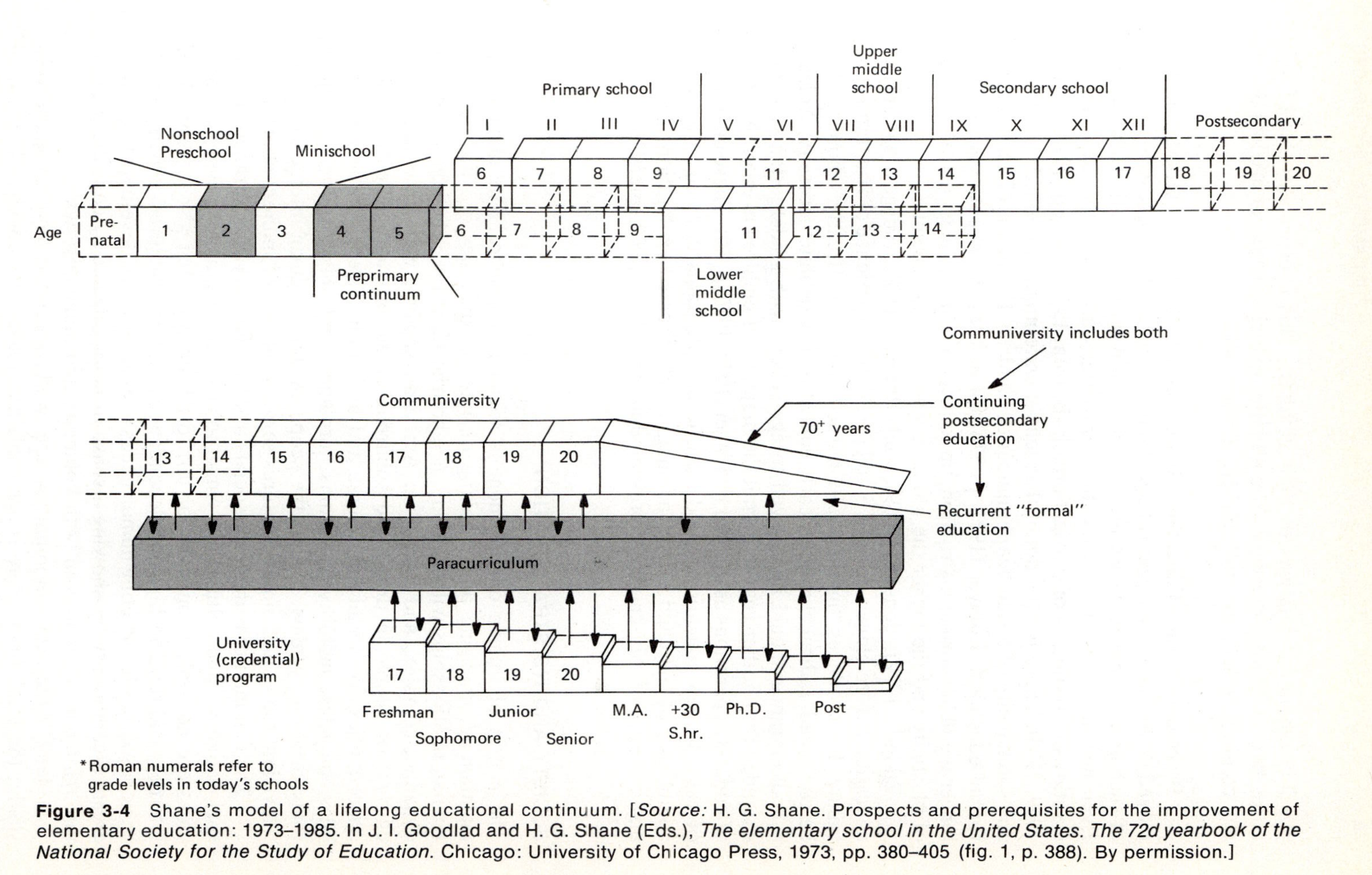

Figure 3-4 Shane's model of a lifelong educational continuum. [*Source:* H. G. Shane. Prospects and prerequisites for the improvement of elementary education: 1973–1985. In J. I. Goodlad and H. G. Shane (Eds.), *The elementary school in the United States. The 72d yearbook of the National Society for the Study of Education.* Chicago: University of Chicago Press, 1973, pp. 380–405 (fig. 1, p. 388). By permission.]

possibly emotional catalysts to their children. This is especially true of any corrective program, otherwise "the situation which creates an original deficit will continue to take its toll" (NIMH, 1970, p. 88).

Decreasing Formal Education In our century the school entrance age has moved downward to six years of age. Men of great erudition question the value of such an early elementary education, indeed of formal education at all levels. Even new and innovative educational programs are currently under serious scrutiny or are targets of outright attack (Binzen, 1972; Haskett, 1973; Jensen, 1969, 1973). There is little doubt that school undermines the *joie de vivre,* zest, and spontaneity of a significant number of children. In support of a late start of formal education, Rohwer (1971, p. 336) formulates the hypothesis that "the longer we delay formal instruction, up to a certain limit, the greater the period of plasticity and the higher the ultimate level of achievement."

Elkind (1972), Jensen (1969), and Rohwer (1971) suggest no formal education before eleven or twelve, to prevent the experience of frustration and failure that produces "intellectually burned" students. If early schooling is to be continued, radical change of early education is a definite necessity. At least school entrance should be delayed to age seven, with kindergarten activities extended for another year to permit the majority of children to enter the phase of concrete operations as defined by Piaget (1970, 1973, p. 30). These opponents of early formal education would then accept programs devoted to enriched stimulation of the child and broadening of self-directed exploration. The role and influence of the school are presented in more detail in Chapters 10, 11, and 17.

SOCIETY AND ITS STRATIFICATION

The matrix of human development is social in character. The making and maintaining, changing and breaking of interpersonal ties are daily affairs. Human relationships within family, neighborhood, school, and community are dynamic, complex, and unfolding. American society, with its concentrated clusters of people in urban and suburban areas, provides almost unlimited opportunities for social intercourse and affiliation. Telephones and automobiles, bicycles, motorcycles, trains, airplanes, and ships, as well as public transportation systems, extend the scope of direct human interaction, while television, radio, and motion pictures serve as means of indirect contact.

The community is the large social structure that serves as a framework for socioeconomic and cultural life. Small towns, suburbs, and segments of large cities are not only geographic or territorial locations but also a network of interpersonal relationships and dependent economic structures, which provide basic resources for meeting human needs at every level of development. People create a community by showing a sense of belonging and by contributing to it through labor and money in both voluntary and mandatory ways.

Many American communities are well integrated and comparatively stable; the total atmosphere is agreeable and reasonably consistent over long periods of time. The directive and corrective influences are sound and extensive. Yet most cities have districts in which the rate of migration is high, disorder frequent, and the sense of belonging weak for most inhabitants. In such places, the morale of people is low and delinquent elements are usually strong (Deutsch, 1973), so that the family receives little assistance in furthering a child's development, from perceptual figure-ground organization to conceptualization of civic obligations.

Socioeconomic subcultures and classes, usually related to ethnic background, have been distinguished throughout the centuries in the United States and elsewhere. There are Jewish, Indian, Negro, Chinese, Chicano, Irish, and Puerto Rican groups, among many others, sharing a particular culture or its distinct versions. *Social class* refers simply to an aggregation of persons and families having approximately the same social status in a society. It denotes a basic stratification of society in terms of similarities along the continuum of the economy, with unequally ascribed prestige and privilege. Basically, class levels are defined in terms of variations in economic assets, education, employment, and residence. Educational attainment and external variables of personality have always played a role in determining prestige status in American society. Social status results from the arrangement of groups of people on a comparative scale, in terms of social distance and prestige as well as of reciprocal rights and duties (Bossard & Boll, 1966, pp. 278–283).

Several objective measures have been devised to establish the socioeconomic status of an individual or family. The Index of Status Characteristics (Warner, Mecker, & Eels, 1960) assigns points for such factors as type of work, income resources, neighborhood, and type of home. For example, a laboratory technician who depends on his salary as a source of income and lives in an apartment building situated in an average residential neighborhood would be placed in the lower middle class.

In American society there is more mobility in the socioeconomic class system and among classes than in most parts of the world. A person born into a low socioeconomic stratum has ample opportunity for change through success in education, athletics, the arts, and business. It is the popular opinion that any man's son can become President of the United States.

Traditionally, three social-class levels are defined: upper, middle, and lower. In most studies each social class is further divided into an upper, a middle, and a lower stratum—for instance, the upper middle class and the lower middle class. General estimates place about 4 percent of the American population in the upper class, about 40 percent in the middle class, and 56 percent in the lower class. In 1975 the population of the United States was made up of about 86 percent Caucasoid, 12 percent Negroid, and 2 percent Mongoloid people (American Indians and Hawaiians are included as Mongoloids). Classes are not only socioeconomic groups but also breeding populations, with more or less separate gene pools from other classes. To a large degree the socioeconomic level is a function of the genetic constitution of its population (Dobzhansky, 1973, pp. 25–26; Eckland, 1971).

The child-management techniques of each of these classes have been analyzed (Becker, 1964; Deutsch, 1973; Kohn, 1968). Lower-class parents tend to be more severe in toilet training, and they resort more often to corporal punishment, than parents in the middle class. Children in the lower class tend to be physically more aggressive and more independent and to have earlier sex experiences. Middle-class parents tend to be more controlling and demanding as parents. Middle-class mothers feel responsible for the behavior of their children and utilize more subtle controls, such as reasoning and appeal, as well as withdrawal of love and attention when the child's behavior falls below the standard set. Many of them are ambitious people who aspire to attain higher status and press their children to raise their status by continuing education and professional training. Middle-class parents regard child rearing as more problematic than working-class parents (Kohn, 1968). Upper-class parents engage in more leisure activities, have been educated in colleges of high reputation, and have usually developed their esthetic appreciation to a higher level than the other classes. Their children are usually encouraged to be self-reliant and to assume a *manipulative* stance toward all aspects of reality. Children of college-educated parents tend to show greater originality and higher scores in verbal creative production than children of less-educated parents (Sheldon, 1968).

The clusters of people in metropolitan and large city areas contribute heavily to the formation of mass society, which, in turn, merges into a mass culture. This process of *homogenization* is accelerated by the dominance of a single educational system and ubiquitous exposure to the competing but similar mass media of communication and entertainment. In the formation of mass culture, diverse political, religious, ethical, and esthetic values and related attitudes are molded one way; the majority of people feel and think almost alike. Since the values and feelings are not deeply anchored, waves of instability occur, as shown by high susceptibility to changing fashions and fads (Wilensky, 1964).

Racial integration of schools and equalization of opportunity for all Americans is a significant factor in blending differences among social classes and in reducing prejudice against blacks and other minority populations. Education for emotional and social maturation is needed to reduce the social conflicts and tensions the current society is subjected to. However, it is worth considering that in a completely open society with full equality of opportunity the future of most children can be as accurately predicted from the status of the biological parents as in a caste society. Most upper-class people remain in the upper class, and most lower-class people circulate within their class for life (Eckland, 1971; Gerbner, 1974; Jencks et al., 1972).

CULTURE

Fundamentally, *culture* refers to the total patterns of a people's way of life seen in terms of artifacts and achievements distinguishing large but similar societies. It encompasses the technology and civilization, law and morality, religion, politics, arts, and recreation, and the training and educational facilities of any specific period of history. Traditional customs and mores and current fashions and fads also constitute part of it. Ruth Benedict's dictum defines culture as "that which binds men together."

Since culture is an end product of centuries of human endeavor and creativity, it cannot be easily revised or abandoned; it maintains a high degree of continuity from generation to generation. Cultural traits, norms, and expectations influence everyday living and special occasions alike. They include not only education, marriage, and family patterns but also feeding. eating, and recreational habits, as well as acceptable and unacceptable means of expressing affection and aggression, sympathy and hostility. Culture focuses on social graces, customs, morals, beliefs, and traditions and on the roles various individuals are expected to play.

With their uniquely structured experiences and personalities, people fashion and modify many aspects of culture but are themselves highly affected by cultural traits and pressures. Scientific breakthroughs further technology, artistic creativity enriches society, political ingenuity changes the laws—and all modify culture. Only pervasive revolutionary destruction disturbs the existing culture seriously by producing a regression that often serves as a basis for its reorganization into a new culture.

American culture is a version of Western culture marked by superior technology, high industrialization, pragmatism, and a rapid rate of change in speech patterns, styles of life, fashions, and fads. Sex and mastery are viewed as leading positive valences. Pursuit of uninhibited pleasure and commission of various forms of aggression and misdemeanor are tolerated to a high extent. Many children are subjected to parental pressures to make the most of themselves. The drive for monetary gain is strongly reinforced by commercial emphasis on economic opportunity and a need for affluence. Middle-class people feel that they are failing unless they are investing or their bank accounts are steadily growing.

American society is like a marathon race in which those in the lead are fearful of those gaining at their heels, while those behind are overtly envious of, if not hostile toward, those ahead. Consequently a large percentage of men and women are tense, jittery, emotionally blunted, and often unsure of their sexual capacity and appeal. Drugs are excessively used to reduce tension and to depress uncomfortable feelings.

Optimism for the future and competence in the present are notable features. Some popular sayings become deeply imprinted in young minds ("you can do anything if you try hard enough"). A dictum attributed to the Seabees describes well this attitude of readiness for technological mastery: "The difficult we do at once, the impossible takes a little longer." Hospitality, helpfulness, and generosity are also marks of the majority of American people, despite the impersonal metropolitan type of society, which tends to encourage detachment and exclusive interest in oneself, one's family, or one's peer group.

Current American culture appears to permit the killing of the unborn, it dislodges the aged, it reveres the child, but it is unduly upset by the teenager's antics. Praise and reward are used as dominant incentives. Vigorous commitment is demonstrated to the rights and dignity of the individual, personal privacy, organizational liberty, and equal opportunity for all—for whites in the past, and for most citizens at present as discrimination continues to decline. Subcultural and countercultural diversification furthers creativity and group esprit de corps but also deepens disorganization of traditional values and mores.

The establishment reinforced by the industrial society presses people into a mold of unidimensionality (Marcuse, 1964) and consciousness II (belief in meritocracy, technology, planning, and the corporate state). A rise to consciousness III—a solution Reich (1970, pp. 217–251, 281–287) suggests—is marked by a general liberation from existing bounds. A person becomes true to himself, open and gentle, and gains in the ethical and spiritual element—one's true potential as a human being. Since American culture is a fusion of many diverse cultural elements, various conflicting tendencies are likely to arise and flare up time and again. The shortness of this presentation of the highly complex structure of American society and culture necessitates the caution that each generalization made not only is tentative but also has notable qualifications.

This chapter has presented genetic, environmental, and related influences. The genetic influences interact with a host of environmental variables, and all of them work together in determining the patterns of growth and behavior of each individual. Parental and peer models are needed for socialization and identity formation. Preschool and school education contribute heavily toward the enculturation of the individual. The culture of the home, the street, and the school provide an atmosphere within which the individual unfolds his or her capacities and becomes a contributing member of the community. While the genetic heritage cannot be abandoned, an individual either fashions his environment in terms of his ideas or moves away to search for a more congenial milieu. Modern American life is full of conflict-producing influences and ensuing struggles. It is not an easy task to grow toward maturity or to age gracefully in a civilization that is full of unsettled issues and contradictions.

QUESTIONS FOR REVIEW

1. What is DNA, and what are its chief functions? How is the genetic code related to DNA?
2. What are the major parental influences on children? Analyze two or three of them.
3. Identify and explain the leading functions of the modern family.
4. Define the father's role and his major functions in the nuclear family.
5. Give some reasons why children need male and female models in the family.
6. With what persons and groups does the older child readily identify himself?
7. What are some of the desirable and detrimental effects of television viewing for children?

8 What are the major functions of the school and of the teacher?
9 In what significant ways do society and social class influence personal behavior?
10 Define *culture* and critically assess several leading traits of American culture.

REFERENCES

Selected Readings

Bogert, L. J., Briggs, G. M., & Calloway, D. H. *Nutrition and physical fitness* (9th ed.). Philadelphia: Saunders, 1973. A specialized source on various aspects of nutrition for infancy and childhood; includes discussions of energy needs, proteins, vitamins, calcium, metabolism, and food preparation and of overweight, diet, and related problems.

Broderick, C. B. (Ed.) *A decade of family research and action.* National Council on Family Relations, 1972. A compilation of fourteen articles reviewing important aspects of the family field published in 1970 and 1971.

Dobzhansky, T. *Genetic diversity and human equality.* New York: Basic Books, 1973. A basic source on genetic influences and subsequent patterns of life-style.

Duvall, E. M. *Family development* (4th ed.). Philadelphia: Lippincott, 1971. A comprehensive and well-written source on family life, stressing developmental tasks and patterns of interaction at different age levels.

McBride, A. B. *The growth and development of mothers.* New York: Harper & Row, 1973. A critical examination of the woman's point of view on growth into motherhood; woman's qualities and traits and oedipal conflict are included.

Thoresen, C. E. (Ed.) *Behavior modification in education.* Chicago (distrib., University of Chicago Press), 1973. Part I of the 72d Yearbook of the National Society for the Study of Education. Includes theoretical and practical exposition of various behavior modification techniques applied to normal, retarded, and other groups of children.

Specific References

Ainsworth, M. D. S. The development of infant-mother attachment. In B. M. Caldwell & H. N, Ricciuti (Eds.), *Review of child development research* (Vol. 3, pp. 1–94). Chicago: University of Chicago Press, 1973.

Anderson, R. H., & Shane, H. G. Implications of early childhood education for lifelong learning. In I. J. Gordon (Ed.), *Early childhood education,* pp. 367–390. (71st Yearbook of the National Society for the Study of Education, Part II.) Chicago: University of Chicago Press, 1972.

Arnold, M. B. *Emotion and personality,* Vol. 1, *Psychological aspects.* New York: Columbia University Press, 1960.

Avery, O. T., MacLeod, C. M., & McCarty, M. Studies on the chemical nature of the living substance including transformation of pneumococcal types. *J. exp. Med.,* 1944, **79,** 137–158.

Ball, S., & Bogatz, G. A. Summative research of Sesame Street: Implications for the study of preschool children. In Ann D. Pick (Ed.), *Minnesota symposia on child psychology* (Vol. 6, pp. 3–17). Minneapolis: University of Minnesota Press, 1972.

Bandura, A., Ross, D., & Ross, S. A. Imitation of film-mediated aggressive models. *J. abnorm. soc. Psychol.,* 1963, **66,** 3–11.

Becker, W. C. Consequences of different kinds of parental discipline. In Martin L. Hoffman & Lois W. Hoffman (Eds.), *Review of child development research* (Vol. 1, pp. 169–183). New York: Russell Sage Foundation, 1964.

Bessell, H., & Palomares, U. *Methods in human development.* San Diego: Human Development Training Institute, 1970.

Binzen, P. Philadelphia: Politics invades the schools. *Saturday Review,* Feb. 5, 1972, pp. 44–49.

Bogert, L. J., Briggs, G. M., & Calloway, D. H. *Nutrition and physical fitness* (9th ed.). Philadelphia: Saunders, 1973.

Bossard, J. H. S., & Boll, E. S. *Sociology of child development* (4th ed.). New York: Harper & Row, 1966.

Bowlby, J. N., Ainsworth, M. B., & Rosenbluth, D. The effects of mother-child separation: A follow-up study. *Brit. J. Med. Psychol.,* 1956, **29,** 211–249.

Bronfenbrenner, U. Is 80% of intelligence genetically determined? In U. Bronfenbrenner (Ed.), *Influences on human development,* pp. 118–127. Hinsdale, Ill.: Dryden, 1972.

Bronfenbrenner, U. Early deprivation in monkey and man. In U. Bronfenbrenner (Ed.), *Influences on human development,* pp. 256–301. Hinsdale, Ill.: Dryden, 1972.

Bruner, J. S. *The relevance of education.* New York: Norton, 1973.

Caldwell, B. M. The effects of psychosocial deprivation on human development in infancy. *Merrill-Palmer Quart.,* 1970, **16,** 260–277.

Cardon, B. W. School psychology for the total school. *Professional Psychol.,* 1972, **3** (1), 53–56.

Chargaff, E. Preface to a grammar of biology. *Science,* 1971, **172,** 637–642.

Cline, V. B., Croft, R. G., & Courrier, S. Desensitization of children to television violence. *Proc. 80th Ann. Convention APA,* 1972, **7,** 99–100.

Crick, F. H. C. On the genetic code. *Science,* 1963, **139,** 461–464.

Culliton, B. J. Reverse transcription: One year later. *Science,* 1971, **172,** 926–928.

Deutsch, C. P. Social class and child development. In B. M. Caldwell & H. N. Ricciuti (Eds.), *Review of child development research* (Vol. 3, pp. 233–282). Chicago: University of Chicago Press, 1973.

Dobzhansky, T. *Genetic diversity and human equality.* New York: Basic Books, 1973.

Eckland, B. K. Social class structure and the genetic basis of intelligence. In R. Cancro (Ed.), *Intelligence,* pp. 65–76. New York: Grune & Straton, 1971.

Eibl-Eibesfeldt, I. *Ethology: The biology of behaviors* (E. Klinghammer, Trans.). New York: Holt, 1970.

Eisenberg, L. The human nature of human nature. *Science,* 1972, **176,** 123–128.

Elkind, D. "Good me" or "bad me"—The Sullivan approach to personality. *New York Times Magazine,* 1972, Sept. 24, p. 18.

Eron, L. D., Lefkowitz, N. N., Huemann, I. R., & Walder, L. O. Does television violence cause aggression? *Amer. Psychologist,* 1972, **27,** 253–263.

Gall, J. G. Chromosomal differentiation. In W. D. McElroy and B. Glass (Eds.), *Chemical basis of development,* pp. 103–135. Baltimore: Johns Hopkins, 1958.

Gallagher, R. E., & Gallo, R. C. Type *C RNA* tumor virus isolated from cultured human acute myelogenous leukemia cells. *Science,* 1975, **187,** 350–353.

Gerbner, G. Teacher image in mass culture: Symbolic functions of the "hidden curriculum." In D. R. Orson (Ed.), *Media and symbols: The forms of expression, communication, and education,* pp. 470–497. (73d Yearbook National Society for the Study of Education, Part I.) Chicago: University of Chicago Press, 1974.

Goodlad, J. I. The elementary school as a social institution. In J. I. Goodlad & H. G.

Shane, *The elementary school in the United States.* (72d Yearbook National Society for the Study of Education, Part II.) Chicago: University of Chicago Press, 1973.

HANES. *Preliminary findings of the first Health and Nutrition Examination Survey, United States, 1971–72.* Washington, D.C.: National Center for Health Statistics, 1974.

Harris, M. Mutagenicity of chemicals and drugs. *Science,* 1971, **171,** 51–52.

Hartup, W. W., Glaser, J., & Charlesworth, R. Peer reinforcement and sociometric status. *Child Developm.,* 1967, **38,** 1017–1024.

Haskett, G. J. Research and early education: Relations among classroom, laboratory, and society. *Amer. Psychologist,* 1973, **28,** 248–256.

Hobart, C. W. Commitment, value conflict, and the future of the American family. *Marriage Fam. Living,* 1963, **25,** 405–412.

Hoffman, L. W. Effects of maternal employment on the child—A review of the research. *Developm. Psychol.,* 1974, **10,** 204–228.

Hoffman, M. L. Moral internalization, parental power, and the nature of parent-child interaction. *Developm. Psychol.,* 1975, **11,** 228–239.

Jelliffe, D. B., & Jelliffe, E. F. Human milk, nutrition, and the world resource crisis. *Science,* 1975, **188,** 557–561.

Jencks, E., et al. *Inequality: A reassessment of the effect of family and schooling in America.* New York: Basic Books, 1972.

Jensen, A. How much can we boost IQ and scholastic achievement? *Harvard educ. Rev.,* 1969, **39,** 1–121.

Jensen, A. The differences are real. *Psychol. Today,* December 1973, pp. 80–86.

Jester, R. Films and television as a source of value judgements. In R. J. Sasnett (Ed.), *Values colloquim III: Value attitudes in a changing society,* pp. 91–102. Pasadena, Calif.: Religion in Education Foundation, 1969.

Kohn, M. L. Social class and parent child relationships: An interpretation. *Amer. J. Sociol.,* 1968, 471–480.

Kornberg, A. The synthesis of DNA. *Scientific American,* 1968, **219,** (4), 64–78.

Landsberg, H. H Low-cost, abundant energy: Paradise lost? *Science,* 1974, **184,** 247–253.

Landsman, T. The humanizer. *Amer. J Orthopsychiat.,* 1974, **44,** 345–352.

Lasker, G. W. Human biological adaptability. *Science,* 1969, **166,** 1480–1486.

Layzer, D. Heritability analyses of IQ scores: Science or numerology? *Science,* 1974, **183,** 1259–1266.

Lynn, D. B. *Parental and sex role identification.* Berkeley, Calif.: McCutchan, 1969.

McBride, A. B. *The growth and development of mothers.* New York: Harper & Row, 1973.

McClearn, G. E. Behavioral genetics: An overview. *Merrill-Palmer Quart.,* 1968, **14,** 9–24.

McGovern, G. The child and the American future. *Amer. Psychologist,* 1970, **25,** 157–160.

Marcuse, H. *One-dimensional man.* Beacon, N.Y.: Beacon House, 1964.

Mesarovic, M., & Pestel, E. *Mankind at the turning point.* The second report to the Club of Rome. New York: Dutton, 1974.

Nash, J. The father in contemporary culture and current psychological literature. *Child Developm.,* 1965, **36,** 261–297.

Newman, D. R. Air pollution. *Curr. Hist.,* 1970, **59,** 13–22.

NIMH. Cognitive and mental development in the first five years of life. Washington, D.C.: National Institute of Mental Health, 1970.

Nirenberg, M. & Leder, P. RNA codewords and protein synthesis. *Science,* 1964, **145,** 1399–1407.

Osborn, D. K., & Endsley, R. Emotional reactions of young children to TV violence. *Child Developm.,* 1971, **42,** 321–331.

Piaget, J. *Science of education and the psychology of the child.* (D. Coleman, Trans.) New York: Orion Press, 1970.

Piaget, J. *The child and reality: Problems of genetic psychology.* New York: Grossman, 1973.

Poleman, T. T. World food: A perspective. *Science,* 1975, **188,** 510–518.

Profiles of children. Washington, D.C.: U.S. Government Printing Office, 1970.

Reich, C. A. *The greening of America.* New York: Random House. 1970.

Rohwer, W. D., Jr. Prime time for education: Early childhood or adolescence? *Harvard educ. Rev.,* 1971, **41,** 316–341.

Rollins, B. C., & Feldman, H. Marital satisfaction over the family cycle. *J. Marriage Fam.* 1970, **32,** 20–28.

Scarr-Salapatek, S. Race, social class and IQ. *Science,* 1971, **174,** 1285–1295.

Scrimshaw, N. S. Early malnutrition and central nervous system function. *Merrill-Palmer Quart.,* 1969, **15,** 375–388.

Shane, H. G. Prospects and prerequisites for the improvement of elementary education: 1973–1985. In J. I. Goodlad & H. G. Shane (Eds.), *The elementary school in the United States,* pp. 380–406. (72d Yearbook National Society for the Study of Education, Part II.) Chicago: University of Chicago Press, 1973.

Sheldon, E. Parental child-rearing attitudes and their relationship to cognitive functioning of their preadolescent sons. *Diss. Abstr.,* 1968, **29,** 4370.

Singer, D. L., Whiton, M. B., & Fried, A. L. An alternative to traditional mental health services and consultation in schools: A social system and group process approach. *J. School Psychol.,* 1970, **8,** 172–526.

Smith, G. R. The role of the schools in cultural, personal, and professional maturity. In J. R. Sasnett (Ed.), *Values colloquium IV: The educational situation as a milieu for maturity,* pp. 84–102. Santa Barbara, Calif.: The Religion in Education Foundation, 1970.

Spock, B. *Infant and child care.* New York: Pocket Books, 1968. (Originally published, 1945.)

Stein, A. H. Mass media and young children's development. In I. J. Gordon (Ed.), *Early childhood education,* pp. 181–202. (71st Yearbook National Society for the Study of Education, Part II.) Chicago: University of Chicago Press, 1972.

Stent, G. DNA. *Daedalus,* 1970, **99,** 909–937.

Sullivan, H. S. *Conceptions of modern psychiatry.* New York: Norton, 1953. (Originally published, 1940.)

Thompson, W. R. Storage mechanisms in early experience. In A. D. Pick (Ed.), *Minnesota symposia on child psychology* (Vol. 6, pp. 97–127). Minneapolis: University of Minnesota Press, 1972.

Thompson, W. R. Some problems in the genetic study of personality and intelligence. In J. Hirsch (Ed.), *Behavior—Genetic analysis.* New York: McGraw-Hill, 1967.

Toffler, A. *The eco-spasm report.* New York: Bantam Books, 1975.

U.S. Bureau of the Census. *Vital Statistics of the United States: 1970* (Vol. 1). Washington, D.C.: U.S. Government Printing Office, 1975.

Walters, J., & Stinnett, N. Parent-child relationships. In C. B. Broderick (Ed.), *A decade of family research and action.* New York: National Council on Family Relations, 1972.

Walters, R. H., & Willows, D. C. Initiative behavior of disturbed and nondisturbed children following exposure to aggressive models. *Child Developm.*, 1968, **39**, 79–89.

Warner, W. L., Mecker, M., & Eells, K. *Social class in America.* New York: Harper & Row, 1960.

White, B. L. *Preschool project: Child rearing practices and the development of competence. Final report.* Washington, D.C.: Office of Economic Opportunity, 1972.

Wilensky, H. L. Mass society and mass culture: Interdependence or independence? *Amer. soc. Rev.*, 1964, **29**, 173–197.

Yarrow, L. J. Maternal deprivation: Toward an empirical and conceptual re-evaluation. *Psychol. Bull.*, 1961, **58**, 459–490.

Yorburg, B. *The changing family.* New York: Columbia University Press, 1973.

Chapter 4

Developmental Aspects and Trends

HIGHLIGHTS

The human individual grows and matures as basic dimensions of organism and personality unfold, each at its own time and pace. Spurred by the genetic code from inside and by nutrition and sensory stimulation from outside, the individual moves uphill toward higher levels of behavioral operation.

Emotional excitement and the cry at birth differentiate into a complex spectrum of socially desirable emotionality, heightened during early childhood and adolescence and lowered after the attainment of adult maturity.

Growth in perceptiveness and cognition takes many avenues of stimulation and exploration before adult logic and symbolic operations become dominant features. Learning to associate sound with meaning in modifiable ways furthers interpersonal communication and broadens the capacity to gain and store knowledge and to use it in novel situations.

From the mother-child dyad to self-assertion in peer groups, the individual undergoes many shifts and refinements as he or she moves toward social maturity marked by responsibility for one's own behavior and a readiness to help others. Learning to see and to do what others expect is a major task of socialization.

Differences in the capacity to learn the three R's and motivational variables affecting various methods of learning baffle many parents and teachers. Intelligence testing sheds much light on cognitive growth; yet darkness covers much of the area of motivational development.

Self-perception rises during the second year of life and generates a lifelong search for models of selfhood and identity. The intricacies of human relationships in our culture are discovered, and acceptable ways to realize some of one's aspirations are usually found during the years of early adolescence and adulthood.

Human sexuality elicits strivings for different degrees of identification with both parents and later with selected peers of the same sex and the opposite sex. Deep identification with others furthers self-identity and humanization.

The human individual is highly integrated in his organismic systems and unfolding behavioral powers. In the prenatal stage, most structures are formed and many physiological functions start. After birth the structures become refined and their behavioral functions multiply, though in a decreasing way, throughout infancy, childhood, and adolescence. Later adolescence and early adulthood are periods of functional specialization and culmination in achievement. While structural differentiation is primarily due to the operation of the genetic code, the upper levels of functional development depend largely on the social milieu in which the individual lives. Figure 4-1 is a schematic model of the general pattern of structural-functional differentiation and decline.

SYNOPSIS OF PHYSIOLOGICAL GROWTH

In its structural aspects physiological growth is a highly uneven process. Extremely rapid prenatal differentiation is closely followed by functional division within the growing organism. Although the genetic curve is progressive

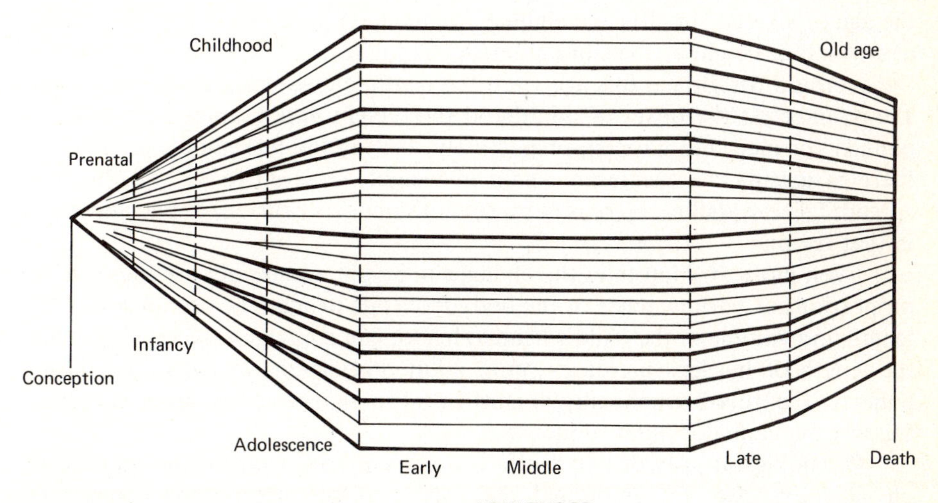

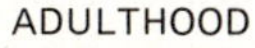

Figure 4-1 A model of growth differentiation. Each line indicates a new structure or function in the total pattern growth. Vertical lines represent demarcations of levels of the life-span.

for most organs and systems until maturity is attained, there are plateaus, even deterioration, for some structures and functions as early as childhood or puberty. The plateau for growth of the sex organs during childhood and the shrinking of the thymus gland after puberty are examples of discontinuity in structural growth. Decrements in most physiological structures and resulting functional losses begin in middle adulthood and intensify during senescence. For example, the mineral mass of the human body rises rather steadily to about thirty-nine years of age and then begins decreasing. How much mineral mass is gained up to this peak age and how much is lost thereafter is crucial to about one-third of our population and to most people over fifty. Many a seventy-year-old woman literally has a subteen mineral mass and is therefore subject to bone fractures or even to vertebral collapse (Garn, 1970).

The prenatal stage, the first and second years of life, and the years of early childhood are periods of tremendous though decelerating physiological growth. Development tapers off during childhood until it reaches its low ebb at about nine or ten years of age. After about two years there is a renewed acceleration of growth, which becomes turbulent during the major pubertal changes. A year later, a final deceleration phase begins, and physiological growth nearly ceases by the age of sixteen or eighteen.

The age at which pubertal changes take place and adult height is attained differs significantly for boys and for girls. For the population of the United States in the early 1970s, the average age at which pubertal changes reached a peak was shortly after twelve-and-a-half for girls and two-tenths over fourteen for boys. Following major pubertal changes the average increase in height for girls is negligible—slightly over 1 centimeter—while boys gain about 6 centimeters (nearly 2½ inches). Increases in weight continue until the late sixties, when a decline begins. Young-adult weight is nearly reached in late adolescence at the age of sixteen for girls and eighteen to nineteen for boys. In the early 1970s, at an average height of 185 centimeters, or 71 inches, boys weighed 148 pounds; the averages for girls at this age were 168 centimeters, or about 66 inches, and 130 pounds. At the onset of adulthood the weight approximates 159 and 136 pounds for men and women, respectively. During the late middle-adult years there is usually a considerable gain in weight, referred to as "middle-aged spread." The waistline increases by several inches as surface fat accumulates in that region.

Even more striking than the changes in height and weight are those in the proportions of various parts of the body from age to age. The proportion of the length and volume of the child's head to his total height keeps decreasing, while the length of his legs increases, until adult proportions are acquired during puberty. Figure 4-2 graphically illustrates these changes as the growth pattern follows the cephalocaudal sequence.

Physiological growth and welfare are highly affected by the amount of physical exercise, by emotional currents, and by dietary provisions. For internal metabolic changes the number of needed nutrients is approximately sixty, most of which are essential for one's welfare (DHEW, 1972, p. 231; Macy

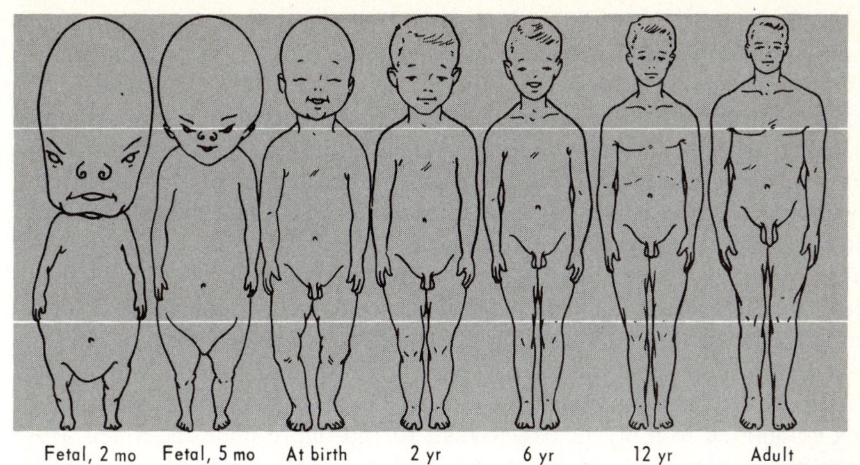

Figure 4-2 Changes in proportions of the human body with age. [*Source:* G. A. Baitsell. *Human biology.* New York: McGraw-Hill, 1940. (after Stratz)]

& Kelly, 1957). For example, a child needs a diet that contains nine amino acids. If proper nutrients are not available, the blood supplies substitutes. When the protein or caloric intake continues to be below the needed amount, physiological growth will be impaired, and with it the capacity to resist disease (Garn, 1970).

Infants and children have different capacities to digest, absorb, utilize, and store the necessary substances from foods making up their diets. Proteins, carbohydrates, fats, water, and oxygen are the raw materials the child needs for growth. The proteins, for example, are broken down by enzyme action into readily transformable amino acids, which are subsequently resynthesized under the direction of DNA and RNA into the particular types of protein characteristic of the tissue cells utilizing them (Patten, 1968, p. 180). Conversion of nutrients into the living cellular matter involves the gastrointestinal system and many other systems and organs of the body. For example, cells divide and grow only if the composition of the blood and the amounts of oxygen and fluids are within rather narrow limits. Under normal dietary provisions, low gains in weight for age indicate organic deficiencies in glandular, circulatory, enzymic, or other functional systems. If food is available, clinical testing is indicated to find the causes curtailing growth and to correct them if medical or related measures are available. Failure to thrive is ascribed to infants whose weight drops to 30 percent or more below the average expected weight for age (Fomon, 1967, p. 11).

Each child has his or her own rate and pattern of physical growth, yet in most aspects these are comparable to those of other children. Among the most striking visible properties of physical growth are height, weight, and the changing proportions of various parts of the body. By the second birthday, children attain about one-half of their adult height. At one year of age most babies are somewhat fat, but at two or three they come into their own—increased physical activity reduces the amount of surplus fat. If excess calories are consumed daily, however, a few of the unused ones are converted to glycogen but the remainder are stored as fat, and fat accumulates over months and years. Obesity in many children originates during infancy. Since babies tend to regulate their food intake by volume rather than by nutritional content, they often consume inordinate quantities of formulas high in calories. This situation promotes the production of excessive numbers of fat storage cells and thereby precipitates the overweight that plagues many persons throughout childhood, adolescence, and adult life. Fat cells are just as demanding of supply as other cells, and the vicious circle is established (DHEW, 1972, p. 26).

It may be noted that the fat four-year-old is usually accelerated in physical growth, but the fat eight-year-old is already "hooked" on fatness, and in most cases committed to a life of overweight, for not more than one in ten obese adolescents securely return to nearly regular weight as adults (Garn, 1970). Overnutrition is a problem for many parents and children alike. A good diet adapted to the activity and metabolic needs of children helps them to grow and at the same time curbs deviations from the optimal pattern of somatic growth.

In the analysis of physiological growth, there are two fundamental questions: How does the individual grow? Why does one grow the way one does? The first question is in part answered by comparing the measurements with the general norms for that age. Height and weight are standard examples. Additional measurements include assessment of head circumference and x-rays revealing the level of bone ossification. The ossification stage of certain bones (hand and wrist, for example) indicates the level of physical maturity of the child x-rayed for that purpose. The calcification and remodeling of hand bones entail many observable changes in the process of maturation. As a result, x-ray analysis serves well as an index to estimate present status as well as future growth, including timing of major pubertal changes (Greulich, 1950). A basic answer to the second question lies in the individualized genetic code, which periodically releases genetic material into the organizing cells. These genetic agents initiate growth changes, provided there are nutritional materials available, to effect the structural changes according to the blueprint of the genetic code.

Glandular functioning is in part the answer to why the child grows in a normal or a deviant way. Malfunctioning of a single endocrine gland—the pituitary or thyroid in particular—produces accelerating, retarding, and other undesirable effects on the total organism. Genetic factors influence endocrine gland growth and activity, but the mechanisms involved are not yet fully understood. A thalamic center acts to stimulate the pituitary gland, but its own stimulation probably comes from a center in the brain. The sympathetic function of the autonomic nervous system affects the output of most endocrine glands. Height and bodily proportion are highly dependent on the pituitary gland, located at the base of the brain. The anterior lobe of the pituitary gland produces growth-stimulating or somatotrophic hormone (STH), which serves as an overall regulator of many metabolic processes in the body. Its influence on growth is the result of this regulation. Low amounts of STH retard physical growth; the child is then small for his age.

In January, 1971, C. H. Li, director of the Hormone Research Laboratory of the University of California, announced in San Francisco that after thirty-two years of hormonal research he had completely synthesized somatotrophin. Li found that the growth hormone consists of a chain of 188 amino acids. STH is the largest protein molecule yet developed by man. The clinical application will be to infants and children who grow slowly because of pituitary deficiency in somatotrophic production. Some forms of dwarfism are likely to disappear as a result of this chemotechnological breakthrough. This discovery also opens the possibility of synthesizing an antigrowth hormone to decelerate growth in fast-growing girls as well as to inhibit tumor-cell growth, including growth of neoplasmic and cancerous tissue.

The pituitary's influence extends to the sexual aspects of child and adolescent growth. Its anterior lobe generates gonadotrophic hormone (GTH), stimulating the growth of testes or ovaries. There are cases of insufficient growth of genital organs, as well as cases of precocious sexual maturation occurring at seven or nine years of age. *Pubertas praecox* is the term used to designate an unusually early structural and functional maturation of the sexual glands. Early sexual maturation produces adjustment problems for many girls. They feel "out of place" with the preadolescents in their classes but usually have difficulty being accepted by older girls. The pituitary gland also produces hormones that regulate other endocrine glands, as well as some functions of the metabolic process. Its posterior lobe generates still other hormones, such as pitressin, which stimulates smooth muscles, and oxytocin, which affects hydration and blood pressure.

The endocrine gland system operates in an interdependent way. Other endocrine glands support many functions primarily controlled by the pituitary. Physical growth also has much to do with the functional capability of the thyroid gland, as protein synthesis and many functions of the metabolic process are controlled by its hormone, thyroxin. As compared with that of the adult, the metabolic rate is highly accelerated throughout infancy and childhood. Slightly below average functioning of the thyroid is related to the rising problem of overweight—too little "burning" of surplus calories occurs, and weight tends to accumulate as a result of any excessive calorie intake over a long period of time. This condition often becomes aggravated by the low level of activity overweight persons tend to prefer.

EMOTIONAL DEVELOPMENT

The human individual is born with a capacity for emotional experience and behavior. Living with parents and siblings and relating to them is a major source for the acquisition of early emotional patterns. If these relationships are pleasant and rewarding, they form a good start for many desirable emotions. As an infant grows, he or she must see and feel affection, sympathy, and love expressed by others before being able to reciprocate. Socially desirable emotions thus need human stimulation for their growth. The expanding amplitude of emotional experience is comparable to a prism: one color shades into another, yet there are highly saturated hues that are obvious to most people. The emotional deficits of some children remind us of partially color-blind persons; they apparently were not provided with some emotion-specific stimulation to unfold a fuller spectrum of human emotionality.

Figure 4-3, based on Katherine M. B. Bridge's empirical assessment, shows the differentiation of excitement (present at birth) into various emotional states during the first two years of life. The diffuse emotional excitement of the neonate, which is a function of the accelerated mass reactivity of the organism in pain, hunger, or discomfort, assumes the quality of a pleasant or unpleasant experience. The resulting delight or disgust are in time refined into affection or anger and subsequently develop into more subtle affective states. Emotions on the right side of the chart are necessary for healthy living and social adjustment. Without a parental show of delight and affection toward the infant, he or she may become an emotionally deficient child, for these and other refined sentiments springing from affection may fail to unfold.

Early childhood is normally the time when the child greatly enlarges the repertoire of emotional states and related behavior. By five years of age, the child is usually capable of nearly the total spectrum of human emotionality, since only moods and some feelings related to mature sexuality appear at puberty. In these early years of life affective states come and go freely. Like a good actor, the child displays contrasting outbursts of anger, jealousy, or temper and intense affection, joy, and love within short periods of time. As the child grows, the ways of expressing emotions are altered. The rising ability to communicate by words replaces in part the use of crying and other motor activity as a necessary means of expressing feelings.

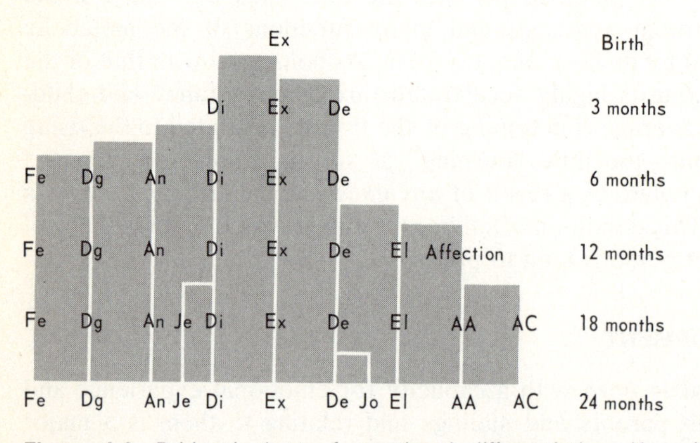

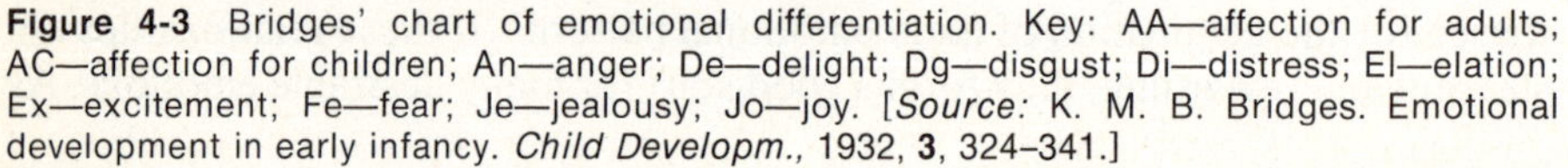
Figure 4-3 Bridges' chart of emotional differentiation. Key: AA—affection for adults; AC—affection for children; An—anger; De—delight; Dg—disgust; Di—distress; El—elation; Ex—excitement; Fe—fear; Je—jealousy; Jo—joy. [*Source:* K. M. B. Bridges. Emotional development in early infancy. *Child Developm.*, 1932, **3**, 324–341.]

Throughout the preschool years, children appear to learn more than they are taught. Under normal conditions of home life, anger, jealousy, and envy but also empathy and love are vividly expressed. During the preschool years, children are typically self-centered and rather demanding. They become aware of the power of their emotional behavior to influence the parents. At the same time they are learning how far they can go in expressing what they feel and know. After being embarrassed by their child's frankness, some parents urge him to subdue his emotions to the point of suppression. The child cannot swallow too much of what he or she intensely feels. Although emotional tone can easily be altered, emotional experiences are often intense and deep enough to penetrate the whole organism and personality of the child. Emotional current does not readily disintegrate; rather, it goes underground. Repeated parental urging to conceal emotion brings about an accumulation of pent-up energies that, when channels of expression are severely curbed, unbalance some bodily systems and in more severe cases produce psychosomatic disturbances and symptoms springing from them (Maccoby, 1971).

Middle and late childhood are periods of further emotional differentiation in the form of sentiment and attitude formation. Anything new the child learns he either likes or dislikes, is attracted or repulsed by. At the same time the child is gaining in ability to control his feelings and emotions and to express them in socially approved ways. Peer associations and group activities are helpful for ventilation of pent-up emotionality (Baumrind, 1972).

Emotional differentiation gains much during the early adolescent years. Before this period ends, all the adult affective states may be experienced. Moods, rarely noticeable in childhood, plague many adolescents. One day the teenager is euphoric and exuberant; the next, he or she finds life bereft of all joy. Middle adolescence is a period of heightened oscillation between polarized moods. In late adolescence, the emotional reactions become subdued but also more mature and adult. An added feature of emotional complexity in adolescence results from the intensification of the sex drive and the experience of deep heterosexual attachment and love. In early adolescence there is usually an unrealistic idea of the love object, which helps to delay fusion of sexual object and love object. It is not until emergence into early adulthood that the majority of persons attain stability in their own identity and readiness for the mature heterosexual love necessary for happy marriage—the development of a lifelong companionship with a single member of the opposite sex (Erikson, 1963, pp. 264–266).

In the periods of adulthood, a considerable increase and maximum refinement of attitudes and sentiments is usually attained. The next significant change in emotions comes in senescence, when social and emotional "disentanglement" occurs. There is generally a drastic reduction in social interaction and emotional involvement. The older person becomes increasingly uninterested, even apathetic at times. He feels that he has had his opportunities; it is too late now. Boredom and depression, irritability, faultfinding, and complaining about one's health problems and painful experiences all rise noticeably with advancing age. Emotional satisfaction is more and more often vicariously derived from memories of past events and from exaggeration and boasting over past achievements. Figure 4-4 depicts schematically Katherine Banham's genetic theory of emotions throughout life.

PERCEPTUAL AND COGNITIVE GROWTH

The process of learning to know the world and oneself begins with perception—an effect of external and internal stimulation of the sense receptors as modified by sense modality and modeling in the brain. *Perceiving* is a process in which sensory input is unified and coded in accordance with an increasing number of operative grouping tendencies and past experience. To a large degree perceiving depends on context, which makes certain aspects of stimuli clear and distinct and masks others. Developmentally, perceiving is marked by new and additional meanings assigned to the previously unattended or undefined sensory data and also by the increase of perceptual constancies, making objects in some respects invariable despite wide variations in viewing conditions. The rising awareness of persons, objects, situations, and relationships is a result of cognitive and emotional developments, highly colored by the effects of original experiences, sets, organismic needs, and other motivational factors.

It was William James who, in *Principles of Psychology* (1890), noted that "to the infant sounds, sights, touches and pains form probably one unanalyzed bloom of confusion." Does the neonate's awareness of reality start that way? Apparently no, since many recent studies demonstrate that the neonate has the capacity to respond distinctly to a considerable variety of stimuli even when asleep (Ashton, 1971a, b; Bower, 1971). During the first five days of life, most neonates can be characterized as being consistently either slightly, moderately, or intensely responsive to stimuli, regardless of their modalities (Birns, 1965; Stechler, 1970).

For Epstein (1967, pp. 260, 283), perceptual development consists of "increasing differentiation, starting from an innate minimum discriminative capacity." The individual learns to respond to new stimuli and new aspects of stimulation that were not responded to previously. The major spurt in perceptual development begins soon after birth and levels off during the second year. Piaget's genetic analysis of perception (1969, pp. 133–134, 137) distinguishes primary field effects perceived in singular fields of centration and perceptual activities as those processes "which occur when centration or their effects have to be related across spatial or temporal intervals," reinforcing coordinations and reducing primary errors of field effect. Field effects are subordinated to perceptual activities, which, in turn, are subordinated to schemata of action (pp. 353–354). Relating previously unrelated elements, perceptual processes of exploration and transposition entail certain deformations or secondary illusions about relations, increasing continuously up to a certain age. The exploratory activity is "the activity which directs eye movements and determines pauses or centrations during the examination of a figure." *Centration* refers to attention on a specific segment of a stimulus,

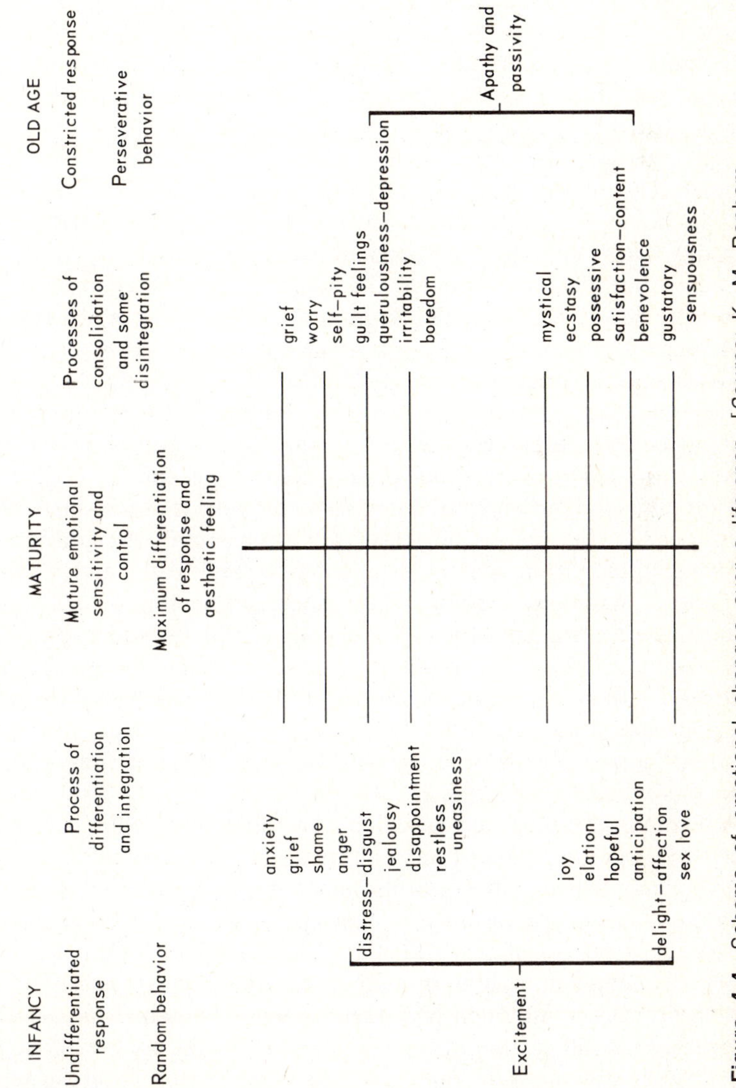

Figure 4-4 Schema of emotional changes over a life-span. [*Source:* K. M. Banham. Senescence and the emotions: A genetic theory. *J. genet. Psychol.*, 1951, **78**, 175–183. By permission.]

leading to its overestimation. Sufficiency of exploratory activity improves with age and practice, the direction of which comes from intelligence—knowing what must be looked at in an object or situation (p. 141).

The child's exploration is syncretic or global, relying on main outlines rather than analysis; he scans and recognizes the overall pattern of a phrase first before he breaks it into words and still later into letters. Piaget (1969, pp. 141–142) accepts Decroly's method of teaching a child to read on the basis of the child's global perception. Generally the child's field of exploration is narrow and dominated by the factor of proximity, which loses much of its importance with age, when fields become more and more extended. Perceptual activities are often followed by behavioral exploration, the patterns of which, when internalized, form a representational system, which ultimately gives rise to formal operations (pp. 283–284).

For Piaget (pp. 284–287), perceiving is egocentric from every point of view, since it is tied to the perceiver's position in relation to the object. This high egocentrism in perceiving is the source of centration and its systematic illusion. Intellectual operations are relatively independent of the ego and its particular point of view; they exceed perceptual data and produce universal knowledge that is common and communicable to others.

The world is naturally and technologically structured, and the growing person attends and reacts to an increasing number of these structures. At the age of four or five the child recognizes destructuring and is ready to provide some parts of a missing structure. As Garner (1966) aptly puts it, to perceive is "to know and comprehend the nature of a stimulus." It is also to organize sets of stimuli never experienced before by the child in terms of one of several alternative modes of perceptual organization. One cannot really know the single stimulus without understanding some properties of the sets within which it is contained. Maturation and learning both improve perceptibility of form, motion and apparent movement, size-distance, and dyadic and group relationship. Critical studies on developmental aspects of perception are presented by William Epstein (1967) and Eleanor J. Gibson (1969). Gibson also expounds her *differentiation* theory of perception, which is stimulus-oriented since the individual continues learning to discriminate the properties, distinctive features, and invariants of a rising host of stimuli. The critical period of visual perceptual training for the development of higher cognitive functions is the age of four to six years (Frostig & Horne, 1964).

Perceptual grouping and modeling, encoding and decoding, underlie the process of cognition. Broad in its function, *cognition* refers primarily to conceiving and dealing with concepts, analyzing and synthesizing, deciding and theorizing. As a generic term it encompasses all processes by which knowlege of an object or relationship is attained, including sensory data and perception, memory and imagination, hypothesizing, and other logical operations. All these cognitive processes, although an integral part of human nature, still must be developed. Like other functions, the developmental course of cognition is marked by several spurts of rapid growth. Concept formation during the third year and questioning during the sixth year of life represent early acceleration of cognitive unfolding.

Progress in verbal skills demonstrates the rise of cognitive power in the child. The speed with which he or she assimilates new words, explains differences and similarities, and increases the length of the sentences used are all fairly reliable indexes of intelligence. When properly reinforced, the child's curiosity is a great spur to cognitive development. Although the frequent "What's that?" or "Why?" of a child at times annoys a busy parent, it is a major means children have for learning about the world around them. The type of questions asked is indicative of the operative level of intelligence. Young children usually want to know "What?" or the name of any new object encountered or brought to their surroundings; they realize that people and things have names. "Why?" or "What for?" is the next question frequently posed, from about four-and-a-half to six. The child is learning that most objects are good for something—to serve human needs and purposes—and so is eager to know the function of any new object. Now he is not only aware of his needs and desires but also of many means to gratify them. Then at eight or nine years of age the child is concerned about causes and effects. When he witnesses a new occurrence, he is interested in its cause; he wants to know how to make it happen again.

What the child will do in various stimulating situations depends greatly on the mother. As a rule, middle-class mothers encourage their children to explore problems on their own and focus on their success, in this way helping them to acquire new learning sets and strategies for future problem-solving situations. By their disapproval and physical intrusion into the problem-solving activity, most lower-class mothers deprive their children of opportunities to work on solving problems, so that their children learn much about what *not* to do rather than *how* to do anything in a systematic way (Baumrind, 1972; Bee et al., 1969).

During early adolescence, adult cognitive functioning is usually attained. Adolescents are capable of formal propositional reasoning, logical theorizing, and questioning about ultimate causes and remote effects. Refinement and polishing of many cognitive functions continues for many years. Special aptitudes and gifts are also vividly shown at this time. As estimated by leading individual intelligence tests, cognitive functioning reaches its peak by the age of nineteen or twenty (Matarazzo, 1972, pp. 105–11; Terman & Merrill, 1973).

With adulthood comes a slackening in the exercise of cognitive functions for many persons who discontinue their educational efforts. The problems of vocation and marriage usually absorb the time earlier spent in developing the cognitive powers. Many persons reduce their interests and activities to practically those of employment and homemaking. If there is a considerable reduction in cognitive stimulation and application, a gradual reduction in cognitive alertness may set in in early adulthood.

In old age, most aspects of perception and some aspects of cognition decline at a moderate rate; the physiological organs and structures decline at a faster rate (Baltes and LaBouvie, 1973; Botwinick, 1967). There are many outstanding exceptions among scientists, philosophers, politicians, clergymen, and professionals who maintain cognitive power and alertness to a very old age, but the generalization holds for most people in American society. The prevailing view of intellectual decline in old age has been challenged by a number of recent studies (Riegel & Riegel, 1972; Schaie, 1973; and Schaie & LaBouvie-Vief, 1974). The issue of cognitive decline is thus open for further research.

Guilford's factor analysis led to a three-dimensional model of intellectual factors, entitled simply the *structure of intellect.* A factor is discovered by virtue of a small cluster of tests that share it—an underlying, latent variable along which persons differ when tested (Guilford, 1967, p. 41). Figure 4-5 presents Guilford's cube model of intellectual operations, contents, and products. For a detailed analysis of this model, the student is referred to the books of Guilford (1967) and Meeker (1969).

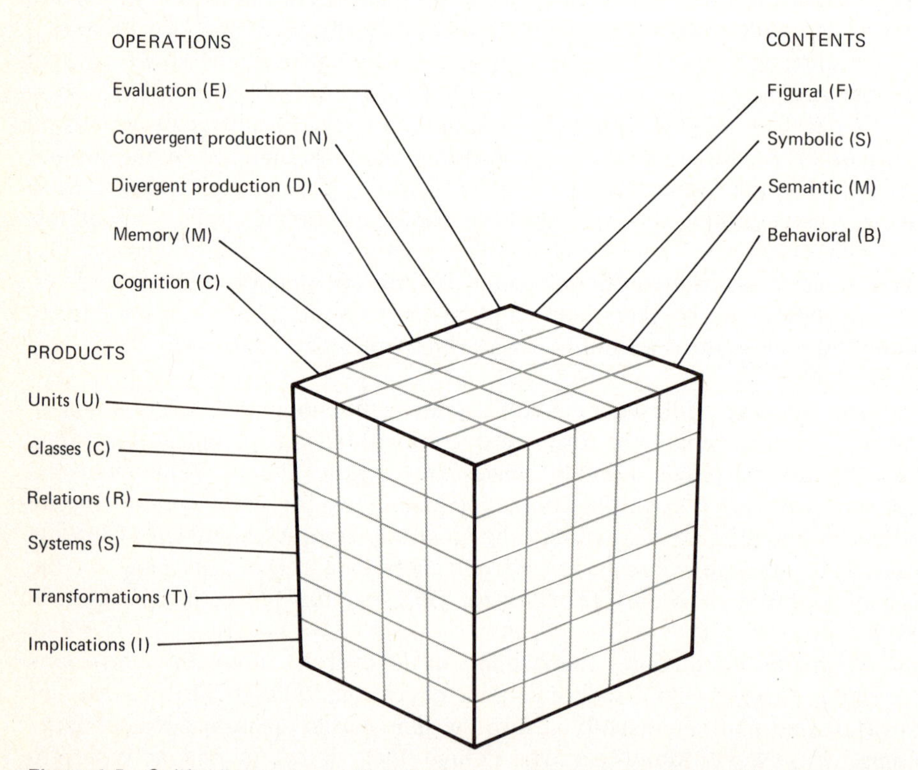

Figure 4-5 Guilford's structure-of-intellect model: 120 related subtypes of intellectual ability derived from the intersections in the cube. The definition of each subtype is derived from three major dimensions—operation, content, and product. Combinations of initial letters provide short names for each subtype or ability. CBU = cognition-behavioral units. [*Source:* J. P. Guilford. *The nature of human intelligence.* New York: McGraw-Hill, 1967 (inside of front cover). By permission.]

Intelligence Testing and Its Applications

One of the controversial problems in the assessment of the cognitive status of a child or adult is intelligence testing and the practical application of IQs. There is considerable uncertainty as to just what intelligence tests measure. Most exponents of testing concede that intelligence tests barely tap the abilities responsible for job proficiency and even less those responsible for accomplishments in the arts, science, or social leadership (Elton & Shevel, 1969). McClelland (1973) argues that intelligence testing in schools leads people "to believe that doing well in school means that people are more competent and therefore more likely to do well in life because of some real ability factor." He suggests "testing for competence in scholastic and other areas to predict grades in life" in the broad theoretical and practical sense. The theoretical foundations and specific rationales of the individual tests offer only limited insight into the probable grades in life.

Do intelligence tests tap the innate cognitive capacity of the individual that could be developed with proper stimulation, or only what is functional at the age of testing? How much can intelligence scores be raised by changes in cultural milieu and by early or enriched education of the child? Extensive surveys by Hans J. Eysenck (1971), Arthur R. Jensen (1969, 1973), and Sandra Scarr-Salapatek (1971) attempt to answer these and related questions. Drawing evidence from a diversity of sources, Jensen (1969, p. 102) concludes that early remedial and compensatory programs produce only small IQ gains and that boosting of the overall cognitive level for life is impossible: "The effort required to boost IQ from 80 to 90 at four or five years of age is minuscule compared to the effort that would be required by age nine or ten." Early efforts apparently only produce "hothouse" effects that largely evaporate within a few years. Providing some evidence, he argues that genetic factors are much more important than environmental ones in producing IQ differences. No doubt extreme deprivation may keep a child from performing up to his genetic potential, but enriched education cannot push the child above that potential. Moreover, Jensen claims that disadvantaged children have an earlier deceleration of intellectual growth as compared with advantaged children, who continue gaining for a longer period of time.

A difference in genetic basis cannot be discovered without evening out a variety of environmental effects. Until many white children are reared in black homes and vice versa, racial differences cannot be definitely measured to close that question. The environmental-disadvantage hypothesis and the genotype-distribution hypothesis are producing two models of search for racial and social-class differences in intelligence. "If all children had optimal environments for development, then genetic differences would account for most of the variance in behavior" (Scarr-Salapatek, 1971).

For purposes of measurement and understanding, cognition must be translated into functional or operational terms. Hence *intelligence* is commonly used to denote the achievements by which an individual uses his knowledge and skills for discriminating stimuli, analyzing and synthesizing, sequencing, and solving verbal and performance problems. A total accomplishment resulting from the application of perceptual and cognitive powers and motor functions, when measured by individual intelligence tests, yields a pattern of the strengths and weaknesses of the functions assessed as well as an IQ, or general index of intelligence. The Stanford-Binet test (Form L-M, 1960, 1973), McCarthy Scales of Children's Abilities (1972), Wechsler Preschool and Primary Scale of Intelligence (WPPSI, 1967), and Wechsler Adult Intelligence Scale (WAIS, 1955) are currently often used for this purpose.

Widely used for several decades, in the 1950s the Stanford-Binet was in part restandardized for the third time in this country. The present Stanford-Binet L-M (1973) will probably retain its position as one of the leading individual tests of child intelligence for many years to come. In testing a ten-year-old child, for example, the examiner establishes his or her basal age (test age at which he succeeds in all subtests) and examines him on successively higher age levels until he fails all the subtests (ceiling age). Giving credits for the lower ages and counting them up to the ceiling age gives the child his mental age (MA). Instead of the earlier-used IQ as a ratio between MA and CA (chronological age), Form L-M utilizes the deviation IQ (that is, a standard score with a mean of 100 and standard deviation of 16 for each age). The deviation IQ improves the consistency from age to age.

Recent rivals of the Stanford-Binet L-M for testing children are the McCarthy (1972) and Wechsler (1974) scales for intelligence testing. The McCarthy Scales of Children's Abilities (MSCA) were designed to test "a variety of cognitive and motor behaviors" by six related scales: Verbal, Perceptual-Performance, Quantitative, General Cognitive, Memory, and Motor. The MSCAs were standardized on a sample of slightly over 1,000 children and "are appropriate for children from two and a half through eight and a half years of age" (McCarthy, 1972). Kaufman (1972) shows that the MSCA is a sensitive tool for diagnosing children with minimal brain dysfunction. The Quantitative and Perceptual-Performance scales contain highly discriminating subtests (Number Questions, Draw-a-Design, Counting and Sorting, Puzzle Solving, Numerical Memory, and Conceptual Group), all showing $p < .001$ for a sample of 44.

From a practical point of view, the psychologist assesses intelligence by requiring the subject to deal with various tasks and problems of increasing difficulty until he or she unequivocally fails on each type of problem administered. The efficiency level reached indicates the functional development of the cognitive powers. A fairly accurate appraisal of intellectual functioning is possible only by use of the individual intelligence tests mentioned before. Even the best group intelligence tests, such as the California Test of Mental Maturity or the Lorge-Thorndike, are only rough estimates of cognitive functioning. Most groups tests of intelligence are based chiefly on examination of verbal skills and perceptionally based analogies.

General distribution of intelligence and other abilities is shown in Figure 4-6. For comparative purposes, distribution of cumulative percentages, sta-

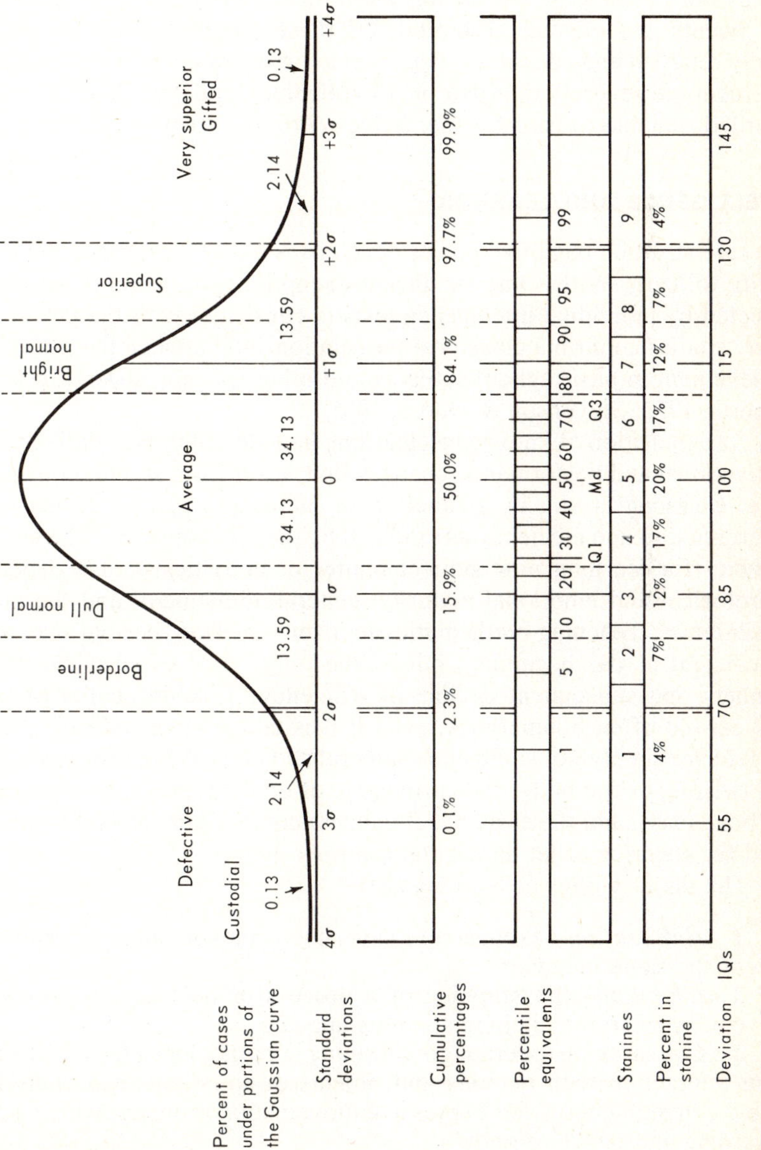

Figure 4-6 General distribution of intelligence and other aptitudes. [*Source:* Adapted from *Test service bulletin* (No. 48). New York: Psychological Corporation, 1955.]

nines, percentiles, and deviation IQs are included. At present, intelligence tests are frequently used for discovering the mentally deficient, retarded, or dull as well as superior and gifted children, in order to place them in appropriate special education programs that will prove stimulating and challenging but not overpowering and frustrating. In a diagnosis of mental deficiency, the individual intelligence test, by classifying the individual according to degree of cognitive deficiency, offers a prognostic differentiation of educable, trainable, and fully dependent cases. In the United States, between 100,000 and 200,000 babies born each year are mentally retarded. About 2,500,000 persons under age twenty are mentally retarded. Of these, approximately 75 percent are mildly retarded and educable, 15 percent are moderately retarded and trainable, 8 percent are severely retarded (many trainable), and 2 percent are profoundly retarded—unable to care for themselves (*Profiles,* 1970, p. 51).

INTELLIGENCE AND LEARNING

The rate at which cognitive power develops is closely related to the increasing ability to learn. With some significant exceptions, the level of intelligence as reflected by individual intelligence tests is consistent with the power to learn. Many studies confirm a close positive relationship between the rise of learning achievement in basic school subjects and other learning tasks (McCandless & Roberts, 1972; Stevenson & Odom, 1965).

As acquisition of knowledge, learning includes all those relatively enduring motivational and performance changes that result from stimulation and experience. Learning is chiefly a function of increasing perceptual and cognitive differentiation and efficiency as modified by the self-concept of the learner. The capacity for learning new subject matter or activities is also dependent on neurological and emotional maturity, general adjustment, and the method of presentation. Teaching mathematics to a four- or five-year-old may appear to be efficient at the beginning, but in the long run it often elicits dislike for numbers and subsequent decline of efficiency. Teaching ballet to a five- or six-year-old often boomerangs, even if it is taught systematically: at this age the feet are simply not sufficiently mineralized and calcified for repeated lifting and twisting of the body, the cartilage is often damaged in its conversion into the bone mass, and the early ballet dancer acquires damaged feet—never again good for superior ballet dancing at a proper age.

The major modes of learning are:

1 *Habituation*—exclusion of superfluous movement as a result of repetition of the same behavior.

2 *Inhibition*—the stopping of a process or behavior in progress or its prevention when the eliciting stimulus appears.

3 *Conditioning* (operant conditioning in particular)—the individual learns from interaction with persons and objects or situations, especially from the verbal communication and expressive movements of others, which actions are rewarding and which are not.

4 *Imitation*—observable late in the first year and avidly practiced at about one-and-a-half.

5 *Identification*—conscious or unconscious assumption of the role or status behavior of a significant person.

6 *Discrimination* (perceptual and conceptual, sign, signal, and symbol)—a major mode for the acquisition of information and experience. Signs, signals, and symbols become modifiers of behavior. This mode of learning includes trial-and-error and trial-and-success behavior.

7 *Logical* or systematic thinking—intellectual and creative facilitation when dealing with tangible and intangible aspects of reality. Through formal analyzing and synthesizing operations of the mind, the person gains in formal knowledge, applicable in many complex reality situations.

After surveying research data on learning, Razran (1965) formulated four levels and four sublevels of learning, coexistent at the higher evolutionary levels (but with higher levels normally dominant). As seen in Figure 4-7, at the highest level, *symbolization* or *Sy(ABC . . . N)* learning configurations become assimilated and integrated to form word symbols, which in turn form sememes (meaning units). This highest level is based on afferent perceptual configuring at the abstract symbolic level. Through the analytic or synthetic operation of the mind, the student continues accumulating abstract and symbolic knowledge that is applicable in many situations of life. Gagné (1968) formulated the *cumulative* learning model. Relating development to learning, he assumes that within limitations imposed by growth, behavioral development results from the cumulative effects of learning. New learning depends primarily on the combining of previously acquired and recalled learned entities or capabilities applied to new situations.

PATTERN OF SOCIALIZATION

Socialization refers to the process of learning to recognize group values and expectations and increasing ability to conform to them. The degree to which a child or adolescent meets the standards of peer groups depends on the amount of his or her social activity. Parent and peer influences change the growing individual from an egocentric child to a socially competent adult. Little is expected from a baby or toddler—a smile, an interested look—but a schoolchild must relate to other people in a social way; he or she must play roles and take positions. Acquisition of communication and interaction skills is a significant part of the socialization process.

Willingly or not, most parents act as representatives of their society and culture. This means that they further the prevailing ethos and cultural traits, at the same time establishing taboos and inhibiting countercultural tendencies. By verbal conditioning and other management techniques, the majority of parents enhance impulse control, responsibility, self-direction, and other positive attributes that usually help children to relate effectively with others. Overpermissive and negligent parents usually damage a child's adjustment when they

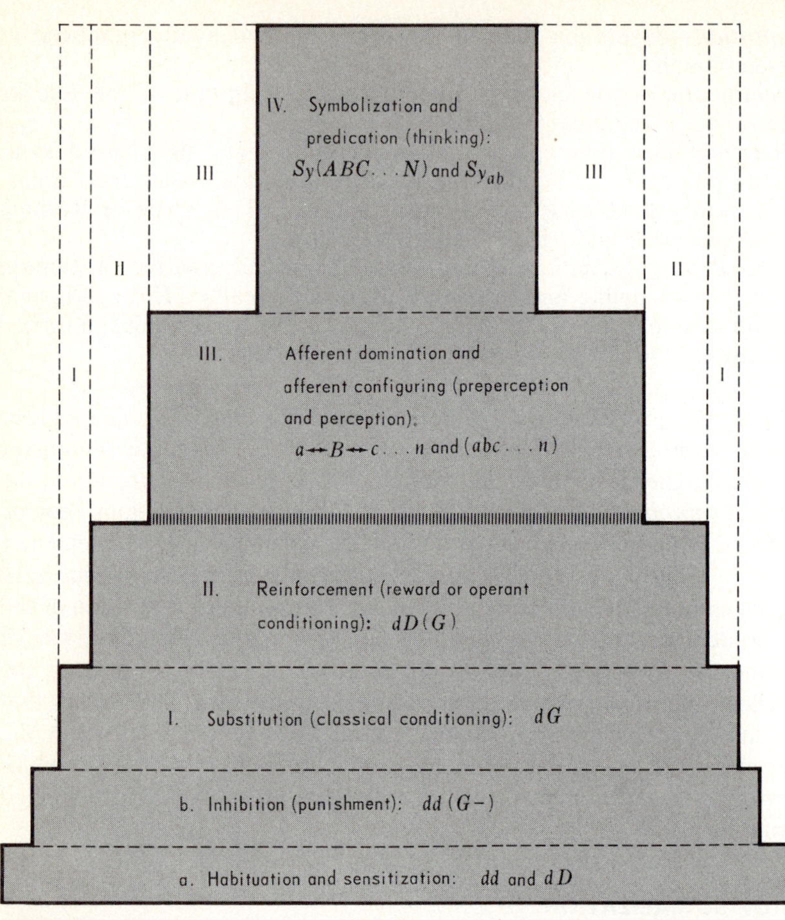

Figure 4-7 Razran's levels of learning. The universality of each level is represented by the width of its parallelogram; its functional efficiency, by the height of the parallelogram; its total efficiency (functional efficiency universality), by the area of the parallelogram. The phyletic antiquity of each level is indicated by its vertical position in the diagram. The dotted lines indicate the coexistence of lower and higher levels. [*Source:* G. Razran. Evolutionary psychology: Levels of learning—and perception and thinking. In B. B. Wolman (Ed.) and E. Nagel (Cons. Ed.), *Scientific psychology: Principles and Approaches.* New York: Basic Books, 1965 (chap. 13, pp. 207–253; fig. 13-1, p. 242). By permission.]

permit him to have his own way too frequently. Later, when their child faces the inevitable frustrations of life, he will be poorly prepared to cope with many of them. As with other functions, the change from egocentrism to mature sociability is neither continuous nor painless. Frequent or not, regressive steps are shown by many children, adolescents, and adults. It is of paramount importance that steps forward be more numerous than setbacks.

The nuclear family is the most influential agent in shaping the child's attitudes, values, and beliefs before the school and peers ascend and become major factors. The peer group often supersedes the family at the age of nine or ten, when the child becomes group-minded and revises his standards to fit the group expectations. In early adolescence there is much dependence on peers and intense loyalty to group leaders and standards. In later adolescence, self-definition and self-reliance increase so that companionships become less passionate than they were earlier (Purnell, 1970).

In late adolescent years the individual is intensely concerned with establishing and furthering heterosexual friendships, but emotional sensitivity and heightened self-consciousness often hinder ease of communication between the sexes. In early adulthood, if not before, the possible choices of a lifetime partner are narrowed down until the decision is made. The necessary adjustments to marriage are a significant developmental task and are indicative of a fairly high degree of socialization and personal maturity. Throughout adulthood, one's social horizon expands through friendships, business contacts, and acquaintances made in the community. With senescence, however, there is an increasingly sharp decline in new friendships and a gradual loss, due to moving, retirement, or death of old associates. Hence senescent persons set the process of socialization into reverse.

SELF-CONCEPT

The twin concepts of ego and self form the core of a controversial area in current psychological theory. The self as the person views it and the self as the agent for activity are probably the leading ways of approaching this concept. In the first sense the ego is an object, in the second a subject. There is an evolving sense of self, and it tends to encompass both knowledge about self and self as an organizer of behavior. What the brain is for the organism, the self is for personality. It is a central system for integrating and directing forces toward various forms of interaction with persons and objects. The sense of *self* refers to the acquired set of feelings and attitudes that a person has toward his own appearance, powers, and behavior. It is the core of each personality. In agreement with George Mead, Harry S. Sullivan, and S. H. Cooley, John W. Kinch (1963) says, "The individual's conception of himself emerges from social interaction, and in turn, guides or influences the behavior of that individual." Kinch feels that consensus could be reached for the following definition: "The self-concept is that organization of qualities that the individual attributes to himself." Since qualities include attributes and roles, this definition is superior to many others. Recently, Boyd R. McCandless and Ellis D. Evans (1973, p. 389) came to a similar definition of the self-concept: it refers to "an individual's awareness of his own characteristics and attributes, and the ways in which he is both like and unlike others." Once the sense of self is experienced, it gains validity for the individual and its behavior-directive power rises. A child of three, for example, feels that he has a mind of his own, and he applies it determinedly. An adolescent wants to be himself and frequently objects to the attempts of others to control or direct his behavior.

In the Freudian sense, *ego* refers to the core of personality that controls impulses and drives in conformity with the demands of reality. Since ego includes nearly the same aspects of the person as the self and since most authors do not make clear distinctions between the two concepts, they may be treated as somewhat equivalent terms, yet some differences are to be noted. While the ego has its roots in unconscious dynamics, the self is anchored in the awareness and experience of identity. Most of the time a person's actions are in accordance with his idea of himself. If a person perceives himself as specially inferior to most peers, whether or not this is objectively true, most of his actions will correspond to this subjectively perceived deficiency.

Seymour Epstein (1973) suggests modifying the self-concept into a *self theory* constructed by the individual about himself as an experiencing and functioning person who deals with the nature of the world, the nature of the self, and their interaction. For Epstein the fundamental purpose of the self theory is to *optimize the pleasure-pain balance of the individual over the course of a lifetime.* Related to this hedonistic purpose are two basic functions: to facilitate the maintenance of self-esteem and to organize the data of experience in a manner that can be coped with effectively. The basic attributes of this self theory are the same as those of all theories: extensivity, internal consistency, usefulness, and testability. Developmentally, the individual acquires first a body self by gaining control over the limbs, by experiencing double sensation when he touches his own body, and by saliency (it hurts when one's own body is pinched but not when someone else's is). Development of a body self facilitates inference of an inner self—a personality identity based on a feeling of continuity of experience and an awareness of one's emotions, drives, and motives.

A moral self is acquired, according to Epstein (1973), when the need to obtain approval and to avoid disapproval rises and the child is capable of reacting in terms of internalized parental values. It is noteworthy that the process of moral learning is a long uphill climb, and the hill seems to get steeper and more slippery as adolescents climb it (Boulding, 1964, pp. 146–147).

When the self-concept is converted into a self theory, a broad base is provided for its study. The self theory is perhaps congenial to highly intraverted adolescents and adults who, by dealing extensively with themselves, construct series of somewhat related hypotheses often organized into hierarchical conceptual systems of some kind worthy of investigation. Lambert (1972) compared ego development with moral development in several age groups as measured by the Kohlberg Moral Development Test and the Loevinger Sentence Completion Test. Findings on six samples ($N = 107$) show that individuals who score high on ego development also score high on moral development and vice versa. High ego strength apparently indicates a high level of moral development.

Consistency or continuity of behavior is largely a result of the self-concept. Although subject to maturation and to environmental influences, modification of the self-concept usually occurs in small steps. Experiences that would radically distort it, even favorably, tend to be denied reference to the self, as when people exclaim "It can't be true," "I wasn't myself," or "I lost my head." The degree of flexibility with regard to awareness differs, of course, from person to person, some persons tolerating a considerable range of modifications and others showing much rigidity. As a rule, flexibility slowly declines with advancing age, older persons showing more rigidity than younger ones. With the advent of adulthood, there arises a natural desire for stability. For many it is rewarding if people continue reacting to them in the same manner as in the past or in a similar manner. For Harry S. Sullivan (1940/1953, pp. 20–23), "The self dynamism is built up out of this experience of approbation and disapproval, of reward and punishment." Later on, the self dynamism "precludes the experience of anything corrective, anything that would be strikingly different."

Maintaining continuity of identity is another major function of the self. In spite of manifold changes in almost all aspects of personality as a result of growth and experience, the individual nevertheless recognizes that he or she is the same person as ten or twenty or more years ago. Many qualities and traits of a person change gradually. This allows self-identity to remain relatively consistent throughout adulthood. Many mentally disturbed persons, however, lose this sense of self-identity and misidentify themselves as extremely powerful or famous personalities.

Genesis of the Self

It is difficult to discover at what point in development the sense of self emerges. There is an increasing awareness of "me" and "not me," "I can" and "I can't"; but exactly when the individual becomes generally conscious of his body, of his various abilities, of his "good me" and "bad me," is still matter for investigation (Elkind, 1972). The fundamental difficulty with the self-concept is that we know about it only by inference. This is true for the person himself as well as for the observer (Combs and Sporer, 1957; Guardo, 1968).

Theodore R. Sarbin (1952) has formulated an interesting set of propositions concerning the genesis of the self-concept. It begins with a "somatic self" at birth, which remains the core of the self-concept. This somatic self consists of percepts pertaining to the neonate's body, such as sensations of hunger, dampness, or indigestion. This self is later supplemented by a "receptor-effector self," which consists of perceptions of the sense organs and musculature. The "primitive construed self," which follows from the previous two, is a rather vague awareness of self as an individual being. The "social self" emerges at approximately two years as a definite sense of self-identity and an awareness of different roles and of relationships with others. This rise in self-awareness, occurring at the beginning of the struggle for autonomy, may be seen as the first manifestation of a genuine self-concept. The young child is led by his or her own desires and wishes. Resistance to parental control at the age of two and three years can be understood in the light of the child's effort to assert himself. The way parents and others react to the child's behavior either encourages or inhibits the child in the process of sensing his or her own selfhood. Acceptance by others leads to a self-accepting attitude and an ability to live comfortably with one's emotions and to stand up for one's own preferences and rights.

Erik Erikson (1963, pp. 235–246) considers the emergence of personal identity an important element contributing to the ego structure. This configuration results from the gradual integration of "constitutions given, idiosyncratic libidal needs, favored capacities, significant identifications, effective defenses, successful sublimations, and consistent roles." Through a synthetic process of the ego, personal identity becomes a part of ego identity sometime during the period leading to adolescence. Erikson's well-known eight life crises, especially "the battle of autonomy," may be seen as milestones for the development of the self-concept.

After surveying studies related to the self-concept and self-identity, Carol Guardo (1968) uncovered the following common dimensions of many theoretical orientations: (1) a sense of singularity, (2) a sense of humanness, (3) a sense of sexuality or sexual identity, and (4) a sense of continuity or sameness from the past to the present and into the future. The sense of self-identity is a developmental phenomenon that unfolds across time. Her later study (Guardo & Bohan, 1971) shows that the sense of self-identity and its postulated dimensions are developmental phenomena demonstrable by empirical results in 116 six- to nine-year-old children. Her results also indicate that these developments parallel Piagetian findings regarding cognitive development of children.

Role of Self-awareness

The growing self-awareness within the child increases his or her behavior regulation. The self-concept becomes a pivot for the integration of past and present experiences. It establishes an order of priority in the response repertoire to various objects, persons, and situations. Slowly the self-concept achieves the place of final arbiter in many conflict situations. Tendencies to act in accordance with the self begin to dominate. The self becomes an architect of the life-style, as the child tends to act more and more in terms of his self-perceived dispositions, abilities, and resources. Self-actualization refers to the gains in ego strength resulting from self-development, self-utilization, and self-identity formation. The resultant selfhood is not a matter of internal maturation; it has to be achieved. Herbert A. Otto (1968) feels that while people make themselves into what they imagine themselves to be, yet they are usually functioning at about 10 percent of their potential, as only a small fraction of their capacities are developed to a high degree.

Stanley Coopersmith (1967, pp. 249, 236) stresses acquisition of high self-esteem as a prerequisite for competence and effectiveness, independence and creativity. Coopersmith's empirical findings suggest three conditions as necessary antecedents of high self-esteem: (1) total or nearly total acceptance of children by their parents, (2) clearly defined and enforced limits, and (3) respect and latitude for individual action within the defined limits.

Throughout middle and late childhood, peers and reference groups begin to play a more dominant role, displacing the parent as the primary modifier of the self-concept. The child more and more comes to identify with those of his or her own age and to adopt a code of behavior from the peer group of the same sex. During later childhood, the self-concept is fairly stable, owing to the relatively even rate of development of the different personality variables. However, with the onset of puberty there is a drastic change in the self-concept. The young adolescent perceives himself as an adult in many ways, yet for most parents he is a child. Though independence from adults is in most cases still impossible for at least several years, the teenager is driven toward self-regulation of behavior. The majority of adolescents become heavily dependent on peers before they gain readiness for a higher mode of self-dependence.

Because of the extensive changes affecting the adolescent in almost all areas, the self-concept is also in a state of flux during this period. The uncertainties of the future make the formulation of definite goals a difficult task. However, it is in the resolution of these adolescent problems and conflicts that the self-concept of the adult is born. The values and attitudes that are part of the self-concept at the end of adolescence are those which tend to remain as relatively permanent organizers of behavior.

In early adulthood, with its new challenges and responsibilities, the self is tested and proved, and by the age of twenty-five or thirty an adult ego is often completely formed. The individual has much autonomy and self-sufficiency. From this age on, the self-concept becomes increasingly resistant to change. In middle adulthood there is generally no major change in the basic qualities of the self-concept, except the modifications that come with age and experience.

The increased longevity human beings are enjoying today has concomitant drawbacks. With the policy of involuntary retirement so prevalent, many persons at this time experience drastic changes in their self-concept. After thirty or forty or more years as breadwinners and contributing members of society, many of these persons find themselves relegated to the position of second-class citizens. The ensuing self-devaluation often disturbs the total personality integration.

DEVELOPMENT OF SEXUALITY

Ascription of gender at birth, followed by parental and social confirmation, impinge on the expanding cognitive abilities and feelings of the child. Perceptual data play a leading role in core gender identity formation, starting in late infancy. Dressed or undressed, one sees oneself consistently as a male or female. Social reinforcement furthers gender identity to a higher extent than later identification mechanisms, castration anxiety, or envy of the male genitals. Preschool childhood is critical for establishing the concept of self as male or female, and perhaps middle adolescence is a critical period for discovering one's sexual attractiveness to members of the opposite sex. Later adolescence serves the function of determining adult identity in its major sexual dimensions and roles. The task of achieving normal heterosexuality is critical for this phase, when being in step with peers matters so much. Once established, these fundamental attitudes tend to persist and are difficult to transform.

Human beings consist of both masculine and feminine components, but the elements of one of them are usually predominant, and the biological givens at birth are usually sufficient to ascribe gender. In the pre-oedipal years of life, masculinity or femininity is furthered by observable differences between boys and girls that call for different parental responses to them (Rosenberg & Sutton-Smith, 1972).

Freud (1962, pp. 97–101) felt that to further its overall maturation, the polymorphous sexual drive needs repression in childhood and for girls in adolescence. The disposition to perversion is a "universal disposition of the human sexual instinct and . . . normal sexual behavior is developed out of it as a result of organic changes and physical inhibitions occurring in the course of maturation." Lack of repression furthers perversions, but continuation of repression feeds neurotic trends of personality. An alternative to these undesirable consequences is offered by Freud's concept of sublimation—redirection of sex energies toward nonsexual goals.

Defining sexual identity constitutes one of the paramount tasks of human development. The relation between boys and girls and between men and women is determined by the biological differences between the sexes and the cultural and religious values ascribed to appropriate sex-role behavior. The sex role is often engendered by autoerotic behavior during the early years of life and expressions of heterosexual interests during the adolescent years. Defining one's sexual role adds much to the formulation of adult sex identity.

The individual's sense of self-identity is largely a result of early identifications with others of his or her sex, parents and selected peers in particular. Biological givens and sexual, emotional, and cognitive differentiation all press the individual to favor certain persons as models for self-development, usually those who express affection and acceptance, respect and approval. Persons used as models who are of the same sex help both males and females to embrace a satisfactory style of life. Sexual adjustment during adolescence and the early years of adulthood is a task of herculean dimensions. All major aspects of development, including the self-concept, are analyzed in the presentation of the levels of life. Chapter 12 presents sexual typing in some detail, while Chapters 13 and 15 deal extensively with the attainment of heterosexuality and adolescent sexual roles.

QUESTIONS FOR REVIEW

1 Describe the pattern of physiological growth and explain the acceleration and deceleration of growth occurring before maturity is attained.

2 Describe somatotrophin and thyroxine and their functions in the organic growth of the individual.

3 Describe Bridge's tree of emotional differentiation. When do the basic emotions, such as anger, fear, and affection appear?
4 What role does human stimulation play in the development of affection, empathy, and other socially desirable emotions?
5 Describe Piaget's theory of perceptual development and explain his concept of centration.
6 Explain cognitive assessment by the use of individual intelligence tests, such as the Stanford-Binet and the McCarthy scales.
7 Why is intelligence testing used for schoolchildren with learning problems?
8 Describe Razran's levels of learning, their universality and efficiency.
9 How does infant socialization begin? When does the child need age-mates for play?
10 Define the self-concept and describe Epstein's self theory.
11 Explain the effects of the child's increasing self-awareness. What basic influence do parents have on the child's self-perception?
12 What factors play a major role in sex-identity formation?

REFERENCES

Selected Reading

Bühler, C., & Massarik, F. (Eds.) *The course of human life: A study of goals in the humanistic perspective.* New York: Springer, 1968. An intensive study of goals in human life by eighteen authors, emphasizing genetic, emotional, and sociocultural factors as well as individual efforts toward integration and self-fulfillment.

Gale, R. F. *Developmental behavior: A humanistic approach.* New York: Macmillan, 1969. An exposition of the various aspects of self, including becoming an authentic person.

Hurlock, E. B. *Personality development.* New York: McGraw-Hill, 1974. A major study of personality patterns and symbols of self, including their persistence and change during life.

Specific References

Ashton, R. The effects of the environment upon state cycles in the human newborn. *J. exp. Child Psychol.,* 1971, **12,** 1–9. (a)

Ashton, R. State and the auditory reactivity of the human neonate. *J. exp. Child Psychol.,* 1971, **12,** 339–346. (b)

Baltes, P. B., & LaBouvie, G. V. Adult development of intellectual performance: Description, explanation, modification. In C. Eisdorfer & M. P. Lawton (Eds.), *The psychology of adult development and aging.* Washington: American Psychological Association, 1973.

Baumrind, D. Some thoughts about childrearing. In U. Bronfenbrenner (Ed.), *Influences on human development,* pp. 396–409. Hinsdale, Ill.: Dryden, 1972.

Bee, H. L., Van Egeren, L. F., Streissguth, A. P., Nyman, B. A., & Leckie, M. S. Social class differences in maternal teaching strategies and speech patterns. *Developm. Psychol.,* 1969, **1,** 726–734.

Birns, B. Individual differences in human neonates' responses to stimulation. *Child Developm.,* 1965, **36,** 249–256.

Botwinick, J. *Cognitive processes in maturity and old age.* New York: Springer, 1967.

Boulding, K. E. *The meaning of the 20th century: The great transition.* New York: Harper & Row, 1964.

Bower, T. G. R. The object in the world of the infant. *Scient. Amer.,* 1971, **225,** 30–38.

Combs, A. W., & Sporer, S. W. The self, its derivate terms, and research. *J. indiv. Psychol.,* 1957, **13,** 134–145.

Coopersmith, S. *The antecedents of self-esteem.* San Francisco: Freeman, 1967.

Department of Health, Education and Welfare (DHEW). *How children grow.* Washington, D.C.: U. S. Government Printing Office, 1972.

Elkind, D. "Good me" or "bad me"—The Sullivan approach to personality. *New York Times Magazine,* 1972, Sept. 24, 18.

Elton, C. F., & Shevel, L. R. *Who is talented? An analysis of achievement* (Res. Rep. No. 31). Iowa City, Ia.: American College Testing Program, 1969.

Epstein, S. The self-concept revisited or a theory of a theory. *Amer. Psychologist,* 1973, **28,** 404–416.

Epstein, W. *Varieties of perceptual learning.* New York: McGraw-Hill, 1967.

Erikson, E. H. *Childhood and society* (2d ed.). New York: Norton, 1963.

Eysenck, H. J. *Race, intelligence, and education.* New York: Library Press, 1971.

Fomon, S. J. *Infant nutrition.* Philadelphia: Saunders, 1967.

Freud, S. *Three essays on the theory of sexuality.* (J. Strachey, Ed. and Trans.) New York: Basic Books, 1962.

Frostig, M., & Horne, D. *The Frostig program for the development of visual perception.* Chicago: Follett, 1964.

Gagné, R. M. Contributions of learning to human development. *Psychol. Rev.,* 1968, **75,** 177–191.

Garn, S. M. Aspects of growth and development. In V. C. Vaughan, III (Ed.), *Issues in human development,* pp. 52–58. Washington: 1970.

Garner, W. R. To perceive is to know. *Amer. Psychologist,* 1966, **21,** 11–19.

Gibson, E. J. *Principles of perceptual learning and development.* New York: Appleton-Century-Crofts, 1969.

Greulich, W. W. The rationale of assessing the developmental status of children from roentgenograms of the hand and wrist. *Child Developm.,* 1950, **2,** 33–44.

Guardo, C. J. Self revisited: The sense of self-identity. *J. human Psychol.,* 1968, **8,** 137–142.

Guardo, C. J., & Bohan, J. B. Development of a sense of self-identity in children. *Child Developm.,* 1971, **42,** 1909–1921.

Guilford, J. P. *The nature of human intelligence.* New York: McGraw-Hill, 1967.

Jensen, A. R. How much can we boost I.Q. and scholastic achievement? *Harvard educ. Rev.,* 1969, **75,** 39, 1–123.

Jensen, A. R. *Educability and group differences.* New York: Harper & Row, 1973.

Kinch, J. W. A formalized theory of the self-concept. *Amer. J. Sociol.,* 1963, **68,** 481–486.

Lambert, H. A comparison of cognitive developmental theories of ego and moral development. *Proc. Ann. Convention, APA,* 1972, **7**(Pt. 1), 115–116.

Macy, I. G., & Kelly, H. J. *Chemical anthropology: New approach to growth in children.* Chicago: University of Chicago Press, 1957.

Maccoby, M. The three C's and discipline for freedom. *School Rev.,* 1971, **79,** 227–242.

McCarthy, D. *Manual for the McCarthy Scales of Children's Abilities.* New York: Psychological Corporation, 1972.

Matarazzo, J. D. *Wechsler's measurement and appraisal of adult intelligence* (5th ed.). Baltimore: Williams & Wilkins, 1972.

McCandless, B. R., & Evans, E. D. *Children and youth: Psychosocial development.* Hinsdale, Ill.: Dryden Press, 1973.

McCandless, B. R., & Roberts, A. Teachers' marks, achievement test scores, and aptitude relations with respect to social class, race, and sex. *J. Educ. Psychol.,* 1972, **63,** 153–159.

McClelland, D. C. Testing for competence rather than for "intelligence." *Amer. Psychologist,* 1973, **28,** 1–14.

Meeker, M. N. *The structure of intellect.* Columbus, Ohio: Merrill, 1969.

Otto, H. A. *Human potentialities: The challenge and promise.* St. Louis: Warren Green, 1968.

Patten, B. M. *Human Embryology* (3d ed.). New York: Blakiston, 1968.

Piaget, J. *The mechanisms of perception.* New York: Basic Books, 1969.

Purnell, R. F. Socioeconomic status and sex differences in adolescent reference-group orientation. *J. genet. Psychol.,* 1970, **116,** 233–239.

Profiles of Children. Washington, D. C.: U. S. Government Printing Office, 1970.

Razran, G. Evolutionary psychology: Levels of learning—and perception and thinking. In B. Wolman (Ed.) & E. Nagel (Cons. Ed.), *Scientific psychology: Principles and approaches,* chap. 13. New York: Basic Books, 1965.

Riegel, K. F., & Riegel, R. M. Development, drop, and death. *Developm. Psychol.,* 1972, **6,** 306–319.

Rosenberg, B. G., & Sutton-Smith, B. *Sex and identity.* New York: Holt, 1972.

Sarbin, T. R. A preface to a psychological analysis of the self. *Psychol. Rev.,* 1952, **59,** 11–22.

Scarr-Salapatek, S. Race, social class, and IQ. *Science,* 1971, **174,** 1285–1295.

Schaie, K. W. Methodological problems in descriptive developmental research on adulthood and aging. In J. R. Nesselroade & H. W. Reese (Eds.), *Life-span developmental psychology: Methodological issues.* New York: Academic Press, 1973.

Schaie, K. W. & LaBouvie-Vief, G. Generational versus ontogenetic components of change in adult cognitive behavior: A fourteen-year cross-sectional study. *Developm. Psychol.,* 1974, **10,** 305–320.

Stechler, G. Autoplastic and alloplastic behavior: How the infant manipulates input. In V. C. Vaughn III (Ed.), *Issues in human development,* pp. 27–31. Washington, D. C.: U. S. Government Printing Office, 1970.

Stevenson, H. W., & Odom, R. D. Interrelationships in children's learning. *Child Developm.,* 1965, **36,** 7–19.

Sullivan, H. S. *Conceptions of modern psychiatry.* New York: Norton, 1953. (Originally published, 1940.)

Terman, L. M., & Merrill, M. A. *Stanford-Binet Intelligence Scale: Manual for the third revision, form L-M.* Boston: Houghton Mifflin, 1960, 1973.

Wechsler, D. *Manual for the Wechsler Adult Intelligence Scale.* New York: Psychological Corporation, 1955.

Wechsler, D. *Manual for the Wechsler Intelligence Scale for Children, Revised.* New York: Psychological Corporation, 1974.

Wechsler, D. *Manual for the Wechsler Preschool and Primary Scale of Intelligence.* New York: Psychological Corporation, 1967.

Part Three

Before and after Birth

Parts One and Two presented the field of human growth science and key influences affecting human beings throughout the life-span. This section will deal with the onset of human life and earliest changes within the living organism, showing the genetic and environmental factors in operation during prenatal life. Both beneficial and detrimental influences are treated, in order to present a coherent and complete picture of this crucial phase of early development.

Although the period within the fluid of the uterus and the first month of postnatal life constitute but a small segment of the human life-span, their significance is tremendous. The viability of the organism (capability of sustaining life outside the mother's uterus) and its very survival after birth are dependent on the fine timing and coordination of a myriad factors. Accidental or not, introduction of any hazardous elements or conditions can have far-reaching effects on the later well-being of the person. Inasmuch as the very bases of behavior, the neuromuscular, glandular, and sensory systems, are formed and integrated during this period, the capacity for adaptation to the environment is largely determined at this stage. In addition to these formative processes—birth and the necessary adjustments to extrauterine life—the individual differences evident after birth and the care provided in the neonatal period are important variables in the future welfare of every human being.

Chapter 5

Prenatal Growth

HIGHLIGHTS

The union of ovum and sperm nuclei defines the hereditary endowment. Embryonic changes bring the first critical period for the growth of the organism and the possibility of external hazards, which depend on the maternal blood chemistry.

After twenty-seven weeks of gestation the fetus has a chance of surviving if born prematurely; gains in viability increase with age. Uniformly throughout the human race, full term comes at the end of the tenth lunar month of pregnancy.

The expectant mother usually has a slightly heightened degree of emotional reactivity during pregnancy and needs support, acceptance, and understanding from her husband and other significant figures in her life to facilitate her total adjustment to the coming motherhood.

When the baby becomes part of the family, each member of the family has some adjustments to make. The baby usually becomes the center of attention and care.

The development of the child begins at conception, about nine months before birth. Two circumstances precede conception or fertilization: ovulation—i.e., the release of one ripened ovum (sometimes two or more ova) from the ovary into the fallopian tube, usually at about the midpoint of the menstrual cycle—and ejaculation of semen fluid of high sperm density and vitality into the vagina near the neck of the uterus within about two to three days before and up to about one day after ovulation. Irregularities in ovulation may cause impregnation at any time during the menstrual cycle. The mature ovum is one of the most massive and probably the most complex of all living cells. Growth of the oocyte into an ovum occurs within the ovarian sanctuary, well protected against any experimental manipulation of its natural differentiation.

Fertilization occurs when one of the millions of spermatozoa migrates (by propelling itself with its long tail) through the cervix to the uterus and up the fallopian tube and penetrates the membrane of the descending ovum. The nuclei of the two gametes approach each other and the chromosomes from the male and the female aggregate. The 23 chromosomes of the sperm join by pairing off with the 23 chromosomes of the ovum. When the ovum is penetrated by a single sperm cell, a wave of chemical changes takes place within the cell, producing a fortified "fertilization membrane." This membrane restrains penetration by additional spermatozoa. As chromosomes and subchromosomal structures become paired and single complementary strands from each of the two parents unite to form a new DNA spiral, the heredity of the human individual is determined.

The sex of the child depends on the sperm's chromosomes. About half of the spermatozoa carry an X chromosome. Their pairing with the female gamete's X chromosome results in female offspring. The other half of the sperms contain a lighter Y chromosome. If a sperm cell with this chromosome fertilizes the ovum, which always carries an X chromosome, the child will be male. These XX and XY combinations of the fertilized cell are also potential carriers of defective sex-linked characteristics, such as hemophilia (a defect in blood coagulation caused by a deficiency of antihemophilic globulin) and color blindness. The fertilized cell may completely divide into two cells. In such a case, each new cell develops into a separate individual, producing identical twins. When the ovary releases several ripened ova, two or more simultaneous fertilizations may occur, resulting in a multiple birth. These fraternal individuals may be of either sex, whereas identical twins must be of the same sex.

Following fertilization the uterine walls, through the action of hormones released by the ovaries, undergo preparation for the reception of the fertilized ovum, the zygote. Upon descent into the uterus, the zygote attaches itself to and becomes implanted in the uterine wall if the lining of that wall is sufficiently receptive. From the onset of the menstrual cycle until menopause, when the cycle of fertility ends, this monthly preparation for reception of the fertilized ovum continues. Thus after nine or ten days of migration and subsistence through nourishment provided by the yolk of the ovum, the zygote begins to derive nourishment from the mother. Within hours after conception, the fertilized cell begins dividing and continuously divides and redivides until a tiny globule of tissue is formed. The globule consists of an outer layer that will support the structures of the growing organism and an inner core from which tissues and organs of the organism itself will develop.

EMBRYONIC PHASE

With the complete implantation of the zygote in the wall of the uterus, the embryonic period begins. The placenta develops rapidly as threadlike structures (villi) of its membrane tap the source of maternal nourishment. The umbilical cord, measuring 10 to 20 inches in length, connects the placenta to the embryo's abdominal wall. The placenta also serves as a filtering device for blood coming from the mother into the umbilical cord. Filtering safeguards the embryo from many bacilli and other nonnutritive contents of the blood. Newly formed blood vessels from the embryo go through the cord to the placenta and back to the embryo, providing nutrition and removing waste materials. The placenta secretes estrogen and progesterone, ACTH, and adrenocorticoid and glucocorticoid hormones; it performs most functions of the lungs, liver, kidneys, and intestines for the embryo as well. The amniotic sac, containing watery fluid, is developed early in the embryonic phase. Attached to the placenta, it surrounds the embryo and equalizes the pressure, thus protecting the embryo. Within the fluid of the amniotic sac the embryo is practically weightless and as it matures has much freedom to move (Ziegel & Van Blarcom, 1972, pp. 84–90).

Human growth proceeds in such a way that during the embryonic period the rate of growth is most rapid. From a minute globule of tissue no larger than a period, the body increases to about ten thousand times that size by the end of the first month of gestation. Along with this rapid growth, much differentiation of cell structure into tissues and organs occurs. Because of the release of genetic information, the dividing cells produce cells of differing structure and functional capacity. Three distinct layers of tissue, each with specific functions, emerge at the onset of the embryonic phase. The outermost layer, the ectoderm, gives rise to the epidermis of the skin, hair, nails, skin glands, sensory cells, brain, and entire nervous system of the body. The middle layer, the mesoderm, produces the deeper skin layers, muscles, bone structure, kidneys, gonads, and circulatory (including blood) and excretory systems. The innermost layer, the endoderm, is the basis for the digestive system, lungs, liver, pancreas, and other internal organs.

By the end of the second lunar month the embryo's features become distinctly human. About 95 percent of the physiological organs and features are already formed. There are rudimentary forms of the nervous system, including the brain and spinal cord and the specialized receptor organs. The ears are just beginning their external formation. After eight weeks of pregnancy, the embryo sufficiently resembles a human being to be recognized as male or female. The EEG activity of the fetal brain reflects an individual pattern. The body proportions, however, differ greatly from those of the newborn: the head is enormously large, the trunk is about as large as the head, and the extremities are very small. The developing embryo is not part of the woman's body, since the amniotic sac and fluid separate the mother's body from that of the embryo, which, by taking nutrition from her through a filtering placenta, develops its own blood supply, circulated by its own heart as it starts beating on the twenty-fourth day after conception.

The systematic formation of the various organs and bodily systems is a notable fact in prenatal development. Each organ has a specific time to emerge. Metabolic activity is intensified in the area where a new organ is being formed. As tissues and organs begin forming, the dividing cells usually start separately but soon begin to clump and to function together, with only a few cells staying apart or straying away. If a toxic substance is introduced at this time—when, for instance, the heart is being formed—the organ is highly likely to be damaged. Once it has acquired its distinct structure, it becomes more capable of withstanding toxic or other disturbance.

Natural and Artificial Attrition

Full-term development is often challenged in both natural and surgical ways. The possibility of miscarriage is one of the great hazards. More than two-thirds of all miscarriages occur by the end of the first trimester, most often during the tenth and eleventh week of pregnancy to women over thirty-five years of age. The causes of prenatal attrition include certain physical conditions of the mother, e.g., deep tears of the cervix inflicted by previous birth or abortion, tumors of the uterus, defective germ plasm and chromosomal aberrations producing embryo defects, toxic and viral agents, glandular or hormonal imbalance, severe malnutrition, and oxygen deprivation. Singly or combined, these and related factors lead to dislodgement of the embryo. It is nature's way of rejecting an unsound pattern of development. Many early embryonic losses commonly go unnoticed (Guttmacher, 1973, pp. 131–135).

In addition to the natural hazards, the rhythm method and various chemical and instrumental contraceptives, including pills for males and "morning after" pills, are used to prevent conception or implantation. The use of many contraceptive methods is chiefly a female responsibility; vasectomy and use of the condom are the male counterpart. The condom helps to prevent spread of venereal disease, while other contraceptive methods do not. Earlier techniques, and some recently devised techniques, increase the choice for couples who want to prevent conception or to control family size.

Diethylstilbestrol, the "morning after" pill, is a contraceptive fraught with danger. The large amounts of estrogen taken over a period of five days can in some cases incite cancerous growth. Any serious health problem, including previous breast cancer and tendency to blood clotting, precludes even one-time use of the pill. Present contraceptive techniques do not offer complete protection against pregnancy. Scattered sampling indicates that about one-third of the couples who use various birth control techniques conceive involuntarily within four to five years. Pohlman and Pohlman (1969, pp. 343–400) present marital and other sexual relations, discuss their significance for husband, wife, and others, and analyze the application of older and newer female and male methods of contraception.

The occurrence of pregnancy can be determined in the physician's office or at home with the Ova II kit, which contains hydrochloric acid and caustic soda and a nine-step direction sheet. The chemicals are mixed with urine specimens as directed. The test indicates pregnancy when the mixtures in the two containers remain nearly the same color.

Technically, an abortion is a spontaneous termination of pregnancy during the first five lunar months, or up to twenty weeks of pregnancy—before the fetus is viable. Surgical abortions are frequently performed during the early months of pregnancy. Dilatation of the cervix and curettage or aspiration evacuation—vacuuming out—is a method frequently used. Late abortions of the fetus (from thirteen weeks on) may use the scooping-out or salting-out technique—injection of hypertonic salt or saline solution into the uterus, which causes death of the fetus and miscarriage within twenty-four hours; otherwise, major surgery similar to cesarean section is performed (Guttmacher, 1973, pp. 147–148). At five to six months of pregnancy, the unborn fetus is probably more functionally alive than many hospital patients who require medically induced stimulation to maintain heartbeat or lung function. For various reasons, including the changing moods and whims of the pregnant woman, birth is denied to an ever-increasing number of human beings during the embryonic or even advanced fetal phase of growth.

Since the Supreme Court decision of 1973 permits abortion up to six months and since length of pregnancy may be miscalculated or falsified, limited infanticide is thus made possible. As Ramsey (1973) observes, future medical knowledge may well move the age of viability into earlier phases of pregnancy, so that application of the present law will decree the death of many viable infants. It is pertinent to say that for prevention of infanticide, the time of viability ought to be qualified by *minus* one lunar month. Indeed, the medical practice of fetology "knows no difference that would warrant less care for the fetus" before or after viability (Ramsey, 1973).

To abort or not to abort is probably an agonizing decision for many women, since induced abortion is linked with hostility and murder (Pohlman & Pohlman, 1969, pp. 398–400). Speaking to a friend or relative is not enough; professional counseling is indicated. The counselor will help the client look into many aspects of the conflict and several alternative solutions.

The availability of contraceptives permits limitation and unnatural spacing of children. Nature's own child-spacing interval of about sixteen to twenty-four months in the breast-feeding mother is noteworthy. The practice of breast feeding is contraceptive for at least three months; beyond that time there is increasing probability of conception. In this natural spread, sibling relationships are likely to be more intimate and less rivalrous. A toddler, who is usually willing to move away from the mother, accepts more readily the addition of a baby to the family than does a one-year-old, whose dependence on the mother is usually very strong. For example, when three children are planned in nature's way, no birth control pills or other contraceptives are needed for about six years during the time of intense male sexuality (Ratner, 1970).

Significant progress has been made in developing special techniques to forecast hereditary diseases of the fetus. For this purpose amniotic fluid containing cells from the fetus is drawn from the uterus after the fourteenth week of pregnancy, a process known as amniocentesis. About one in five of the cells survive and multiply in a prepared test-tube fluid. Analysis of the chromosomal and subchromosomal structures reveals the presence or absence of hereditary aberrations (Friedmann, 1971). By 1975, about 60 genetic diseases could be diagnosed by amniocentesis, including Down's syndrome, Tay-Sachs disease, Huntington's chorea, cystic fibrosis, and sickle-cell and I-cell diseases. Karyotype analysis is also used, usually before conception, to discover the parents' probability of bearing a child with muscular dystrophy or certain other genetic diseases. Genetic testing and counseling can prevent defective children and heartbroken parents.

After three months of gestation, the sex of the offspring can be determined by withdrawing some of the fluid surrounding the fetus, spinning the fluid, and then analyzing cells shed from the fetus for chromosome composition. If the spun cells are placed in a cow-serum culture and allowed to grow, the method of chromosome analysis becomes definite. Sex determination is done not to satisfy the parents' curiosity but mainly for medical reasons, especially when certain genetic diseases that pass only to male or only to female offspring are being considered.

Embryonic growth consists chiefly of structural differentiation and organ formation. Much refinement and relocation of the organs and systems occurs during the fetal period. Gains and improvements in the functional efficiency of most structures and physiological systems constitute the key task of fetal development.

FETAL PERIOD

The fetus is a rapidly developing male or female. The fetal period extends from about two months after fertilization to the time of birth, which normally occurs at the end of the tenth lunar month, or about 266 days after conception. Since the differentiation of tissue into various bodily organs and systems occurs during the embryonic phase of development, the fetal period is largely one of structural refinement and proper organ positioning for most body systems. Cartilage formed in the late phase of the embryo is now in part converted into the bone mass. Primary centers of ossification—protein matrices absorbing calcium and other minerals from the bloodstream—are set up in various areas of the body. X-rays of the bones are occasionally used to determine the biological age of children. The greater the number of ossification centers and the greater the degree of epiphyseal closure, the greater is the biological age (or bone age) of the child tested. The rate of growth may also be accurately assessed by gauging fetal size by means of an ultrasonic sound device, which yields a three-dimensional "map" of the unborn child. Moreover, amniocentesis allows a precise measure of the duration of the pregnancy. One or two of these clinical tests enables an obstetrician to discover whether a fetus is of a size appropriate to its age (DHEW, 1973, pp. 73–166).

Fetal growth prepares the organism to sustain itself after birth. One by one the organs and systems become partially or fully functional and serve the organism. At the present time, *viability,* or ability to survive premature birth, starts at about the end of the sixth month of gestation: by twenty-seven weeks of age the fetus has a chance of surviving if it is born prematurely (Hooker, 1969, pp. 76–77).

During the fetal stage, physiological growth occurs at a rapid rate. About fifteen weeks after conception, the pituitary gland begins secretion of somatotrophin—the growth hormone regulating many differentiation processes in the body. Thyroxin and cancitonin produced by the thyroid, parathormone excreted by the parathyroids, and insulin secreted by the islets of Langerhans are other significant hormones regulating metabolic and growth patterns. By the end of the third month, an average human fetus weighs only 1 ounce and is 3 inches long. By the end of five lunar months, the fetus weighs 9 to 10 ounces and is about 9 inches long. At eight months it weighs 4 to 5 pounds and is 17 to 18 inches long. At full-term birth the infant is about 20 inches, or 50 centimeters, long and on the average weighs 7 pounds 4 ounces, or about 3,300 grams. The extensive physiological changes are illustrated in Figure 5-1, which shows the actual size of the embryo in the early weeks of growth and the fetus in its formative stages.

Newborn infants weighing less than 2,500 grams (5½ pounds) are defined by international standards as *premature.* Many prematurely born infants need incubators and special care in order to survive. Most hospitals retain the neonates until they reach either 5 or 5½ pounds. The nearer the infant is to full term, the better are the chances of survival. Birth weight alone is not a satisfactory indicator of gestational age, which is more closely related to proximity to full-term maturity. Many fetuses simply weigh less than expected at any age of gestation, while others weigh more. Hence accurate determination of prematurity is a matter of considerable complexity.

There is a growing tendency to define prematurity by a gestational age of less than thirty-seven weeks. In this case, too, the reliability of the last menstrual period is not high, since about 20 percent of women have episodes resembling a menstrual period during pregnancy (Battaglia, Frazier, & Hellegers, 1966). For this reason, charts relating the specific age at which various neuromuscular reflexes begin responding are now available in delivery rooms and nurseries in many countries. Growth-retarded neonates usually exhibit the same reaction patterns as full-term infants. They usually need an injection of glucose, because their glycogen stores are depleted before birth. The genuinely premature baby lacks the expected neuromuscular responses; he is subject to various respiratory complications, especially hyaline membrane disease—progressive collapse of the thin air sacs that form the lung. Decreased oxygen

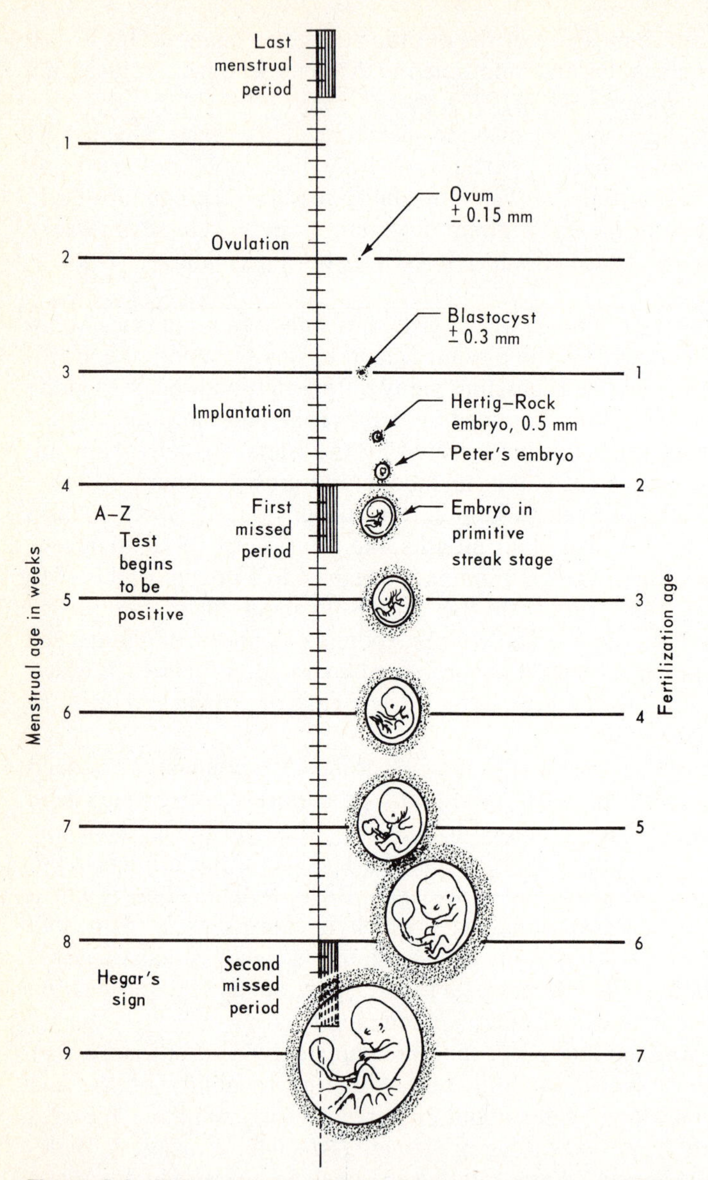

Figure 5-1 Early prenatal mitosis. Diagrams show actual size of embryos and their membranes, based on the mother's menstrual history. [*Source:* B. M. Patten. *Human embryology.* (3d ed.) New York: McGraw-Hill, 1968, p. 145. By permission.]

supply renders the baby vulnerable to brain damage (DHEW, 1973, pp. 73–166). Infants of low birth weight who survive the neonatal phase usually grow well and attain a normal growth pattern (Warkany, 1966).

PHYSIOLOGY AND BEHAVIOR

The internal organs are structured early in prenatal life. By the fifth week, the principal structures of the brain—midbrain, cerebellum, medulla—can be distinguished. The peripheral nerves, spinal ganglia, and related structures necessary for reflexive or autonomic activity are largely acquired in the third month of gestation. Before the embryonic phase ends, the brain becomes partially functional and conveys impulses aiding coordination of the other bodily systems. The heart is formed early and, as already noted, starts beating on the twenty-fourth day of life (Gilbert, 1963, p. 39). At about the fifteenth week, the fetal heartbeat can be detected with a stethoscope. The heartbeat of female fetuses is more rapid than that of males, but the difference is not significant enough to identify the sex of the individual with a high degree of reliability. The internal organs move to approximately mature positions as early as the fifth lunar month. By the end of the fifth month of gestation, the complete quota of neurons that will eventually form the mature brain is already present. Myelenization (protective-sheath formation) of the neurons progresses rapidly and increases their functional capacity.

Motor activity begins at a late stage of embryonic development. Progressing in a cephalocaudal direction, reflexes occur in the region of the head before they occur in the trunk and later in the extremities. On the basis of motion-picture sequences, Humphrey (1970) reported fetuses from seven and a half to ten weeks having widespread reactivity of total pattern reflexes: first the head bends, then the hands uncover the mouth and move backward, and finally trunk and pelvic movement develops. Avoidance and protective reactions, such as moving away from a stimulus, and feeding reflexes—opening the mouth and swallowing—begin to appear at the eleventh week. Humphrey notes a relationship between fetal and infant movements and exemplifies it by the partial finger closure seen in both fetuses and infants.

Type and amount of activity differs from fetus to fetus. Some are active as much as 75 percent of the time, others as little as about 5 percent. Some fetuses almost constantly kick, turn, and squirm; others maintain the same position but thrust and kick their hands and feet. Squirming or writhing movements are most frequent during the third to fourth month before birth and then decline steadily until birth.

Maternal emotions such as fear, grief, or anxiety cause an immediate and profound increase in the activity level of the fetus, especially during the last trimester of pregnancy (Sontag, 1966). Some fetuses have hiccups (quick jerks of shoulders and trunk) almost every day, while others do not have them at all. Spelt's experiments (1948), based on the mother's perception of a conditioned stimulus, point to the possibility of establishing a conditioned response in the human fetus within the womb during the last two months of gestation. At the eighth month of pregnancy, for example, violent kicking occurs as a response to a bell. No other supportive studies of prenatal conditioning have been reported (Stevenson, 1970, p. 854). After reviewing the conditioning reports for three- to ten-day-old infants, Sameroff (1971) concluded that the occurrence of conditioning is apparently dependent on the development of sensory and motor schemata, since only after differentiation of these schemata can connections between stimulus and response be formed and strengthened.

Sensitivity of the fetus develops later than mobility. Cutaneous reactivity and especially the pain reaction are slow to unfold during the prenatal period. Prematurely born infants, when stimulated until blood comes, exhibit little, if any, reaction. The temperature senses are well developed. There is a stronger reaction to stimuli that are warmer than the body than to colder stimuli. Occasionally the obstetrician uses hot and cold water to stimulate crying and breathing in the neonate. The gustatory sense is well developed, and the taste buds are more widely distributed in the fetus than in an adult, being found not only on the tongue but on the hard palate, the tonsils, and parts of the esophagus. The olfactory sense is well developed before birth and functions immediately after birth. Both full-term and prematurely born infants react differently to sweet and salt and to sour and bitter.

Although the senses of sight and hearing are less developed before birth than most other senses, pupillary reflexes, such as reactions to a beam of light, occur in the prematurely born. The adult arrangement of the retina and the fovea is complete before birth. Infants begin to acquire impressions of the world around them as soon as they are born. Newborn infants fixate objects and patterns in their surroundings with their eyes, and they fixate some patterns more often and for longer periods of time than others. Infants are remarkable habituators to stimulation of many sorts (Lipsitt, 1970).

INFLUENCES ON PRENATAL GROWTH

For sound prenatal development, the mother's condition is paramount. What is her nutritional status? Does she use drugs or alcohol? What is her developmental status, diet, and overall health? These are some of the questions that need answers, since the period of gestation puts considerable demands on the expectant mother.

Diet

The mother supplies all the nutrients for the rapidly growing embryo and fetus. However, the old saying that a mother must "eat enough for two" is not true. Overeating often places a burden of additional weight on the prospective mother. If her nutritional condition before pregnancy was satisfactory, she has a reserve supply for the embryo's needs. As compared with the mother, the fetus is supplied on a priority basis. In cases of malnutrition the expectant mother will have to augment her nutrition during pregnancy to supply the growing fetus satisfactorily. In cases of serious malnutrition both mother and fetus are likely to suffer. Insufficient calcium in the mother's system, for instance, will impair ossification of the bones and teeth of the fetus, and the mother's teeth are also likely to become more decalcified than before. A greatly lowered amount of protein will reduce the rate of physiological growth of the fetus.

Malnutrition is a major determinant of serious defects in children. "If the caloric intake is too little, growth is impaired and with it, the capacity to resist disease" (Garn, 1971). A long-term survey by Pasamanick and Knobloch (1966) showed that major prenatal deformities, as well as premature birth, neonatal death, and small size of the newborn infant, were highly related to the mother's diet during pregnancy. Intrauterine mortality tends to vary directly with the rate of malformation. What about those who survive the effects of prenatal malnutrition? Prenatal exposure to the Dutch famine of 1944–1945 seemed not to be related to mental performance at the age of nineteen. The study population comprised 125,000 males born in selected famine and control cities in the three-year period 1944–1946. The findings support the hypothesis of the priority of fetal supply to the mother's own nutrient supply (Stein, Sasser, Saenger, & Marolla, 1972).

During the second and third trimesters of pregnancy, the quantities of protein, calcium, and iron as well as vitamins A, B_6, B_{12}, C, D, and E have to be moderately increased to meet the demands of fetal growth (Hepner, 1958; Williams, 1974). Increase of milk and vegetables in the expectant mother's diet is likely to maintain this necessary supply. The added requirement is about 600 to 700 calories per day during these two phases of pregnancy.

Drugs and Illness

Limiting use of drugs to those prescribed (with prescriptions reexamined by the obstetrician, especially if drugs were prescribed by several medical practitioners) is a necessity for the expectant mother in order to avoid intoxication of the fetus. Depending on the quantity, overindulgence in alcohol or nicotine produces adverse effects on the fetus. Statistics show that mothers who smoke give birth to smaller babies. Slightly to moderately higher rates of prematurity for the offspring of mothers (and fathers) who smoke than for those of nonsmokers has been repeatedly confirmed. A significant rise in the number of infants with low birth weights occurred only when both parents were heavy smokers (Joffe, 1969, pp. 237–241; Yerushalny, 1964). A 1971 report by the U.S. Surgeon General reveals that the infants of more than 100,000 smokers studied showed a mean birth weight of about 6 ounces less than that of babies of nonsmokers. At the end of the first year, however, all the infants tended to weigh about the same. If an expectant mother cannot quit smoking, she can reduce its growth-decelerating effects by smoking less and by picking a brand with lower nicotine and tar content.

The most critical prenatal period is the first trimester of pregnancy, which includes the embryonic stage and the early phase of fetal development. Any diseased or toxic condition of the mother significantly affecting her metabolic rate and blood composition is detrimental to the growth of the child, because most toxins penetrate the placenta and invade the fetus. The nature of the teratogenesis (organ malformation) is dependent on the specific embryonic or fetal phase during which viruses, bacilli, drugs, or other detrimental agents act (Joffe, 1969; Pasamanick & Knobloch, 1966; Wilson & Warkany, 1965). It is difficult, however, to determine the exact point at which detrimental effects begin taking their toll. Besides drugs, normal growth can be disrupted by syphilitic spirochetes, bacilli, and viruses, especially the causative agents of infectious diseases such as rubella, measles, whooping cough, and mumps. During the first trimester of gestation and also later such agents can produce effects ranging from fetal death to severe organismic damage, resulting in neurological disorders and subsequent behavior deficiencies and problems in childhood. In designating such dire consequences, Pasamanick, Knobloch, and Lilienfeld (1956) use the concept of a "continuum of reproductive casualty," since a wide variety of prenatal factors are related to the early growth deficiencies.

Metabolic deficiencies add significantly to the disturbances of growth. Phenylketonuria (PKU) results when the enzyme that metabolizes phenylalanine is defective. Hyposecretion of some endocrine glands, such as the pituitary, thyroid, adrenal, and islets of Langerhans, can be very serious unless timely diagnostic examination leads to proper medication. Undersecretion of the thyroid gland, for instance, can lead to a form of severe physical and metal retardation called *cretinism.* As medication, thyroid pills have been available for many years.

Blood incompatibility is another serious condition afflicting some fetuses whose blood type is Rh+ while the mother's is Rh−. Rh incompatibility usually occurs when an Rh− mother is pregnant with an Rh+ fetus for the second time. During her first such pregnancy, Rh+ in the fetal blood, inherited from an Rh+ father, sensitizes the mother, and she produces antibodies usually several weeks after delivery. These antibodies become activated about thirty weeks after the next conception of an Rh+ fetus. Many of them penetrate the placenta and cause a breakdown of the fetal blood cells. In the past, massive blood transfusions after birth saved about half of the afflicted infants. In 1968 a vaccine, Rhogan, was introduced which when administered once after every Rh+ childbirth prevents complications due to this blood incompatibility.

Emotional Adjustment

Pregnancy brings new bodily sensations and new feelings, occasionally toward the fetus. Maternal stress increases fetal movements beyond the normal amount, to the point where the fetus is strained. Joyful and welcome as pregnancy may be, there is a growing burden to be carried, which often intensifies some regular discomforts, such as nausea, heartburn, fatigue, and shortness of breath because of the squeezed lungs. The expectant mother may have doubts about her potential as a mother and wife. She is aware of children being born with a variety of defective conditions and may fear that this will happen to her child. Some strain and distress is experienced throughout the period of pregnancy, though for different reasons. In one of the early empirical studies at the Fels Research Institute, Sontag (1941) demonstrated that strong emotional reactions are irritating to the fetus. The movements of fetuses increased several hundred percent while their mothers were under emotional stress. Even when the disturbance was brief, heightened behavioral irritability of the fetus lasted as long as several hours. An unusually large amount of activity seems to correlate with a "hyperactive, irritable, squirming, crying infant after birth"—a "neurotic" baby (Sontag, 1961).

The heartbeat of the fetus is accelerated for nearly one hour when the mother smokes a cigarette. If she smokes frequently, the little heart may be strained to the point of damage. Although the sensitivity to toxins, hormones, and viruses differs greatly during various phases of prenatal growth, Scott (1962) speaks of marked congenital vulnerability from conception to infancy. Prenatal growth constitutes a major critical period in the total cycle of human development. It determines the soundness of the foundation for postnatal development.

An expectant mother should live her life as any healthy, active woman would but should avoid excesses of any sort: eating, drinking, and smoking all need moderation. The mother's psychological adaptation to her pregnancy is influenced either favorably or unfavorably by her own adjustment to family life. An overall feeling of gratification results when both parents have desired and planned for the offspring. When an unwanted pregnancy occurs, fears and anxiety usually intensify.

Since gonorrhea is widespread among young females and many are not aware of this affliction, there is a possibility of gonococcal infection at the birth of their children. As the fetus passes through the birth canal, germs often get into the eyes, making them swollen and red. If not treated, gonococcal infection often results in blindness. Penicillin ointment or a silver nitrate solution put into each eye after birth prevents complications. Since this is an effective preventive measure, most states require ointment for all newborn children.

About two-thirds of pregnant women experience heightened irritability, the "blues," and crying spells for no apparent reason other than pregnancy. Moods swing more powerfully during this phase of life that at other times. If the expectant mother's health is precarious and if conflicts over pregnancy become severe, early consultation with the family physician or obstetrician should be sought to help her in making decisions and avoiding possible complications. It is the family's, and more directly the expectant mother's, responsibility to do all that can be done to provide a climate for sound prenatal growth of the child.

The expectant mother needs support and frequent approval and reassurance, especially from her husband; additionally she needs lines of communication with other significant persons in her life. If they express approval and understanding of her condition, this helps to raise her sense of worth and dignity as a woman, fulfilling her unique function in the family and society. Opportunities to talk about pregnancy or impending birth when she wishes to are deeply gratifying and reassuring. It reduces the emotional tension resulting from questions pregnancy raises for her way of life, especially if this is her first pregnancy.

TRENDS IN HUMAN DEVELOPMENT

Predictability of Human Development

The prenatal pattern of human growth gives many leads to the formulation of regularities and constancies in human development. Since development consists of orderly sequences of differentiation in growth and behavior, an expert in human development, assisted by laboratory and psychodiagnostic assessment, is in a good position to specify where the individual stands and how fast he or she is likely to move. A modern, thorough biochemical examination is likely to show major aspects of organismic growth and health as well as physiological retardation and illness. Psychological testing, both metric and projective, enables the psychologist to estimate with considerable accuracy a person's aptitudes, needs, defenses, and cognitive abilities and thus infer the direction of the driving forces. Persons found to be highly intelligent, for example, will use their intelligence in a variety of activities—success in school may be predicted. The major aspects of adjustment and psychopathology are usually revealed in projective testing. Thus comprehensive testing is likely to give many useful leads in predicting the motivational trends and achievements of children or adolescents. Since each person is fairly constant in rate of development, the expert can chart the further course of growth and maturation, learning and adjustment. When growth retardation is due to poor nutrition or lack of stimulation, recovery is often substantial when these needs are gratified for a long period of time (Jones, 1970; Kagan & Klein, 1973; Stein, Susser, Saenger, & Marolla, 1972).

Individuality of Each Person

Though to an inexperienced eye many newborn infants and older babies appear to be much alike, observation reveals that quite the opposite is true. There are striking and minute differences in all aspects of appearance, growth, and behavior. Moreover, individual differences increase with age, even though, for example, identical twins may remain very similar throughout childhood. Various forms of analysis—such as fingerprints, handwriting, free drawing, and voice prints—show individualized features even in identical twins. If a teacher asks her pupils to write on any selected theme, each child will begin, continue, and finish it as no other pupil will. Not differences but similarities are likely to be a surprise. Allport (1955, p. 19) catches this emphasis on individual distinction: "Each person is an idiom unto himself, an apparent violation of the syntax of the species. An idiom develops in its own peculiar context, and this context must be understood in order to comprehend the idiom." Bayley's study (1956) of interindividual variations of developmental patterns is supported by various measurements. Many factors contribute to individual differences. Biological identity based on biochemical distinctness in genetic material and metabolism is always a part of the individual and can never be lost or even completely homogenized (Wald, 1965). The rate of maturation, position in the family, and parental backgrounds, as well as the sequence of experiential events, all vary in some significant way from person to person. All these variations add their share toward the distinctness of each child, adolescent, and adult.

QUESTIONS FOR REVIEW

1. What constitutes conception or fertilization? How does it occur?
2. Which chromosomes determine the sex of the individual? How can the sex be determined during prenatal life?
3. Discuss the implantation of the zygote. What developments follow implantation?
4. List and clarify the functions of the placenta. What are the functions of the amniotic sac?
5. Give the major causes for abortion and explain two of them.
6. Explain the concept of viability. When does viability begin? Relate it to late abortion and to premature birth.
7. When does mobility begin and what sequence does it follow? What are the individual differences in kind or amount of activity?
8. Describe the functioning of the several senses at premature and full-term birth.
9. Discuss the expectant mother's diet and some effects of malnutrition.
10. List the major hazards to fetal development and explain some of them, including drug use.
11. What are the basic growth regularities that facilitate prediction? What measures can be used to predict developmental probabilities?
12. What factors contribute to individual differences and how? What happens to these differences with age?

REFERENCES

Selected Reading

Guttmacher, A. F. *Pregnancy, birth, and family planning: A guide for expectant parents in the 1970s.* New York: Viking, 1973. A practical source on pregnancy, birth control, birth complications, and some major aspects of family adjustment.

Patten, B. M. *Human embryology* (3d ed.). New York: McGraw-Hill, 1968. A comprehensive reference work on fertilization and the prenatal growth of various somatic systems.

Prenatal care. Washington, D. C.: Office of Child Development, Department of Health, Education, and Welfare, 1970. Fundamental information about pregnancy and related problems, prenatal growth, birth, and care of the neonate.

Ziegel, E., & Van Blarcom, C. C. *Obstetrical nursing* (6th ed.). New York: Macmillan, 1972. A general source on the female reproductive organs, development of the fetus, and birth and care of the baby.

Specific References

Allport, G. W. *Becoming: Basic considerations for a psychology of personality.* New Haven, Conn.: Yale University Press, 1955.

Battaglia, F. C., Frazier, T. M, & Hellegers, A. E. Birth weight, gestational age, and pregnancy outcome with special reference to high birth weight–low gestational age infants. *Pediatrics,* 1966, **37,** 417–422.

Bayley, N. Individual patterns of development. *Child Developm.,* 1956, **27,** 45–74.

DHEW, National Institutes of Health. *How children grow* (DHEW Publ. No. 18). Bethesda, Md.: National Institutes of Health, 1973.

Friedmann, T. Prenatal diagnosis of genetic disease. *Sci. Amer.,* 1971, **225** (5), 34–42.

Garn, S. M. Aspects of growth and development. In V. C. Vaughn, III (Sci. Ed.), *Issues in human development,* pp. 52–58. Washington, D. C.: U.S. Government Printing Office, 1971.

Gilbert, M. *Biography of the unborn.* New York: Hafner, 1963.

Guttmacher, A. F. *Pregnancy, birth, and family planning: A guide for expectant parents in the 1970s.* New York: Viking, 1973.

Hepner, R. Maternal nutrition and the fetus. *J. Amer. Med. Assoc.,* 1958, **168,** 1774–1777.

Hooker, D. *The prenatal origin of behavior.* New York: Hafner, 1969. (Originally published, 1952).

Humphrey, T. Development of the fetal activity and its relation to postnatal behavior. In H. W. Reese and L. P. Lipsitt (Eds.), *Advances in child development and behavior* (Vol. 5, pp. 1–51). New York: Academic Press, 1970.

Joffe, J. M. *Prenatal determinants of behavior.* New York: Pergamon, 1969.

Jones, M. R. *Miami symposium on the prediction of behavior, 1968: Effects of early experience.* Coral Gables, Fla.: University of Miami Press, 1970.

Kagan, J., & Klein, R. E. Cross-cultural perspectives on early development. *Amer. Psychologist,* 1973, **28,** 947–961.

Lipsitt, L. The experiential origins of human behavior. In L. R. Goulet and P. B. Baltes (Eds.), *Life-span developmental psychology: Research and theory* (chap. 12, pp. 285–303). New York: Academic Press, 1970.

Mace, D. R. *Abortion: The agonizing decision.* Nashville, Tenn.: Abingdon, 1972.

Pasamanick, B., & Knobloch, H. Retrospective studies in the epidemiology of reproductive casualty: Old and new. *Merrill-Palmer Quart.,* 1966, **12,** 7–26.

Pasamanick, B., Knobloch, H., & Lilienfeld, A. M. Socioeconomic status and some precursors of neuropsychiatric disorders. *Amer. J. Orthopsychiat.,* 1956, **26,** 594–601.

Pohlman, E., & Pohlman, J. M. *The psychology of birth planning.* Cambridge, Mass.: Schenkman, 1969.

Ramsey, P. Feticide/infanticide upon request. *Child Fam.,* 1970, **9,** 257–272.

Ratner, H. Child spacing: II. Nature's subtleties. *Child Fam.,* 1970, **9,** 2–3.

Sameroff, A. J. Can conditioned responses be established in the newborn infant? *Developm. Psychol.,* 1971, **5,** 1–12.

Scott, J. P. Critical periods in behavioral development. *Science,* 1962, **138,** 949–958.

Sontag, L. W. The significance of fetal environmental differences. *Amer. J. Obstetr. Gynec.,* 1941, **42,** 996–1003.

Sontag, L. W. Implications of fetal behavior and environment for adult personalities. *Ann. N.Y. Acad. Sci.,* 1966, **132,** 782–786.

Spelt, D. K. The conditioning of the human fetus in utero. *J. exp. Psychol.,* 1948, **38,** 375–376.

Stein, Z., Susser, M., Saenger, G., & Marolla, F. Nutrition and mental performance. *Science,* 1972, **178,** 708–713.

Stevenson, H. W. Learning in children. In P. H. Mussen (Ed.), *Carmichael's manual of child psychology* (Vol. I, pp. 849–938). New York: Wiley, 1970.

Wald, G. Determinancy, individuality, and the problem of free will. In John R. Platt (Ed.), *New views of the nature of man,* pp. 16–46. Chicago: University of Chicago Press, 1965.

Warkany, J. Solved and unsolved problems of intrauterine growth retardation. In R. M. Blizzard (Ed.), *Human pituitary growth hormone,* pp. 16–23 (Report of the Fifty-fourth Ross Conference on Pediatric Research). Columbus, Ohio: Ross Laboratories, 1966.

Williams. S. R. *Essentials of nutrition and diet therapy.* Toronto: Mosby, 1974.

Wilson, J. G., & Warkany, J. *Teratology: Principles and techniques.* Chicago: University of Chicago Press, 1965.

Yerushalny, J. Mother's cigarette smoking and survival of infant. *Amer. J. Obstetr. Gynec.,* 1964, **88,** 505–518.

Ziegel, E., & Van Blarcom, C. C. *Obstetrical nursing* (6th ed.). New York: Macmillan, 1972.

Chapter 6

The Human Being at Birth

HIGHLIGHTS

When the fetus descends into the lower abdominal cavity, birth is likely to occur in less than two weeks. When labor begins and uterine contractions become harder and regular, the process of birth is taking place. Medical assistance helps to relieve pain and provides assurance that the vital needs of both mother and infant will be met.

The introduction of "the pill" and other contraceptives has significantly reduced the fertility rate. Children born during 1976 are expected to live on the average seventy-two years.

The neonate is a highly receptive individual whose orienting reflex maximizes the stimulus value of the surroundings—learning sets in within the first days of life.

Individual differences mark all aspects of the neonate's appearance and behavior. Any mother recognizes her baby's crying because each infant cries as no other does; careful biochemical and behavioral examination shows that no two infants are exactly alike in genetic, structural, or behavioral makeup. Individuality is present in children from the earliest days of life.

Birth is the process during which the developing infant is expelled from the liquid uterine environment of weightlessness into the external world. Although it is impossible to set the exact date for an expected birth, it usually occurs at the end of the tenth *lunar* month, or about 266 days after conception. Whether prenatal development is complete or not, birth is followed by early infancy, a period of less direct dependence on the mother. Birth brings the necessity of breathing, followed shortly by self-feeding and elimination.

About two weeks before birth, *lightening* takes place: the fetus descends into the lower abdominal cavity; the head of the fetus sinks into the pelvis, and its body falls a little forward. For the mother, in most cases lightening is a barely perceptible process of changing appearance and relief. After lightening, the expectant mother can breathe more easily, because the upper abdomen and chest are relieved of the pressure caused by the previous position of the fetus. The occurrence of lightening shows that the fetal head is not too large for the pelvic inlet and that the fetal position is favorable for birth.

PROCESS OF BIRTH

Light uterine contractions initiate labor; the onset of regular contractions leads to birth. The onset of labor is usually preceded by one, two, or all of the following occurrences: (1) false labor pains caused by irregular, intermittent contractions of the uterus, (2) cervical dilation and pink, blood-tinged discharge from the vagina, and (3) labor pain and rupture of the membranes containing the fetus and amniotic fluid.

Birth is accomplished by rhythmic contractions of the uterine muscles at diminishing intervals, which increase pressure, causing the expectant mother to feel pain and become more aware of the pending birth of her child. The successive and increasingly more powerful and frequent contractions rupture the membranes containing the fetus, and in several phases force out the infant into a new state of air-surrounded existence. At a full-term birth, prenatal growth is complete and the plasticity of the fetal body is heightened.

Three phases of labor are usually distinguished. The first phase on the average lasts about twelve hours and is characterized by harder and more regular uterine muscle contractions lasting for longer periods of time than at the beginning of the phase. During this stage of labor the cervix dilates and the birth canal widens in preparation for the passage of the fetus. The second stage of labor, lasting about eighty minutes, brings about the actual birth of the child. It is estimated that 95 percent of all children are born head first. In the third phase, within a few minutes, the placenta with its amniotic and chorionic membranes is expelled from the uterus.

During the second phase of labor the mother is advised to make repeated efforts to relax and rest between contractions, to breathe deeply, and to bear down with her abdominal muscles while contractions are occurring. To facilitate the baby's respiration after birth, it is desirable that she control her fear so that analgesic (pain-relieving) medication is not required at an early stage of labor. As the second phase advances, the contractions usually become painful and severe. The frequently used spinal anesthetic affords much relief from pain yet leaves the mother sufficiently conscious to follow the progress of the birth. Although the average time of total labor is about fourteen hours for the first birth and eight hours for subsequent ones, it may last much longer or be much shorter. Hospitalization for childbirth is a necessity if all emergencies are to be met most effectively, since only a hospital has all the equipment necessary to facilitate the process of birth. Hospitals are well equipped to meet any emergency that may arise during birth, including any special care or medication a newborn infant may need.

In cases of difficult labor, fetal monitoring devices are often set to check on an infant's heartbeat and other signs of strain. If strain is severe, a cesarean section may be ordered to facilitate and shorten the process of birth. Right after birth, the obstetrician holds the baby on his palm, head and nose down, so that the remaining birth fluids can be sucked out with small rubber syringes; then the umbilical cord is clamped in two places and cut in between. After a close inspection and rating of the baby's health, premeasured silver nitrate is dropped into its eyes to prevent infection, in case the mother had undiagnosed gonorrhea. Unless some emergency measures need to be applied immediately, an identifying bracelet is placed on the infant's wrist and then the nurse brings the baby to the mother. Other regular procedures follow, including bottle-feeding with about $^1/_4$ ounce water with vitamin K to reduce any bleeding, then a bath, and frequent checks of body temperature.

The presently prevailing horizontal position of the mother during delivery disregards natural movement and female anatomy. The pressure of the heavy uterus on the body's largest vein (vena cava) reduces the blood flow through the heart and in long deliveries cuts down significantly the blood and oxygen supply to the fetus. The vertical or semivertical position for delivery is the apparently natural one taken by mothers in primitive tribes in all parts of the world. In 1971 Tucho Perrusi, an Argentinian physician, constructed a delivery chair adjustable to the comfort of the mother. In the nearly vertical position gravity prevents the fetus from sliding back between contractions, as frequently happens when the mother is in a horizontal position. (At full-term birth, the fetus is covered by vernix—a thick greasy material that helps the child slip easily down through the birth canal.) The obstetrician's role in vertical birth is lessened, while the mother's role is increased but facilitated, since her pain in this position is less severe. The sporadic change to vertical delivery is another significant step along nature's route—one for which even the most complicated techniques cannot easily be substituted.

Occasionally *hypnosis* is used to facilitate the mother's relaxation and cooperation. In some cases with complications, such as heart weakness or a preceding spinal disease, that make the use of anesthetic or even analgesic drugs inadvisable because of side effects, hypnotic trance seems to be a proper nondetrimental substitute. To assure hypnotizability, one or two preparatory sessions are necessary for checking and raising the expectant mother's susceptibility to a trance state. No detrimental side effects have been reported from hypnosis under professional guidance, and it is safe for the baby.

The ancient Chinese technique of *acupuncture* is another alternative to regular anesthesia during childbirth. Several long but very thin needles are used in this form of anesthetic, stimulating the nerve-connecting centers related to the painful area. Relaxation and numbness result from twirling needles or supplying them with a mild electric charge. The experimental use of acupuncture in childbirth involves the use of two or four needles—two in the ankle and another near the wrist to relax the abdominal area. Additionally, a series of wired electrodes are often attached to the abdominal area. Increase of pain is moderated by an increase in pulsating electric current applied by the anesthetist or obstetrician. Light electric shocks are experienced by the mother. Self-regulation of the current is a possibility, since only the mother knows the intensity of her contractions and how much pain she is willing to stand without diffusion. Deliveries through cesarean section using acupuncture as the only anesthetic have been reported since 1973. Patients report much less postoperative discomfort than from the use of other anesthetics. Only a limited number of experts are presently available for the professional use of either hypnosis or acupuncture.

Several studies have reported the effects of analgesic medication administered during labor. Stechler (1964) showed that twenty infants two to four days old whose mothers had received depressant drugs during labor fixated several visual stimuli for significantly shorter periods of time than neonates of mothers who did not receive pain-relieving medication; moreover, the higher the amounts of medication, the less the total attentiveness of the infants in this experimental situation. A single intravenous injection of 200 milligrams secobarbital during labor significantly reduced the amount, rate, and pressure of sucking in ten babies two, three, and four days old as compared with control neonates of mothers who received no sedation (Kron, Stein, & Goddard, 1966). These two experimental studies support the concept of greater vulnerability in the immature organism. A newborn infant is left with the anesthetic in his or her blood that was present in the mother's blood at the time the cord was severed. The immature liver and kidneys detoxify this medication at a slower rate than the mother's, so the immature brain is likely to be more and longer affected by any sedative medication in the blood (Brazelton, 1972; Kraemer, Korner, & Thoman, 1972).

Many hospitals and some obstetricians offer expectant mothers an opportunity to attend prenatal and postnatal classes or other forms of instruction. One of the objectives of prenatal classes is to teach the expectant mother about natural childbirth in order to help reduce her fear. The mother is taught proper physical exercise and deep abdominal breathing that will help her to relax. At the time of delivery, the deep abdominal breathing will not only decrease the mother's awareness of pain but also give the unborn infant better oxygenation. If the abdominal muscles are not tense, there will be less resistance to the actual progress of birth. It has been estimated that this practice has reduced the total time of labor by as much as about two hours. After childbirth the mothers are taught how to feed, bathe, and dress their babies and how to make a formula before they leave the hospital. Actual demonstrations and audiovisual aids are often used in these postnatal teaching programs. The skills in infant care acquired give the mother the feeling of security she especially needs for her first child.

One of the newer hospital arrangements permits the baby to live with its mother. "Rooming in" allows an early start in infant care, which is usually a great advantage for both mother and child. The close relationship helps to elicit affection, care, and concern—exactly what a helpless newborn infant needs. Parke, O'Leary, and West's (1972) findings on nineteen *S*s suggest that the father is a very active participant in the new family triad. In this study he was twice as likely as the mother to hold the infant and equaled her in the extent to which he looked at and touched the infant and vocalized. Indeed, smiling was the only behavior in which the mother was more active. Both mothers and fathers touched male babies significantly more than they did females.

Except for cases of serious postpartum (afterbirth) bleeding, most mothers leave the hospital in three or four days after childbirth and assume nearly normal activities at home within a week. A number of mothers need assistance with housework for two or three weeks in order to have daily rest periods. This way they regain their regular strength more easily and heighten their resistance to infection. Life with the baby binds the mother to home; it is good for her if she loves the baby and appreciates this temporary and partial isolation from social and vocational life.

Many mothers go through a short period of mild to moderate depression within the first week after the baby is born. The process of bodily readjustment is not smooth—some glands overshoot their normal targets, and "after-baby blues" occur. A mother may feel that she will never be as beautiful as she was before pregnancy or that her husband is neglecting her. Heightened emotionality marks each major event of life, and giving birth to a baby is no exception. Opportunities to get ample rest and support and encouragment from the husband help to overcome the emotional excitability within a week or two.

NEONATAL PHASE

The beginning of life after birth is primarily a time for the infant to relate and adjust to the variable situations encounted outside the mother's uterus. The immense stimulation of the skin and deeper regions of the body produced by the contracting muscles of the uterus helps to activate the respiratory system as well as the gastrointestinal and genitourinary tracts (Montagu, 1965, p. 40). These are the vital functions the newborn infant must be able to control (respiration, digestion, and elimination).

The birth cry is a reflexive inhalation and exhalation of air across the vocal cords, which are tightened by the shocking stimulation of the birth and the change of air pressure in the extrauterine environment. The cry after birth is a sign of live birth. Usually holding the neonate by the feet and gently patting the buttocks is sufficient to elicit crying. As birth excitement declines, the baby goes into a long period of sleep. Everything will be new to the baby when he or she wakes up after a day of sleep. Never before has he felt his own weight or free air space around him. He has never been hungry before or felt cold or hot. Time and again, these sensations will elicit crying.

The birth cry marks the beginning of vocal communication of the infant's needs and emotional states. The newborn infant's utter helplessness and high vulnerability readily elicit assistance and ego involvement from people in the environment. "Defenseless as babies are, there are mothers at their command, families to protect the mothers, societies to support the structure of families and traditions to give a cultural continuity to systems of tending and training" (Erikson, 1961, p. 151).

The early phase of infancy is marked by a rapid succession of developments in the physiological and perceptual aspects of growth. Sensory receptivity is active, and the orienting reflex (OR) plays the role of "tuning" the sensory receptors to a stimulus of low or moderate intensity and to its change. Operating soon after birth, the OR refers to a complex set of unconditioned and autonomic motor reactions dependent on peripheral vasoconstriction and cephalic vasodilation (Sokolov, 1963). This is why neonates respond to relatively subtle changes in stimulation; for example, their responses to flashing chromatic or achromatic lights are similar to adult reactivity (Lodge, 1969). Reaction to visual stimuli is marked by the closing of the eyes when a flash of light is applied to either one eye or both and by expansion or contraction of the pupils in response to decreased or increased light intensity. Many auditory stimuli of ordinary intensity cause a momentary suspension of the neonate's activity.

During the first phase of postnatal life the infant copes with the environment by means of the senses and motor reactions. He or she is sensitive to tactual stimuli and to changes in position. Opening and closing of the mouth is a part of the baby's "search" movements and efforts to suckle practically anything that comes in contact with the mouth. Salty, acid, or bitter solutions usually make the infant stop sucking, while sweet solutions elicit and maintain the reaction. The neonate is "remarkably sensitive to environmental events, perhaps particularly to those relating to oral stimulation and to food intake" (Lipsitt, 1967, p. 241). After a number of conditioning experiments, Papoušek (1967) found some evidence to conclude that "discriminative" learning occurs during the first days of life.

The physiological development before birth laid a satisfactory foundation for the vital processes, yet for several days, if not weeks or months, their internal organization remains precarious. The organismic equilibrium is readily upset—temperature may rise and blood composition change for no apparent reason. Discomfort in the digestive system—indigestion and colic pain—are among the main causes for excessive neonatal crying.

In terms of weight, physical growth shows a loss for several days. During the first three to four days of life, the neonate loses 6 to 9 percent of the birth weight, or as much as 6 or 7 ounces, before beginning to gain and recovering it in seven to nine days. This is not surprising when one considers the large amount of energy the infant expends in responding to various stimuli and in crying. This early plateau is followed by a period of rapid growth and manifold gains beginning during the second week and continuing afterward and to a lesser degree throughout the total span of infancy.

In addition to random mass motion, the neonate exhibits a considerable repertoire of behavior, such as fair adaptive "search" movements, as well as sucking and rooting, rejecting substances from the mouth, yawning, sneezing, vocalizing, and holding the breath. Hands and feet also produce several reflex patterns: the Moro reflex, a clutching movement of the arms and legs, resulting from a blow to the surface on which the baby is lying or intense sounds; the Babinski reflex, an upward and fanning movement of the toes as a response to stroking of the soles; the palmar reflex; rubbing the face; and the knee jerk. Nursing posture, adjustment to the shoulder when carried, and the startle response involve considerable coordination of many parts of the body (Birns, 1965; Dennis, 1934).

The infant's acquisition of control over various muscle groups follows the cephalocaudal sequence. This means that the muscles of the head region are the first to respond to internal or external stimuli and later to voluntary control. Before the age of one month, the baby is able to lift the head from time to time and hold it erect for a moment when supported, to focus the eyes on moving objects, and to make simple crawling movements when laid prone on a firm surface.

Parents or other observers also notice the neonate's capacity for emotional response, marked by pervasive excitement and is usually accompanied by crying and signs of displeasure. Relaxation signifies lack of emotionally exciting stimuli. This restful state can be induced by picking the baby up and carrying him on the shoulder, by rocking, and by feeding. In this early postnatal phase, Schaefer and Bell (1957) distinguish two types of affect: embeddedness affect, described as the diffuse discharge of tension caused by a need arousal or an external stimulus interrupting quiescence; and activity affect, characterized as directed, sustained, enjoyable, and motion-oriented. In her classic study of emotions and personality, Magda B. Arnold (1960, p. 209) recognizes four emotions at this time. The infant "*wants* food, *enjoys* it, is *angry* when restrained or disturbed, and can be *startled* or alarmed." In her view all these emotions continue throughout life. Since there are no recent substantial studies on the subject, the issue of early emotionality cannot be resolved at this time and is open for future empirical research.

SURVIVAL AND ADJUSTMENT

Since newborn infants face many new tasks, some of them have serious difficulties early in life. In some cases parental care and stimulation frustrate the intrinsic biogenic and emergent psychogenic needs. One of the frequent problems related to the satisfying of biogenic needs is feeding. In the hospital or at home, the scheduled feeding hours are often unsatisfactory, since many babies are likely to get hungry early, cry themselves to sleep, and show little initiative at the scheduled time. At the beginning, self-demand feeding (feeding whenever the baby appears to desire food) usually does a better job, although it does not guarantee either sucking satisfaction or the intake of a sufficient amount of food. Some infants show poor response to breast feeding and formula feeding alike; diarrhea and vomiting are frequent. For many of them it takes weeks for such feeding difficulties to subside.

Setting up a gratifying feeding situation is a task for mother and infant alike. The emotional tone with which the baby accepts food is often traceable to the mother's attitude and her level of relaxation while she is feeding him. Quietly talking to the baby, handling him with slow and deliberate movements, and attempting to gain the infant's attention are all of considerable significance in avoiding feeding problems. Breast feeding contributes to the affective bond between mother and child. It is her natural way of giving "self" in an intimate way, if she is internally undivided in her desire to breast-feed her child. For success in breast feeding, consultation with a pediatrician about two months before the infant's birth is necessary. The pediatrician will examine the expectant mother's nipples. In many cases breast massage and specific daily exercises may be necessary to get a more protractible nipple posture. Good nipple protractibility assures proper nursing grasp by the baby right at the start of breast feeding—about one hour after delivery (Applebaum, 1970). The best modern formulas are only imitations of the mother's changing supply from colostrum to mature milk, the antibodies of which protect the baby against many infections and allergies. Moreover, breast feeding accelerates the return of the mother's figure to normal.

During the first three days there is just a little colostrum in the mother's breasts, but what the baby really needs during these days is sucking practice. He is better off with little food until he is a bit older and his stomach is really ready for food. Usually the mother's breasts need considerable stimulation to produce milk in sufficient amounts, and most babies do this job superbly (Bakwin, 1964). Odenwald (1958, pp. 16–17) considers the periods of feeding the most important moments of the infant's life. It helps to satisfy the sucking urge and offers pleasure and emotional gratification for both the baby and the mother. The mother's holding the baby in her arms while he is taking nourishment, whether at the breast or from a bottle, gives the bodily contact from which is derived the sense of security that every infant needs.

Although the average food intake per day during the first year of life is about 2 ounces to every pound of body weight, no definite amount can be prescribed: the quantity depends on the neonate's level of physical maturity, which is difficult to estimate. Prematurely born infants are capable of digesting only much smaller amounts of food than full-term babies. Too often, however, parents believe that a certain amount of food must be taken in order for the child to grow adequately. As a result, they use various techniques to increase the amount and by doing this magnify discomfort in the infant's digestive system. Both the amount of milk and the timing have to be flexible to adjust to the changing physiological and experiential states of the infant.

A condition known as *colic* can cause new parents much anxiety. Colic in infants is characterized by a sudden onset of loud and persistent crying with the knees drawn up to the abdomen. The abdomen is tense and distended. An attack may last one-and-a-half or two hours, but in some cases it may continue many hours. The cause is frequently a digestive disturbance. It may be air in the intestines that the baby swallowed during feeding and has not eliminated by burping. It is usually relieved when the baby is able to expel the air by rectum. Other causes of colic are overfeeding or a formula that is too rich or otherwise disagreeable to the baby. Even the casein in milk can cause a digestive disturbance, and an allergy to a certain element in the diet can lead to much distress, in which case the type of feeding formula should be changed. Some babies who are allergic to cow's milk have to be fed goat's milk or milk made from soybeans.

The expectant mother's adherence to medical regulations, combined with the prenatal education in infant care offered by high schools, hospitals, and other institutions, has made major contributions to the prevention of diseases of early infancy. This also guards the biological welfare of the infant against the "experimental" trial-and-error approach in the managment of the first child during the first and most crucial year of life. Some dangers, however, still exist. Epidemic diarrhea in newborn infants is a serious condition that can result in death. Pneumonia in neonates is also serious, but with modern methods of treatment, barring other complications, chances of survival have increased.

Attention has recently been given to babies born of narcotic-addicted mothers. It has been observed by obstetricians and pediatricians that babies born of addicted mothers manifest withdrawal symptoms after delivery, the severity of which depends on the amount of drug taken and the time of the mother's last dose before delivery, as well as the frequency with which the mother was accustomed to take the drug. Many babies of addicted mothers have died as a result of withdrawal reactions, especially if the baby's difficulty was unnoticed and therefore untreated (Kunstadter, 1958). With treatment, the chances of survival are fairly good, and the baby can be cured of the addiction. Whether a baby cured of addiction will later in life be emotionally unbalanced and readily resort to narcotics is yet to be determined by empirical research.

Premature births are often prolonged and catalysmic and subject the infant to greater than normal stress. The organism of the premature infant is somewhat fragile and generally not ready for birth (CPRP, 1963; Drillien, 1963). The lungs may not be developed enough to allow for an adequate exchange of gases, or the digestive system may not be sufficiently developed to utilize milk and break it into the proper compounds for use in cellular life. Susceptibility to infections of various kinds is considerable in the premature infant, since the immunity level is low. Secondary infections, such as pneumonia or diarrhea, seriously threaten the infant's life, even at several months of age. The mortality rate is inversely proportional to the birth weight. If a neonate weighs less than 2,500 grams (5 pounds 8 ounces) or has had a gestation period of less than thirty-seven weeks, he is usually considered premature. About one-third of infant deaths under one year of age are due to prematurity, and almost one-half of the infants who die in the first month of life are premature babies (Ziegel &

Van Blarcom, 1972, p. 698). The psychological effects of prematurity are also numerous. The prematurity per se arouses the concern and anxiety of parents and stirs up feelings and attitudes of overprotectiveness. The difficulties in feeding the premature infant and the small amounts of fluid taken reinforce parental anxieties. Yet studies on prematurity suggest that if the corrected chronological age (CCA) is used to estimate growth and behavior, the premature baby develops satisfactorily (Drillien, 1963; Gesell, 1952; Illingworth, 1972, pp. 37–40). The CCA is counted from the time of conception rather than birth. Thus a premature infant born at thirty-five weeks of gestation has an age allowance of five weeks whenever his behavioral maturity is appraised in terms of age norms.

INDIVIDUAL DIFFERENCES

Watching babies in a nursery helps one realize the great variability among infants in their first days of life. They differ in all observable aspects, and as they grow older the differences are magnified. After two or three days the mother recognizes her baby. Why does a mother recognize the crying of her baby when many other babies are around? The answer is clear—she recognizes it because her baby cries as no other child does. Cry prints are at least as distinctive as fingerprints for identifying newborn babies (Wolff, 1969).

Among newborn infants individual differences have been observed for centuries, yet there is a dearth of systematic studies investigating the consistency of this variation over long periods of time (Horowitz, 1969). Much significance is usually attributed to the appearance and weight of the neonate. Many infants are large-boned and tend toward a husky structure, while others tend toward a slim, fragile, delicate structure. Some have a large face but small hands and feet; others have small faces but large stomachs. A parent may be frightened by the neonate's coloration or slanting eyes. Often this is but a sign of some prematurity; in some cases it points to Down's syndrome, in which many Mongolian features are present. Following full-term birth or at the corrected chronological age for birth, the observer gets a preview of the baby's body build. It is a source of either joy and pleasure or anxiety and concern to many parents. Temperamentally, some infants are placid; others are highly excitable.

The spectrum of normal individual differences is broad indeed. Inborn differences predetermine the particular styles of development for high, average, or low activity (Brazelton, 1972, p. 281). Kagan (1971, p. 4) aptly puts it this way:

> There are quiet and active babies, irritable and dour babies, heavy and light babies, premature and term babies, hypoxic and full breathing babies. Everyone acknowledges the obviousness of these differences, but only a few believe they play a permanent role in guiding the psychological organization of the older child.

Pattern and amounts of activity, thresholds of sensory reactivity, amounts of vocalization, rest and sleep, adaptability and sucking patterns—all differ in a variety of ways from one infant to another. Many young infants are highly reactive and awake at the slightest touch or glimmer of light, while others sleep unperturbed. Some babies are highly irritable and cry easily, while others are not readily aroused to crying. One infant cries for feeding at intervals less than two-and-a-half hours, while another baby easily waits for more than four hours. Infants also differ in their overall reactivity to stimulation. They vary in their sensitivities to different stimuli—some being more responsive to sound as compared with tactile stimulation. Their capacity to discriminate between qualities within a given sensory modality is apparently a congenital characteristic (Lipton & Steinschneider, 1964). The patterns of cardiac acceleration in the startle reflex and other defensive responses, as well as cardiac deceleration in the orienting reflex and related behavior, show individual variation among infants from the very early days of life (Graham & Jackson, 1970). These reactions are a forecast of temperamental and emotional variation, while sensory acuity and threshold of responsiveness probably foreshadow cognitive differences at a later age.

Applying a stream of air of five seconds' duration on the abdomen, Lipton and Steinschneider (1964) found consistent differences in autonomic function among sixteen neonates who were studied between the second and fifth days of life. The neonates showed individual patterns of autonomic behavior. Polygraph records indicated a rapid startle response to a stimulus while the respiration rate increased and amplitude decreased. The cardiotachometer rose to a peak, then returned to prestimulus level. EEG tracings showed less discernible changes. Timing and magnitude differed on all measures for all neonates at this age as well as at two and five months of age. At this later age the responses were recognized as more "mature." It was also found that the responses of individuals maintained their relative ranking within a group.

Measurements of fetal heart rate, a ten-beat sample of twelve fetuses eight months old, and the pattern of the resting heart rate of the same individuals at twenty years of age show a high positive correlation. Prediction of adult autonomic functions thus appears possible (Sontag, 1966). Porges, Arnold, and Forbes (1973) assessed heart-rate patterns of twenty-four full-term neonates responding to a thirty-second moderately intense auditory stimulus one to three days after birth. The pattern was marked by acceleration to onset, deceleration to offset, and decreased variability during the stimulus of a 250-hertz sine wave. It is noteworthy that the high-variability subgroup exhibited response patterns resembling those of mature adults.

In Prague, Czechoslovakia, Papoušek (1965) showed individual differences in higher nervous activity in the first half-year of life. The rapidity and course of establishing the conditioned food-seeking reflex showed marked individual differences in the neonatal phase. When eleven infants aged four to six months received sweet milk from the left with a bell signal and a bitter solution from the right with a buzzer signal, they showed different reactions to the two stimuli in terms of facial or vocal responses. Soon the infants began to react with the same emotional responses to the acoustic signals alone. Then *E* switched the signals, and "the infant was offered bitter solution with the bell and milk with the buzzer. He rejected the milk and spit it out, but accepted the bitter solution eagerly as long as the bell was ringing. Only gradually did the infant learn to readjust his responses." On the basis of his experimental data, Papoušek claims that "around the end of the second and beginning of the third month of life, changes occur that suggest a marked qualitative change in higher nervous function."

Behavioral differences between male and female infants have been observed from the early weeks of life. Using a sample of thirty first-born babies and data collected under natural conditions, Moss (1967) found some behavioral sex differences at three weeks of age. When the mother initially responds to the infant's behavior, she is guided by her notion of a boy or a girl, so that her consistent behavior reinforces and regulates the infant's reactivity. This study also showed that the mothers responded more frequently toward the female infants, thereby facilitating social responses in female children from the very early weeks of life. The infant is dependent on the mother and others for reinforcement, and the amount of stimulation provided by adults is one of the major determinants of an infant's attachment behavior. It is common for the father, on returning home from work, to initiate actions aimed at stimulating the child, and these actions are usually responded to with smiles and clear pleasure (Goldberg & Lewis, 1969; Moss, 1967; Schaffer & Emerson, 1964).

Different biological givens, variation of parental approach to each child, and distinct rates of learning—all magnify individual differences as infants grow into the diverse constellations of their families and face varied environmental settings from the earliest days of life. Considering learning alone, Harold W. Stevenson (1970, p. 919) concisely states: "Learning occurs from the time of earliest infancy and at all ages there are large individual differences in the rate of attaining the correct response." It is only natural that as children grow older they should show more and more distinct individual differences.

TRENDS IN LIFE DATA IN THE UNITED STATES

Selected birth and death statistics for the United States will help to illustrate recent trends in the nation's vital data. During the decade 1954–1964, the annual number of live births varied within the range of 4.0 and 4.3 million, but the decade 1965–1975 showed a general drop to 3,760,358 in 1965 and 3,717,910 in 1970. Children born during the 1970s will live, on the average, nearly seventy-two years. The median weight at birth for infants born alive in 1960 was 3,310 grams; in 1965, 3,270 grams; and in 1970, 3,300 grams—about 7 pounds 4 ounces for the decade. As seen in Figure 6-1, statistics show annually decreasing rates of infant and maternal deaths Infant mortality rates per 1,000 live births dropped from 68 in 1930 to 29.2 in 1950 and to 19.8 in 1970. An impressive decrease has also occurred in the maternal death rate per 10,000 childbirths: 1930, 68.2; 1940, 37.4; 1950, 8.3; 1960, 3.7; and 1970, 2.5. The rate in 1970 was equivalent to one maternal death for every 4,000 live births. This spectacular decrease in the death rates of mothers and infants is due particularly to the

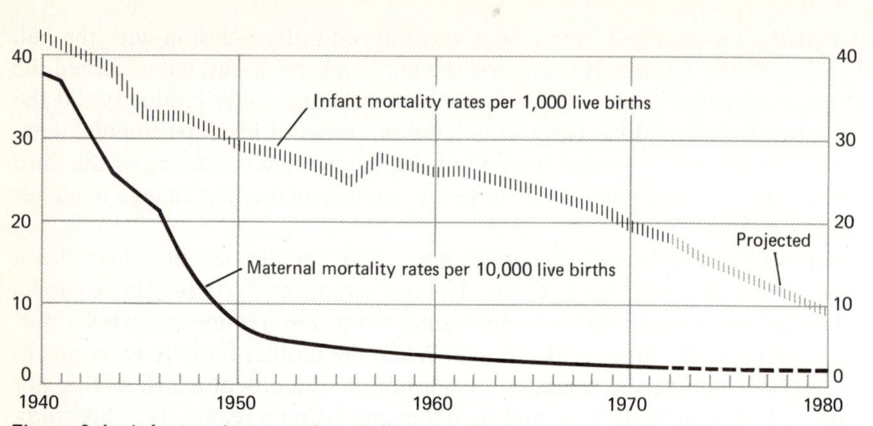

Figure 6-1 Infant and maternal mortality rates: United States, 1940–1980. [*Source: The facts of life and death: Selected statistics on the nation's health and people.* National Center for Health Statistics (No. 600, rev. 1965); and *Monthly vital statistics report, annual summary for the United States, 1970, 1971, 1972* (Vols. 19, 20, 21, No. 13).]

availability of and manifold improvements in medical service. Table 6-1 shows total *N* of births, method of delivery, median birth weight, birthrate, *N* of fetal and infant deaths, and related life data for four decades, 1940 to 1970, in the United States.

As Figure 6-2 indicates, in 1947 nearly four million babies were born in the United States, the largest number in the history of the country until then. During that year the birthrate soared to its highest point in decades. The predicted drop in the birthrate since 1947 has been smaller than expected, and

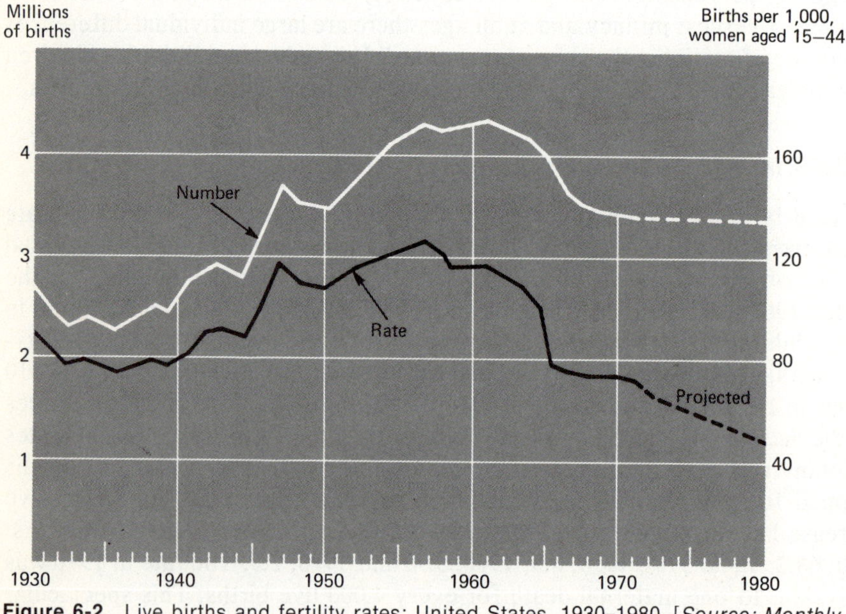

Figure 6-2 Live births and fertility rates: United States, 1930–1980. [*Source: Monthly vital statistics report: Final nativity statistics, 1966,* **15**, (No. 8), and 1974.]

Table 6-1 Life Data: United States, 1940–1970

	1940	1950	1960	1965	1970
Total births:					
Registered	2,360,399	3,554,149	4,257,850	3,760,358	3,717,910
Adjusted for underregistration	2,558,647	3,631,512			
Live birthrate per 1,000:					
Registered	17.9	23.61	23.7	19.4	17.8
Adjusted for underregistration	19.4	24.1			
Fetal deaths:					
Ratio per 1,000	31.3	19.2	16.1	16.2	
Total *N*	73,802	81,300	94,452	93,200	
Infants under one year, deaths:					
Ratio per 1,000	47.0	29.2	26.0	24.7	19.8
Total *N*	110,894	103,825	110,873	93,200	73,735
Delivery:					
Physician in hospital	55.8 percent	88 percent	96.6 percent	97.4 percent	N.A.
Midwife	N.A.	N.A.	2.0 percent	1.5 percent	N.A.
Median birth weight	N.A.	7 lb. 5 oz.	7 lb. 5 oz.	7 lb. 4 oz.	3,300 grams
Immature birth (2,500 grams or less)	N.A.	7.5 percent	7.7 percent	8.3 percent	7.9 percent

Sources: Vital statistics of the United States: 1965, Vol. 1, 1967; *Monthly vital statistics report,* Vol. 14, No. 12, 1966, and Vol. 22, No. 5, 1973; United Nations, *Demographic yearbook 1965,* 1966; *Statistical Abstract of the United States: 1973,* 1973.

the following decade was marked by a slight rise. The 1960s and early 1970s witnessed a moderate decrease, resulting chiefly from the availability of the birth-control pill and other newly manufactured contraceptive devices. Federal legalization of abortion in 1973 added another factor of significance in the decline of births. Since the early 1960s, the population of the United States has been growing at an irregular but generally decreasing rate. The estimated rate for 1975 is less than one-half the high rates of about 16 per 1,000 population in the mid-1950s. Since the death rate has been quite stable, the decline in the rate of natural increase of the population has been due almost entirely to the decline in the birthrate.

QUESTIONS FOR REVIEW

1. What are the main signs indicative of pending birth?
2. Describe the first and second phases of labor, including delivery of the child.
3. What is the chief means for control of pain in childbirth, and what means are often used to lessen the awareness of pain?
4. Why is the hospital the most advantageous place for childbirth?
5. What favors the vertical or semivertical rather than the horizontal position for delivery?
6. What skills does the mother need when she leaves the hospital and begins caring independently for her baby?
7. Explain the significance of the birth cry and give several reasons for infants' crying.
8. List the kinds of behavior the neonate is capable of. Analyze their effects on postnatal adjustment.
9. How do major patterns of feeding affect the baby? Consider the mother's attitude toward holding the baby during bottle feeding.
10. Explain how premature birth and full-term birth are defined, and discuss the effects of prematurity on postnatal adjustment.
11. What makes the first day of life critical for the neonate's survival?
12. What constitutes an Rh hemolytic condition, and what preventive measures are necessary at what times to avoid Rh incompatibility?
13. Discuss observable individual differences among newborn infants, including some differences between males and females.
14. What do statistics for the United States show about the decline in the birthrate and the decline in maternal and infant deaths during recent decades?

REFERENCES

Selected Reading

American Academy of Pediatrics. *Standards and recommendations on the hospital care of newborn infants.* Evanston, Ill.: The Academy, 1964. An authoritative presentation of sound and desirable hospital practices.

Collaborative Perinatal Research Project. *Five years of progress.* Bethesda, Md.: National Institute of Neurological Diseases and Blindness, NIH, 1963. A major investigation by fourteen medical institutions of the long-term effects of pregnancy and delivery, following up 40,000 cases until 1972, when the offspring would be age seven.

Illingworth, R. S. *The development of the infant and young child: Normal and abnormal.* Edinburgh and London: Livingstone, 1972. Prenatal and perinatal factors are presented at length, including assessment of the newborn baby and patterns of development.

Specific References

Applebaum, R. M. The modern management of successful breast-feeding. *Child Fam.,* 1970, **9,** 61–84.

Arnold, M. B. *Emotion and personality* (Vol. I). New York: Columbia University Press, 1960.

Bakwin, H. The secular change in growth and development. *Acta Pediat.*, 1964, **53**(1), 79–89.

Birns, B. Individual differences in human neonates' responses to stimulation. *Child Developm.*, 1965, **36,** 349–356.

Brazelton, T. B. *Infants and mothers.* New York: Dell, 1972.

CPRP (Collaborative Perinatal Research Project). *Five years of progress.* Bethesda, Md.: NIH, National Institute of Neurological Diseases and Blindness, 1963.

Dennis, W. A description and classification of the responses of the newborn. *Psychol. Bull.,* 1934, **31,** 5–22.

Drillien, C. M. *The growth and development of the prematurely born infant.* Edinburgh: Livingstone, 1963.

Erikson, E. H. The roots of virtue. In J. Huxley (Ed.), *The humanistic frame,* pp. 145–166. New York: Harper & Row, 1961.

Gesell, A. *Infant development: The embryology of early human behavior.* New York: Harper, 1952.

Goldberg, S., & Lewis, M. Play behavior in the year-old infant: Early sex differences. *Child Developm.,* 1969, **40,** 21–31.

Graham, F. K., & Jackson, J. C. Arousal systems and infant heart rate responses. In H. W. Reese & L. P. Lipsitt (Eds.), *Advances in child development and behavior* (Vol. 5). New York: Academic Press, 1970.

Horowitz, F. D. Learning, developmental research, and individual differences. In L. P. Lipsitt & H. W. Reese (Eds.), *Advances in child development and behavior* (Vol. 4, pp. 83–126). New York: Academic Press, 1969.

Illingworth, R. S. *The development of the infant and young child: Normal and abnormal.* Edinburgh and London: Livingstone, 1972.

Kagan, J. *Change and continuity in infancy.* New York: Wiley, 1971.

Kraemer, H. C., Korner, A. F., & Thoman, E. B. Methodological considerations in evaluating the influence of drugs used during labor and delivery on the behavior of the newborn. *Developm. Psychol.,* 1972, **6,** 128–134.

Kron, R. E., Stein, M., & Goddard, K. E. Newborn sucking behavior affected by obstetric sedation. *Pediatrics,* 1966, **37,** 1012–1016.

Kunstadter, T. H., et al. Narcotic withdrawal symptoms in newborn infants. *J. Amer. Med. Assoc.,* 1958, **168,** 1008–1010.

Lipsitt, L. P. Learning in the human infant. In H. W. Stevenson, E. H. Hess, & H. L. Rheingold (Eds.), *Early Behavior: Comparative and developmental approaches,* pp. 225–347. New York: Wiley, 1967.

Lipton, E. L., & Steinschneider, A. Studies on the psychophysiology of infancy. *Merrill-Palmer Quart.,* 1964, **10,** 103–117.

Lodge, A., et al. Newborn infants' electroretinograms and evoked electroencephalographic responses to orange and white light. *Child Developm.*, 1969, **40**, 267–293.

Montagu, A. *The human revolution.* Cleveland: World Publishing, 1965.

Moss, H. A. Sex, age, and state as determinants of mother-infant interaction. *Merrill-Palmer Quart.*, 1967, **13**, 19–36.

Odenwald, R. P. *Your child's world from infancy through adolescence.* New York: Random House, 1958.

Papoušek, H. The development of higher nervous activity in children in the first half-year of life. *Monogr. Soc. Res. Child Developm.*, 1965, **30**(2), 102–111.

Papoušek, H. Experimental studies of appetitional behavior in human newborns and infants. In H. W. Stevenson, E. H. Hess, & H. L. Rheingold (Eds.), *Early behavior: Comparative and developmental approaches*, pp. 249–277. New York: Wiley, 1967.

Parke, R. D., O'Leary, S. E., & West, S. Mother-father-newborn interaction: Effects of maternal medication, labor, and sex of infant. *Proc. 80th Ann. Convention, APA*, 1972, **7**, 85–86.

Porges, S. W., Arnold, W. R., & Forbes, E. J. Heart rate variability: An index of attentional responsivity in human newborns. *Developm. Psychol.*, 1973, **8**, 85–92.

Schaefer, E. S., & Bell, R. Q. Patterns of attitudes toward child rearing and the family. *J. abnorm. soc. Psychol.*, 1957, **54**, 391–395.

Schaffer, H. R., & Emerson, P. E. Patterns of response to physical contact in early human development. *J. child Psychol. Psychiat.*, 1964, **5**, 1–13.

Sokolov, E. N. Higher nervous functions: The orienting reflex. *Ann. Rev. Physiol.*, 1963, **25**, 545–580.

Sontag, L. W. Implications of fetal behavior and environment for adult personalities. *Ann. N.Y. Acad. Sci.*, 1966, **134**, 782–786.

Stechler, G. Newborn attention as affected by medication during labor. *Science*, 1964, **144**, 315–317.

Stevenson, H. W. Learning in children. In P. H. Mussen (Ed.), *Carmichael's manual of child psychology* (Vol. 1, pp. 849–938). New York: Wiley, 1970.

Wolff, P. H. The natural history of crying and other vocalizations in early infancy. In B. M. Foss (Ed.), *Determinants of infant behavior.* New York: Barnes & Noble, 1969.

Ziegel, E., & Van Blarcom, C.C. *Obstetrical nursing* (6th ed.). New York: Macmillan, 1972.

Part Four

Infancy

Although many developments during infancy are preconditioned by intrauterine growth, a great deal of behavioral differentiation results from external stimulation and from the parents' approach to child care and management. Unlike other mammals, human beings have a prolonged infancy which exposes them to the hazards of extreme dependence on the parents. Throughout infancy, rapid growth occurs in the physical, emotional, language, and cognitive areas. Middle infancy is marked by gains in control over large muscle groups and various parts of the body; later infancy is characterized by the acquisition of locomotion and speech, abilities that serve as communicators of personality qualities and traits. Increasing self-awareness and ego formation are usually accompanied by marked resistance to parental control.

Developmental tasks of infancy include learning new modes of locomotion, learning to take solid foods, establishing daytime toilet control, and learning to communicate by gestures, signals, and words, as well as learning to relate emotionally to parents, siblings, and other persons in the home environment. During the years of infancy, foundations are laid for further behavioral differentiation in most aspects of personality—the fundamental growth fabric is now woven for life.

Chapter 7

Middle Infancy

HIGHLIGHTS

The middle phase of infancy is marked by orderly unfoldings in most aspects of human development. The quantity and quality of maternal care is a crucial stimulating power in furthering growth of the unfolding behavioral system.

Despite socioeconomic and cultural stratification, there is great uniformity of psychomotor and cognitive growth during the first twelve to fifteen months of life. Physiological, perceptual-motor, and emotional factors come into display according to the individual genetic timetable. Social and personality developments, however, show significant difference among one-year-olds.

The infant's activity from scanning to touching and biting and the contingent discrimination of the stimuli surrounding him or her generate patterns for the lifelong career of learning. Cries for parental interaction, when properly responded to by parents, further social learning. Improper handling of the infant may include monotonous routine at one extreme and overstimulation at the other. Consideration of the changing physiological state and feelings of the infant is fundamental to his or her welfare.

Immunization programs and protection from infections and high body temperatures help to maintain health and avoid breaks in the metabolic and growth processes of the individual.

Middle infancy extends from the time when the infant begins to make satisfactory adjustments to the tasks of gratifying biological needs until the time when he or she takes the first steps independently and begins to use words as a means of communication. For the majority of children, this happens at the thirteen- to fifteen-month level; for some, later; for others, earlier. Throughout middle infancy, the developing infant depends almost wholly on parental care for gratification of needs. The middle stage of infancy is marked by vivid physiological, emotional, and sensorimotor developments and by rapid and substantial increments in height and weight. It is also the phase of intense unfolding of molar, or purposeful, behavior and the foundation of manifold learning experiences.

PSYCHOMOTOR DEVELOPMENTS AND CONTROL

The infant's control of the large muscles increases to a significant extent in middle infancy. Since muscle control is governed by cephalocaudal as well as proximodistal sequence, head movements are achieved before trunk and extremities reach the same level of development and control. Crawling allows for some movement at five months; creeping and standing are often acquired by ten months. At five or six months, the baby is able to grasp, hold, and manipulate small objects and put them into the mouth. At this age "infants regularly respond to the nearer of two objects when their separation in depth is considerable" (Wohlwill, 1960). Infants scan and observe objects and persons and like to see them in motion. They like exploring their surroundings. They are pleased to reach what they see and become upset when their efforts fail. Motor activities increase the ability to satisfy curiosity and enhance rising cognitive and social interests, the advancement of which is a major characteristic of the later phase of infancy. As simple behavioral sequences become organized into higher-order skills, competence rises and exploration techniques multiply (Bruner, 1973).

Achieving the power to move by creeping is a significant accomplishment for a baby. Now he can reach the objects and persons attracting his attention. His helplessness decreases and his perceptual field is enlarged as he observes persons and objects in his surroundings from many possible points of view. If he is kept constantly in a crib and later in a playpen, perceptual deprivation is likely to occur. By eleven or twelve months, exploratory activities increase considerably and consume about two-thirds of the infant's waking time, unless environmental changes are so simple and limited that they present insufficient challenge for the growing mind of the infant. A stimulating home provides ample opportunities for developing eye-hand coordination. It is beneficial for the infant to have toys and other safe objects available for manipulation. From the early days of life, the infant orients himself so as to maximize the stimulus value impinging on his sense organs. He turns his head, scans and positions his eyes, moves toward most new stimuli, reaches out his hands and fingers, touches and manipulates objects extensively. Then he readies himself for other stimuli. It is good for the infant to have new stimuli introduced from time to time and to have his environment expanded with age.

A major comparative study of motor and mental development (Bayley, 1965), based on 1,409 infants aged one to fifteen months, from twelve metropolitan areas, showed no significant differences in revised scores on Bayley's scales between males and females or between firstborn and later-born infants. Nor was parental education, socioeconomic status, or geographic location a significant determinant. Figure 7-1 illustrates Bayley's findings and indicates no significant differences between whites and Negroes on the mental scale, but during the first twelve months the Negro infants scored consistently

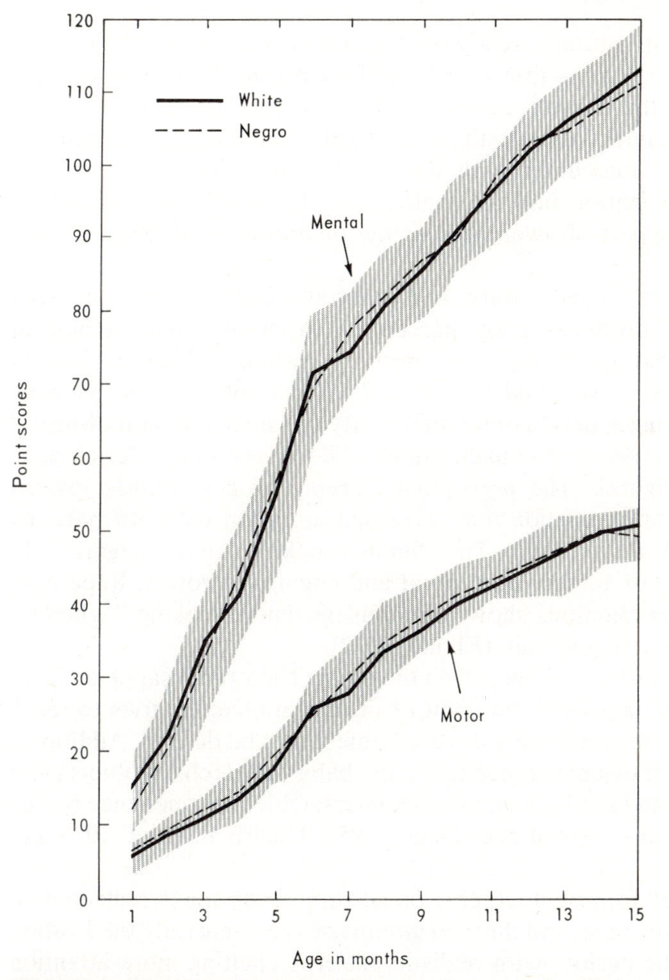

Figure 7-1 Mental and motor point scores: means and ± SD for months 1 to 15. [*Source:* N. Bayley. Comparisons of mental and motor test scores for ages 1-15 months by sex, birth order, race, geographical location, and education of parents. *Child Developm.*, 1965, **36**, 379–411. By permission.]

above whites on the motor scale. As in prenatal development, great uniformity in early postnatal growth is suggested. Early forms of behavior, whether perceptual and adaptive abilities or motor skills, are very similar for the majority of infants of any background. Earlier, Knobloch and Pasamanick (1953) reported that Negro infants showed no significant deficit in functioning and were slightly accelerated in motor development. These studies point to a general uniformity of rate in early postnatal development among various population groups in the United States. Prenatal uniformity of growth has been attested to by many studies.

LEARNING NEW RESPONSES

For Piaget, the newborn infant is ready for the lifelong career of learning. The protoplasmic sensitivity of the first weeks of life, when no distinction is made between the self and the nonself, constitutes the morphologic-reflex organization—a basis for accommodation to the object and assimilation to the subject. The progress of this encounter with the surroundings works in the dual directions of externalization and internalization, that is, the acquisition of perceptual experience and of awareness of the operation itself (Piaget, 1954, pp. 355–357).

Sensitivity to light, sound, odors, and other stimulation is present from birth. Sense activity produces early perceptual experience with a built-in coding mechanism that is the basis for primary learning. Newborn infants fixate objects with their eyes, and they attend to some objects and patterns more often and for longer periods than others. By the second month of life, if not earlier, infants discriminate configurational differences and prefer circular (facial) over linear figures. The perceptual learning process moves toward increasing stimulus differentiation and increased attention to novel patterns (Fantz, 1967; Fantz & Nevis, 1967). The stimulus-seeking activity begins early and forms a baseline of further perceptual and cognitive growth. Repetitive behaviors and circular reactions show "stimulus nutriment seeking," whereby the baby ignores distracting stimuli (Elkind, 1972).

At seven or eight months of age, when the object followed disappears from view the infant's gaze lingers at the point of disappearance. He tries to reach for, and usually picks up, a partially covered object that he desires. Additional encounters with object disappearance incite the baby to search for objects lost by removing covers and later by opening containers. Object permanence is thus established at about one year of age (Piaget, 1954; Užgiris & Hunt, 1975, pp. 34–35).

Lewis (1969) used four photographic facial stimuli varying in realism with 120 infants at three, six, nine, and thirteen months of age. Generally the fixation data declined over the period, with realistic patterns eliciting more attention than schematic ones, possibly as a function of form schema formation. There was an increase of vocalization and smiling with age and a decrease in fretting or crying behavior. Regular faces elicited more vocalization than other faces. No fear responses to facial stimuli were observed, but the male experimeters elicited fear at nine and thirteen months of age. Possibly the long age intervals did not give the full story of change in infant reactivity to facial stimuli. Related studies suggest that as early as six to eight weeks of age, infants show reliable preferences for familiar over novel stimuli; after eight weeks of age, however, they consistently prefer novel to familiar stimuli. Apparently when perceptual schemata of familiar objects are formed, attendance to novelty or to discrepant patterns increases (Weizmann, Cohen, & Pratt, 1971; White, 1972).

Infants four to six months of age show more cardiac orienting behavior to changes in stimulus intensity than adult subjects (Berg et al., 1971). They need stimulation to nurture their unfolding cognitive structures. Therefore, one should not "leave the baby alone to grow up," as John B. Watson suggested many years ago, but carry him around to provide sufficient amounts of sensory data for information processing and the perceptual organization of reality. The sensorimotor experience becomes more meaningful as the sensory input is more adequately translated into perceiving and feeling, leading to preference for activities and experiences that more directly serve biological and emotional ends. As experiential retention increases, the infant recognizes his mother by her way of handling him. He recognizes foods by their color and smell. Gradually likes and dislikes are stabilized and lay the foundation for the development of attitudes. Distinct reactions to all members of the family are acquired during the second half of the first year.

Perceptual refinements further differential responsiveness—interpretation of what has been perceived increases. The perceptual differentiation depends on the amount of exposure to learning materials. Accumulated experience forms the basis for conditioned and refined responses to persons, objects, and situations. By approximately ten months of age, the infant should have a repertoire of movements, sounds, and other self-expressive devices to enable him to imitate others with some success. Imitation of others, especially of the mother, father, and older siblings, serves as a dynamic incentive to further development of language, emotions, and interests. These early achievements lay the foundations for other peculiarly human activities.

With progress in this stage of development, the sensorimotor coordination improves, and the newly established motor patterns allow for a thorough exploration of the surroundings and expansion of the infant's world. As this phase progresses, the need for stimulation intensifies and greater curiosity is shown toward novel and complex objects and situations. Stimulus change is a potent factor in early learning, because infants are more reliable discriminators of stimulation variability than past studies suggested (Fantz, 1967; Fitzgerald et al., 1967; Horowitz, 1969). After surveying infant learning studies, Frances D. Horowitz (1969) defined the infant rather clearly:

> The young human infant is a learning organism, subject to a variety of environment-

> al contingencies, able to behave reliably in relation to associated events, and, within a range of response capabilities, his behavior is modifiable, trial by trial, over a somewhat generally predictable conditioning course.

For Eleanor Gibson, perceptual learning occurs from the early days of infancy. The child pays attention to the distinctive features of persons, objects, and situations, and also to the invariants of events that occur over time. Distinctive features and invariants are discovered sequentially one by one in the flowing array of stimulation—a filtering process attenuates and suppresses the less relevant. Orienting improves, and successful strategies of attending are developed and used against distraction from the less significant features. Gibson (1971) assumes a built-in need to get information about one's environment, which she calls the "search for invariance." The resultant reduction of uncertainty reinforces various forms of perceptual learning. For Gibson, words, like other entities in our environment, are perceived by detecting their distinctive features, classifiable as phonological, graphic, semantic, and syntactic. Of course, the young infant's attentive strategy deals primarily with phonologically distinct features.

Visual-cliff experiments show that human infants are able to perceive and avoid an apparent chasm as soon as they can move about, when creeping is established (Walk & Gibson, 1961). Using Walk and Gibson's visual-cliff model III, Schwartz, Campos, and Baisel (1973) explored cardiac responses of twenty infants five months old and twenty infants nine months old. They found that on the deep side of the cliff, heart rate shifted to acceleration with the nine-month-olds as compared with the five-month-olds, and visual attention was greater on the deep side, indicating a shift with age from attentiveness to fearfulness. A playful attitude and curiosity mark the infant's approach to reality in most of its tangible aspects. Play is a medium of converting dormant powers and endowments into various abilities and skills. Simple forms of play serve as a preparation for complex activities. With progress in sensorimotor coordination, simple forms of object manipulation are supplemented by more complex and goal-related activities. Regular ties in manipulation patterns begin to appear toward the end of middle infancy and set the stage for advancement in the infant's differential approach to his or her surroundings. At the ten-month level, for example, many babies make the important observation that objects fall down and produce noise. For some time they enjoy this discovery by allowing their toys, spoons, and foods to fall. Various play activities train children's senses and initiate and promote insight into the fundamental laws of nature. Within the first year of life the child accumulates the bulk of preverbal and unconscious experience; i.e., behavioral experience that cannot be verbalized.

Drawing the mother's attention to his needs by means of vocalization is one of the first actions the child learns to perform quite well. When he is in the mother's arms, crying stops and many discomforts become tolerable. Crying for attention is often observed at about six months of age, and sometimes as early as the three-month level. Cries for attention increase as the attachment to the mother becomes firm. The baby seeks contact and physical proximity with her in order to be exposed to her "relatively high (but also accommodating) arousal value, and protests when he is prevented from achieving this end" (Schaffer, 1963, p. 194).

Visual, tactile, and kinesthetic stimulation in a nursery for premature infants at low birth weight (1,800 grams) for six weeks and weekly home visits until twelve months of age aimed at improving maternal care were found to result in significantly higher Cattell IQ scores at one year of age than the control *S*s achieved (Scarr-Salapatek & Williams, 1973). The early stimulation program was effective in furthering behavioral development of the experimental group of fifteen premature infants. The study suggests that early professional assistance is required for low-income families, especially those with high-risk infants, "if their development is to reach socially adaptive levels." Otherwise the biological vulnerability and nonstimulative social circumstances usually have "disastrous effects upon later intellectual functioning" (Birch et al., 1970; Scarr-Salapatek & Williams, 1973).

White (1972, pp. 92–97) attributes special importance to the ten- to eighteen-month age period for the acquisition of overall ability and general competence. At this age the infant is primed for expending enormous amounts of energy exploring and learning about his world. His curiosity is immensely magnified by his newly-acquired ability to cover space by crawling, creeping, and walking in any direction the mother approves. Within this age range the toddler begins to engage seriously in interpersonal activities and contests. He is learning a great deal about his mother and her reactions, since his life is usually centered on her because of a strong attachment to her. This age of increased demand for exploration and locomotion tests the mother's ability to rear children. The basic foundations of language and general education are shaped by the quality of experience during this period of about eight months. At its end, predictive developmental divergence becomes clear when Bayley's, Griffiths', or Gesell's tests are applied.

EMOTIONS AND NEEDS

Emotional growth during middle infancy is rapid and manifold. Its developmental trend is based on innate temperamental sensitivity and environmental stimulation. As the infant begins to relate to persons and objects more clearly, he or she is able to react with an emotional tone proportional to the depth of perception of pleasant and threatening stimuli. Diffuse emotional excitement and massive motion in response to any strong stimulation, which is present at the neonate level, now unfolds a variety of modifications of feeling. The basic feelings, such as anger, rage, fear, delight, and affection, arise in the early months of life and assume important self-expressive and self-assertive roles.

Some specific emotions can be seen early in life. One of these is distress, which may be provoked at an age as early as three weeks in some infants. Annoyances caused by bathing and dressing arouse distress and crying in the infant. As the baby grows older, he meets with attempts by people to restrain him, and being unable to comprehend all aspects of such situations, he is quickly aroused to anger. By the time the infant is nine months old, he becomes angered by his own limited ability to do as he wishes. An older infant who succeeds in pulling himself to a standing position may become angry and cry when he is not capable of doing it again. A young baby expresses his anger in a generalized way by crying and kicking, but as his comprehension develops, his anger is directed at whatever object he perceives as the source of his annoyance.

Fear is an emotion experienced less often than anger by babies, because the incidents that cause fear usually occur less frequently than do the daily annoyances that incite anger. Generally, fear is caused by stimuli that startle the baby. Sudden loud noises, strange objects or persons, objects associated with pain, or sudden removal of support tend to arouse the emotion of fear, depending somewhat on the child's comprehension of the situation. A baby reacts to fear by crying, withdrawing, or seeking refuge in the arms of the mother. Fear is learned, often by conditioning. A classic study by John B. Watson and R. Raynor (1920) gives an example of this in a conditioning experiment in which an eleven-month-old baby was exposed to a furry animal. At first, "Albert," the baby, had no fear of the animal, but when he reached for it, the experimenter made a loud noise behind him. This sequence of events was repeated several times, and each time the baby showed greater signs of fear. It was further observed that the fear was transferred to other furry animals and objects. ("Albert" was later reconditioned, with the help of ice cream, to remove this fear.)

Pleasant emotions are also seen in babies, two of them being delight and joy. In the small baby, delight is the result of physical contentment. As the baby grows older, he is delighted by being tickled or talked to or played with. The young baby expresses his pleasure or joy by cooing or smiling, while the older baby laughs aloud. The older baby derives great satisfaction and joy from accomplishing a difficult task, such as turning over, rising to a standing position, or climbing on a piece of furniture. He often expresses his pleasure by vocalizing and smiling.

Affection is another of the pleasant emotions, and it can be seen by the eleventh month. At this time the baby gazes at another person's face, touches it, smiles, waves his arms, and shows in other ways that he is not indifferent to those whom he loves. From about the twelfth month on, the baby becomes more selective in giving his affection. Adult members of his family are the prime objects of his love, though he may be readily attracted to some strangers after he becomes accustomed to them. At about this age the baby stretches his arms toward the loved one and pats and plays with his or her face. In the second year other children and some inanimate toys also become objects of the baby's affection. The child's affections grow the more often he comes in contact with people who are kind and affectionate to him. The important principle concerning the development of affection is that an individual must be loved in order to learn to love. Usually the mother's love is unconditional; she loves the baby because he is *her* child. She loves her baby for what he is. Exceptions are not infrequent, however. Maternal indifference and even hostility are often stimulated by a lack of happiness and adjustment in marriage. Lack of affection from parents often causes a child to withdraw into himself; but too much affection, "smothering," furthers the child's extreme dependence on the mother.

FOUNDATIONS OF LANGUAGE

Language refers to vocal forms and other means used in interpersonal communication. *Speech* designates that segment of language in which articulated sounds are used as means of communication; it is the deliberate, rule-defined vocal behavior of a human being. It involves assimilation of many speech sounds through observation and practice. Language is the chief vehicle by means of which human beings communicate their needs, ideas, feelings, and desires to other persons. Speech is a stimulus to behavior; it evokes responses in others, and it is a response to many interpersonal situations. Speech is helpful in making and maintaining contacts with others. A silent situation is usually uncomfortable and anxiety-provoking. Except for some infrequent cases, the neonate possesses all the equipment and dispositions needed for the acquisition of the various forms of language and speech communication. The laryngeal activity of sound-making is pleasurable to infants and children alike. There is experimental evidence that infants as young as one month of age are "not only responsive to speech sounds and able to make fine discriminations, but are also perceiving speech sounds in a manner similar to categorical perception of adults" (Eimas, Sizueland, Jusczak, & Vicorito, 1971). Apparently, a biological makeup for speech perception operates at a very early age, since young infants "sort acoustic variations of speech sounds into categories."

During the first month of life, vocalization consists of cries. At about four or five weeks of age, the infant begins to gurgle and coo. The maternal voice is a typical stimulus that furthers vocal activity at this age and later. Gurgling and cooing sounds are caused by spontaneous movements of air through the vocal chords. As a rule, during the first month the infant achieves some voluntary control over the flow of air through these chords and begins to produce "explosive" sounds. As soon as control over the speech mechanism is improved, rhythm is introduced. This form of vocalization is referred to as *babbling,* or lalling. Like crying, babbling is a preliminary to speech and is used extensively from six to ten months of age. The infant seems to derive great pleasure from listening to his own sounds, the quantity and quality of which increase until at about the one-year level *baby talk* begins.

Babbling, as well as baby talk, is produced by combining vowels with consonants, as in "wa," "la," "mah," "bah," "hah," and "ugh." Baby talk implies a certain refinement of these sounds, including a clearer duplication and greater, although inaccurate, imitation of adult articulated sounds, closely resembling certain words and phrases. Crude replicas of adult words are often recognized, but these are a long way from phonetic identity. Since what is first acquired is patterns and structure, constituent phonemes come later (Lenneberg, 1967). The accurate imitation of words produced by others is a major step toward speech. Two other major steps include the understanding of gestures and the association of the articulated sound with its meaning. Considerable progress in both steps is usually observed by the eleventh or twelfth month.

At present it is generally agreed that the infant responds to and understands gestures long before he or she is able to comprehend words. Pantomimes and various expressive movements often accompany vocal expressions and supplement them. The infant points to objects long before he can ask for them. Usually it is not until the age of about four years that he learns to combine words into complete sentences. The need for expressive movement is then progressively reduced, but it can never be fully abandoned. It remains a modifier that amplifies oral communication.

For the purpose of articulation and diction, speech organs continue to improve in efficiency throughout infancy and childhood. Ability to modify the passage of air, using movements of tongue, lips, teeth, and soft palate to produce most sounds, is sufficiently advanced by about the one-year level. As a result, rapid development in speech occurs early in the second year. Piaget (1954) has observed: "The infant with a good phonemic storehouse is likely to become the child with a good lexical storehouse."

Watson (1971) summarizes many aspects of perceptual-cognitive development in early infancy and examines means to stimulate attention, memory, and sensorimotor schemata. He suggests that "a young infant needs to be exposed to a variety of stimulation in order to adequately adapt and accommodate his basic sensorimotor schema." Since presentation of meaningful speech utterances is more attention-arousing than gibberish, a mere playing of speech records would provide infants with additional vocal material to gain and hold their attention. Also, since violation of expected sequences gains more attention, it may be used for that purpose. Apparently language has a natural salience for infants. Yet this artificial speech stimulation would not reward the infant's interactive behavior, as siblings and parents do. Therefore, if used excessively, it might depress the infant's adjustive behavior. It is an educated guess whether or not young infants would profit from construction of sensorimotor stimulation systems, such as mobiles responding to the infant's head movements by means of a special pillow. In addition to speech, it cannot easily be determined what other stimulation is genuinely desirable for the enhancement of early cognitive development.

Early forms of communication, such as crying and movements of the eyes and face precede the formal steps to socialization, but to a large extent they are intertwined. At two months of age, the baby turns his head in the direction of a human voice and shows relaxation when soft music is played. At three months of age, the infant responds with a smile to his mother, and he may cry in order to secure her attention. At about eight to ten months of age, the infant makes another distinct step in relating himself to others: he reacts appropriately to friendly, affectionate, angry, or scolding expressions. At this age he recognizes familiar persons and welcomes their approach; he exhibits signs of fear and eventually cries when strangers approach or watch him closely. At the one-year level many babies respond when spoken to. This is a source of joy to parents, who are frequently surprised by new types of oral reaction. It has to be kept in mind, however, that the recipient of a message decodes and transforms it according to his own life experience and readiness to respond (Frank, 1966, p. 176).

Bowlby (1969, p. 178) feels that an infant's clinging to humans implies an "in-built propensity to be in touch with and to cling to a human being." At four months of age, most infants smile and vocalize readily and follow the mother with their eyes for longer periods. Selective social smiling begins at this age. Attachment behavior is clearly present at six months. The mother is the principal attachment figure; the infant clings to her more strongly than to other persons. Fully established by the end of the first year, attachment behavior persists and means for achieving it become increasingly diverse (pp. 350–351). This in-built disposition or drive to be in touch with and to cling to a human adult, the mother in particular, is attested by Ainsworth and Bell's (1970) report of the behavior of fifty-six one-year-olds in a strange situation in a laboratory with some novelties in the setting. The presence of the mothers was found to encourage exploratory behavior; their absence heightened the search for attachment and crying and lessened exploration. Reunion episodes evoked contact-maintaining behaviors and to a lesser degree contact-resisting behaviors. Generally, infants used their mothers as secure bases from which they ventured to explore; the mother's absence was intolerable to most infants.

In surveying the development of early skilled action, Bruner (1973) discusses the use of the hand-eye-brain system for solving problems through the growth of manipulatory skill. He recognizes quantum jumps in performance due in part to the progress of the feedback processes. Bruner distinguishes three processes in feedback operation: (1) internal feedback, which signals an intended action; (2) feedback proper, coming from the effector system during action; and (3) feedback coming after action is completed—awareness of results. Anticipatory or internal feedback has been described by Evarts (1971). Bruner (1973) assumes that skilled behavior is not selected by contingencies of reinforcement—a Skinnerian position—but is virtually "released" by appropriate objects in the infant's environment, presented under appropriate conditions of arousal.

HEALTH AND ADJUSTMENT

During middle infancy most of the homeostatic mechanisms increase in efficiency, and their original lability and fluctuation following birth tend to decrease. Trivial and expected variations in temperature, diet stimulation, or other external conditions become less disturbing to basal metabolism, hydration, or rate of heartbeat. Throughout infancy the somatic equilibrium continues to show considerable fluctuation. The same applies to diet.

An infant's appetite often varies greatly from meal to meal; however, the healthy infant is likely to compensate for any nourishment loss at another feeding. Nutrition is an aspect of health that is still misunderstood by some parents. If any baby is above standard weight, the parents conclude that he is growing and is healthy, whereas in reality he may be in need of certain nutritive materials, such as iron and vitamins. The infant takes nutrition by volume rather than by richness. It is frequently the overweight babies who succumb to upper respiratory tract infection, pneumonia, and cardiac deficiencies.

If the mother's attitude is right, she is ready to show about the right amount of attention, care, and love the baby needs, just as her breasts are usually ready to give the right amount of milk at the right time. When a young infant is breast-fed by a healthy mother, he or she receives an adequate caloric intake, and practically all the requirements for most specific nutrients are fulfilled. Iron, fluoride, and vitamin D often need supplementation from the early weeks of such feeding. An additional bottle at 6:00 P.M. is practical for most breast-fed babies. It is important to note that breast feeding transmits many antibacterial and antiviral bodies that provide significant protection against bacteria and viruses that invade the body through the gastrointestinal tract (Fomon, 1967, pp. 245–246). Hence breast feeding tides the immunocologically sterile baby over until his own immune system becomes operative and efficient. This is a vital function of breast feeding that is often overlooked.

Infancy is noted for various disturbances in health and difficulties in adjustment. Both, but especially the latter, depend on the ability of the parents to protect the baby and to gratify his needs. A mother who is tense and anxious while feeding her baby transmits this emotional disturbance to the baby, who in some manner senses his mother's uneasiness and also becomes tense. This tension may cause the baby to have digestive disturbances so that at times he cries extensively and even refuses to eat. Sensitivity to colds and fever at times takes serious forms and leads to eczema, earaches, pneumonia, prolonged vomiting, diarrhea, or gastritis. Rapid physiological development, excessive crying, or unsatisfactory care often result in various skin rashes. Heat rashes and allergy rashes are also frequent in middle infancy. Apart from cases of infant neglect and abuse, sudden infant death syndrome (SIDS), or crib death. strikes apparently well-cared-for babies. The syndrome of crib death is the leading cause of infant mortality after the age of one month. Research on SIDS is only starting, and little is known about its etiology at this time.

During infancy, body temperature is somewhat unstable and the baby's temperature may rise as high as 105° to 106°F (40° to 41°C) with an infection. This is often seen when the baby has an upper respiratory tract infection. Hence it is important for the mother to know how to take the baby's temperature and how to reduce it. Sponging the baby's body with tepid or lukewarm water, a limb at a time, is recommended in order to maintain the baby's temperature under 104° until a physician can be reached. High temperatures can cause convulsions and, when prolonged, even brain damage. Babies are especially susceptible to infection and should be protected from cold drafts. Proper clothing helps control perspiration and dehydration. Infants should be dressed according to the changing temperature of their surroundings. Overdressing or overheating the baby can be just as bad as not dressing him warmly enough.

Vaccinations, especially for smallpox, diphtheria, whooping cough, poliomyelitis, and tetanus, are usually given to babies after they are three months old. Booster shots to continue protection usually begin at about three years of age. These vaccines have done much to guard babies and young children from the more serious communicable diseases. When a baby or child has been exposed to one of the other communicable diseases and the physician fears the disease may be too harmful, he can administer gamma globulin, which will give the child temporary protection and either prevent the disease altogether or lessen its severity. In the United States, the Food and Drug Administration and public health departments do much to protect residents against unsanitary food and water, which might be a source of diarrhea or other disease.

Babies will attempt to creep or stand at about ten months of age, and possible movements should be anticipated to protect the baby from injury. The older baby needs more supervision, because he is now capable of creating dangers for himself. The need for protection continues throughout infancy and childhood, although the type and amount will vary from year to year.

QUESTIONS FOR REVIEW

1. Identify some of the motor activities achieved during middle infancy and explain their significance.
2. How does an infant maximize the stimulus value of various objects in his or her surroundings?
3. What are the emotional and social responses that emerge during this stage of life? Discuss fear and anger.
4. How do infants react to normal and deviant faces and to visual-cliff experiments at different ages?
5. Identify several needs of the infant and explain one of them.
6. What are cooing, babbling, and baby talk? Explain the differences among them.
7. Define play and indicate its role in middle infancy.
8. How does the infant's exploratory behavior relate to the mother's presence or absence?
9. What marks the termination of middle infancy? Do other observers distinguish this level of maturity from the preceding and following ones?
10. What do parents have to exhibit and provide during middle infancy in order to stimulate genetic dispositions?
11. What can parents do to protect infants from harm in late babyhood?
12. Explain what a mother should do when her baby has a high temperature.

REFERENCES

Selected Readings

Bowlby, J. *Attachment and loss.* Vol. 1, *Attachment.* Vol. 2, *Separation: Anxiety and anger.* New York: Basic Books, 1969 and 1973. A major treatise on attachment and separation and their effects on the infant and young child. Primary object clinging and attachment behavior, its lability and stability, and its ways of interaction are analyzed.

Frank, L. K. *On the importance of infancy.* New York: Random House, 1966. A comprehensive work on infant development and related theories, in a paperback edition.

Horowitz, F. D. Learning, developmental research, and individual differences. In F. D. Horowitz (Ed.), *Advances in child development and behavior,* Vol. 4, pp. 83–126. New York: Academic Press, 1969. A major chapter on the topics indicated, emphasizing research in learning.

Provence, S. The first year of life: The infant. In L. L. Dittman (Ed.), *Early child care: The new perspectives,* pp. 27–39. New York: Atherton Press, 1968. A practical survey of first-year developments and the fundamentals of infant care.

Specific References

Ainsworth, M. D. S., & Bell, S. M. Attachment, exploration, and separation: Illustrated by the behavior of one-year-olds in a strange situation. *Child Developm.,* 1970, **41,** 49–67.

Bayley, N. Comparisons of mental and motor test scores for ages 1-15 months by sex, birth order, race, geographical location, and education of parents. *Child Developm.,* 1965, **36,** 379–411.

Berg, K. M., Berg, W. K., & Graham, F. K. Infant heart rate response as a function of stimulus and state. *Psychophysiology,* 1971, **8,** 30–44.

Birch, H. G., et al. *Mental subnormality in the community.* Baltimore: Williams & Wilkins, 1970.

Bowlby, J. *Attachment and loss.* Vol. 1, *Attachment.* New York: Basic Books, 1969.

Bruner, J. S. Organization of early skilled action. *Child Developm.,* 1973, **44,** 1–11.

Eimas, P., Sizueland, E., Jusczak, P., & Vicorito, J. Speech perception in infants. *Science,* 1971, **71,** 303–306.

Elkind, D. Cognitive growth cycles in mental development. In M. J. K. Cole (Ed.), *Nebraska Symposium on Motivation.* Lincoln: University of Nebraska Press, 1972.

Evarts, E. V. Central control of movement: Feedback and corollary discharge: A merging of the concepts. *Neurosciences Research Program Bulletin,* 1971, **9**(1), 86–112.

Fantz, R. L. Visual perception and experience in early infancy: A look at the hidden side of behavior development. In H. W. Stevenson, E. H. Hess, & H. L. Rheingold (Eds.), *Early behavior: Comparative and developmental approaches,* pp. 181–224. New York: Wiley, 1967.

Fantz, R. L., & Nevis, S. Pattern preferences and perceptual-cognitive development in early infancy. *Merrill-Palmer Quart.,* 1967, **13,** 77–108.

Fitzgerald, H. E., Lintz, L. M., Brackbill, Y., & Adams, G. Time perception and conditioning an autonomic response in human infants. *Perceptual motor skills,* 1967, **24,** 479–486.

Fomon, S. J. *Infant nutrition.* Philadelphia: Saunders, 1967.

Frank, L. K. *On the importance of infancy.* New York: Random House, 1966.

Gibson, E. J. Perceptual learning and theory of word perception. *Cogn. Psychol.,* 1971, **2,** 351–368.

Horowitz, F. D. Social reinforcement: Effects on child behavior. In *The young child: Reviews of research,* pp. 27–41. Washington, D.C.: National Association for the Education of Young Children, 1969.

Knobloch, H., & Pasamanick, B. Further observations on the behavior development of Negro children. *J. genet. Psychol.,* 1953, **83,** 137–157.

Lenneberg, E. H. *Biological foundations of language* (chap. 7). New York: Wiley, 1967.

Lewis, M. Infants' responses to facial stimuli during the first year of life. *Developm. Psychol.,* 1969, **1,** 75–86.

Piaget, J. *The construction of reality in the child.* New York: Basic Books, 1954.

Scarr-Salapatek, S., & Williams, M. L. The effects of early stimulation on low-birth-weight infants. *Child Developm.,* 1973, **44,** 94–101.

Schaffer, H. R. Some issues for research in the study of attachment behavior. In B. M. Foss (Ed.), *Determinants of infant behavior,* Vol. 2, pp. 179–199. London: Methuen, 1963.

Schwartz, A. N., Campos, J. J., & Baisel, E. J., Jr. The visual cliff: Cardiac and behavioral responses on the deep and shallow sides at five and nine months of age. *J. exp. Child Psychol.,* 1973, **15,** 86–99.

Užgiris, I. Č., & Hunt, J. McV. *Assessment in infancy: Ordinal scales of psychological development.* Urbana, Ill.: University of Illinois Press, 1975.

Walk, R. D., & Gibson, E. J. A comparative and analytical study of visual depth perception. *Psychol. Monogr.,* 1961, **75** (15, Whole No. 519).

Watson, J. B., & Raynor, R. Conditioned emotional reactions. *J. exp. Psychol.,* 1920, **3,** 1–4.

Watson, J. S. Cognitive-perceptual development in infancy: Setting for the seventies. *Merrill-Palmer Quart.,* 1971, **17,** 139–152.

Weizmann, F., Cohen, L. B., & Pratt, R. J. Novelty, familiarity, and the development of infant attention. *Developm. Psychol.,* 1971, **4,** 149–154.

White, B. L. *Pre-school project: Child rearing practices and the development of competence. Final report.* Washington, D.C.: Office of Economic Opportunity, 1972.

Wohlwill, J. F. Developmental studies of perception. *Psychol. Bull.,* 1960, **57,** 249–288.

Chapter 8

Toddlerhood: The Dawning of the Child

HIGHLIGHTS

The infant who walks, climbs, uses some words, and understands a number of fundamental laws of nature and social relationships is a very different individual from a five- or ten-month-old baby. The toddler is ready for autonomous exploration of many aspects of the environment: people, objects, situations, and relationships. He or she has a powerful urge to convert inner experiences into behavioral modalities.

Toddlerhood is a delightful age when many new meanings emerge as the infant discovers new ways of search for semiotic representation of reality. For many toddlers it is a highly frustrating period, as their explorations are cut off too often and behavioral controls are unduly pressed upon them. Usually the toddler shows a strong preference for the mother, unless somebody else knows better how to gratify the gamut of physiological, emotional, and social needs.

As self-awareness grows, the child discovers his or her own mind and asserts it eagerly. Various forms of noncompliance and active resistance follow one another. From about two years of age, features of childhood begin to emerge at a rapid rate—a sense of self gains in directive power to become an architect of the child's growth in personality traits and strategies of adjustment. The mold of many formative experiences takes effect at this phase of life.

During the last phase of infancy, from approximately fifteen months to two-and-a-half years of age, the child considerably expands his environment through increased movement, new forms of language communication, and understanding of fundamental relationships. During this phase the infant achieves greater control over bodily functions and home situations, and infantile helplessness decreases. Newly acquired abilities also play a major role in assisting the development of individual initiative and assertiveness. Awareness of individuality occurs early in this stage. From about two years of age it is marked by expressions of self-reference. Many features of childhood emerge and gradually overshadow babyhood characteristics. Toward the end of toddlerhood, the infant looks more like a child. His maturity level and personality organization are still predominantly infantile, but will be reorganized at some time during the year. From this time on, facility and speed in various motor and play skills will depend chiefly on perceptual maturation and practice. Cognitive development involves perceptual growth, which depends on the adequate functioning of all the human senses, especially those for evaluating distance and depth. Perception is a process of cognitive interpretation that includes both the sensory experience and its meaning and value for the individual. The decreasing structural growth in late infancy permits the child to make more rapid advances in the organization of new behavior traits and the acquisition of many human abilities and skills, furthered by the society and the culture (Murphy, 1968: Piaget, 1937/1954).

IMAGINATION AND UNDERSTANDING

Interest in and partial understanding of pictures and stories now emerge and develop at a rapid rate. Whenever books are available, the infant makes frequent attempts to turn the pages and seems to enjoy their contents. Many delightful experiences result as he makes associations between pictures and previous observations of his toys and television characters. The first spurt of imaginative growth occurs at this midphase. The qualities and activities of living individuals are readily attributed to inanimate representations of reality. Television programs designed for children, in which birds, dogs, and other animals assume the roles of people, appeal to infants at this stage and serve as stimuli for imitation, however imperfect this may be. Through character modeling the child assimilates language, social behavior, and other forms of complex human activity as presented through interpersonal communication and mass media. Most children now have many opportunities to learn much and early (Bandura & Barab, 1971).

The neonate shows rudimentary forms of perceptivity but no intellectual behavior for some time after birth. Indeed, he lacks the proper means for it, since speech is the vehicle par excellence for distinctly cognitive functioning through representation of stimuli. However, intelligence can occasionally be assessed at three to six months of age. Seemingly, the more intellectually gifted infant tends to develop faster in various aspects of growth; perceptivity and motility patterns are therefore used in rating intelligence. In mental testing of young infants, Nancy Bayley (1969) begins with such test items as "responds to sound of bell," "quiets when picked up," "responds to sound of rattle and click of light switch," and "momentary regard of red ring." Age placement for these items is 0.1 month. At the age of six to seven months the following test items are used: "looks for fallen spoon," "playful response to mirror," "retains two or three cubes offered," "manipulates bell," "interest in detail," and "vocalizes four different syllables." All infant growth is included in Gessell's Developmental Schedules (1949), based on normative summaries of motor, adaptive, language, and personal-social behavior. The resulting developmental and intelligence quotients (DQ and IQ) are useful in estimating an infant's progress in psychomotor, cognitive, and social growth. It may be added that until the second year of life there is relatively poor prediction from infant tests to IQ assessments in middle or late childhood. At this early age parental socioeconomic status alone is the best single predictor. Empirical testing data on a large infant and child sample do not support simple continuity of general precocity at one age with general precocity at another age. A constant *g* factor is probably not tenable as a model for infant cognitive development (McCall, Hogarty, & Hurburt, 1972).

The toddler discovers the world through his mother and father, but he adds quite a bit of magic as his imagination begins to provide ideas and answers of his own. Though he imitates his parents, he also imitates nonhuman, even nonliving, objects as he sees them. He discovers new ways of playing by using fantasy: "Let's pretend" appears in a variety of situations. Beyond that, he starts imitating absent persons and objects once they have been "interiorized," to use Piaget's term, and the pattern of activity is worked out in his head before it is portrayed (Piaget, 1945/1951, p. 65). The infant does not yet know how to discriminate between effects of his own actions and those of other objects or persons. He lives in a world with a few permanent objects and with only a shadowy awareness of himself as a person. His cognitive behavior organization is both preverbal and prelogical—not represented by any symbol system. The use of speech accelerates and symbolic representation begins toward the end of this phase, and the search for names and identity commences (Piaget, 1937/1954; Užgiris, 1973).

EXPLORATIONS OF THE ENVIRONMENT

Throughout late infancy, the child eagerly engages in the process of exploring and becoming familiar with his or her own environment in most of its aspects and many of its vicissitudes. Curiosity is readily aroused. Toddlers make repeated efforts to get anywhere and everywhere, in and out, up and down. They make use of chairs and other household furniture in order to climb and reach high places in their homes. Drawers, boxes, cans, and bottles are, whenever possible, opened and their contents examined. Eagerness to manipulate external objects in every way possible increases as late infancy advances;

it helps toddlers recognize the invariance or constancy of people, objects, and relationships (Piaget, 1937/1954; Užgiris, 1973).

The object-grouping activity observed at the beginning of this phase turns into selective ordering—the toddler classifies and separates small objects on the basis of their differences. At about two years of age toddlers begin to make simple graphic collections without any verbal instruction or presentation of any pattern. The mere presence of different small objects is a sufficient stimulus for this discrimination play. Additionally, occasional opportunities for free play to mess with sand, mud, and even leftover food is a good way to let toddlers explore textures (Pavenstedt, 1969).

In touching, grasping, pushing, pulling, sucking, throwing, and banging, the infant stimulates his senses, engages his muscles, and gains much enjoyment, fun, and surprise. He has a tendency to keep a variety of familiar objects in his possession and use them for old and new activities as his ingenuity inspires him. Depending on opportunities, neighborhood exploration also advances to a significant extent. Toddlers need and profit from this type of stimulation. As a rule, they try hard to imitate older children's behavior.

Using his power of locomotion, the infant slowly increases his ability to estimate distance and depth. For example, he readily observes a change of line or color, but he needs an accumulation of experience in order to relate these details to distance. Also, he must learn that in many instances a surface continues unchanged even though a color or line has changed. The many difficulties young children have in perceiving depth on the basis of minor changes in line, shading, and object interposition are puzzling to adults, since perception of distance and depth seem so natural to them.

At this stage the infant willingly engages in play with other children if they are ready to contribute the lion's share of cooperation. He learns to enjoy infants of his own age, especially when he has frequent contacts with them. If parents are alert to the likes and dislikes of their children, if they know what they can take and tolerate and what helps to further their interest in people and children, they can help young children to unfold their powers and talents at an early age (Murphy, 1968).

Parallel play and independence in play activity are the first signs of a growing desire for autonomy and greater self-expression in proportion to the infant's own needs and desires. Engaging in a great variety of exploratory activities shows that the toddler age is well advanced and his reservoir of knowledge greatly enriched. The toddler spends a great deal of time seeking new stimuli, the significance of which cannot be fully understood until later (Berlyne, 1966).

SEARCH FOR NAMES

Some ability to communicate through speech usually appears during the late stage of infancy, when the infant begins to attend to the language communication of others for increasingly longer periods of time. The toddler partially or completely articulates sounds, attaching either the usual meaning or his own to them. He first learns several common nouns, often the names of a few people and things; then several verbs are added. Later he begins to include simple adjectives, adverbs, conjunctions, and prepositions, and much later personal pronouns and articles. The single-word sentence is sometimes well established before the infant is one-and-a-half years of age. At this age he begins using phrases and brief sentences. Infant speech is self-centered and is chiefly used to communicate more fully the infant's own needs, emotions, and desires.

The terms *language* and *speech* are frequently used interchangeably, even though they denote distinct concepts. First of all, not all sounds made by humans are classified as speech. Vocalizations in the form of cries and explosive or babbling sounds are preliminary forms of linguistic communication. Two basic criteria should be used to determine the extent to which the infant at this stage is capable of speech: (1) clarity in pronouncing words and (2) ability to associate specific meaning with the pronounced sound. Clarity in articulation results from the power to interrupt and modify the sound waves as they pass through the throat, larynx and pharynx, mouth, and nose by intricate movements of the vocal cords, tongue, soft palate, teeth, and lips. Proper pronunciation involves fine coordination, precise timing, and a delicate interaction of these and other organs forming the so-called speech mechanism. It is not surprising, therefore, that rhythm difficulties are frequent at this stage, when basic speech developments are taking place.

Though part of the baby talk is usually understood by parents and others who are in frequent contact with the infant, baby talk often does not meet the above criteria. Thus, if the baby says "mama" to every woman, or if his associations with "bird" or "toy" are very general, the specific sound-object association has not yet taken place. The actual association of articulated sound with a particular object or person is fundamentally different from other forms of language shared with certain other mammals. The development of true speech out of jargon is subtle and can be readily misinterpreted by parents and outside observers, including those who are aware of the main criteria. The infant does not readily respond to adults in terms of their requests. This makes it difficult to tell whether or not he possesses certain speech abilities. Many fifteen-month-olds use five or six words, but there is great variability among infants at this age.

In the late phase of infancy, the naming stage usually begins. It is marked by the frequent question, "What's that?" The toddler begins to realize that various objects and persons have names, and he or she wants to know them in order to promote familiarity with them. As Piaget often points out, this development represents the first major step in a realistic approach to various environmental factors (Piaget, 1937/1954). Berlyne (1966) speaks of "epistemic" curiosity, since the child's responses are aimed not only at dispelling the uncertainties of the moment but also at acquiring knowledge, that is, "information stored in the form of ideational structures and giving rise to internal symbolic responses that can guide behavior on future occasions."

Many studies (Golden & Birns, 1968; Katz, 1963) show that language skills facilitate learning. Golden, Bridger, and Montare (1972) provide evidence that two-year-old middle-class children are superior to lower-class children in their ability to use language as a tool for learning. This explains in part why middle-class children continue gaining superiority in utilizing verbal instructions when responding to many learning tasks. Lower-class children, however, do not show significant differences on the nonverbal or sensorimotor learning tasks.

DEVELOPMENTAL TASKS

The late phase of infancy is marked by many developments and by acquisition of new abilities and skills as the infant becomes more familiar with the home environment and more effective in meeting its demands. Certain fundamental tasks must be mastered at this time:

1 *Taking solids.* Early opportunities to eat a few solid foods in small quantities and a gradual increase in their variety help the one-year-old infant to develop enjoyment of new textures and tastes. Otherwise, some children will restrict their choices of foods to such a degree that a balanced diet becomes an impossibility. An increasingly competent use of chair, spoon, cup, and dishes is expected, provided the infant has ample opportunity to learn. Limited use of baby foods is helpful, but extreme reliance on them occasionally results in fixation on certain foods and rejection of many others.

2 *Physical control.* The improvement of coordination and of the neuromuscular system is an important developmental task at this stage, since it gives the child confidence in handling himself in a variety of home situations. A well-furnished home and age-related play materials provide opportunities for eye-hand and hand-foot coordination. Experimentation with stair climbing, running, jumping, and free rhythm dancing is part of this. Now the infant has to engage in much activity in order to improve control over the fine muscle groups.

3 *Understanding communication.* Learning to interpret and use speech is the chief way the child progresses in self-expression and adjustment to other people. It is good for the child if parents take pains to use simple words and phrases distinctly. A correct interpretation of basic concepts, such as "yes" and "no," "come" and "go," and "take" and "give," helps the infant greatly in recognizing the do's and don'ts of his environment. It represents the first step toward further levels of discrimination and personal use of these and related concepts.

4 *Learning toilet control.* Acquiring toilet control is a task of late infancy. Understanding basic concepts and gestures facilitates cooperation in terms of need, procedure, and place. Increased awareness of the function, stemming from neuromuscular readiness, and favorable emotional ties with the mother are key factors in toilet-training success. In this phase, some infants resist regulation of bowel or bladder movement for some time, but complete sphincter control is frequently attained during the third year of life.

5 *Promoting self-assertion.* Self-awareness is a task pertaining more to preschool years than to late infancy. It begins with the discovery of one's individuality, including likes and dislikes and preferences for and relationships with parents, siblings, and others. Acquiring suitable forms of self-assertion is important in fostering the concept of the rights of the individual. Parents,

through careful guidance, can help the child attain mastery over obstacles in an efficient manner without displaying an impatient or domineering attitude toward him. An insufficient number of contacts with other people and children at this age often results in shyness and withdrawal during the preschool years.

6 *Recognizing limits.* At present there is a general recognition that toddlers need definite limits beyond which they cannot venture on their own (Coopersmith, 1968). As running and climbing develop, the toddler is likely to create occasional dangers for himself unless protective measures are taken and limits are explained by gestures and words. When the toddler is pressed to adapt himself to a few consistently defined limits, his reality testing within accepted limits is strengthened (Pavenstedt, 1969).

SELF-AWARENESS

The toddler exhibits many signs of his self-awareness and of personal choice. He readily makes up his own mind and takes initiative in planning play, locating desired food, and getting the attention of the desired person. Self-initiative rises toward the end of late infancy, and parental corrections are objected to vigorously. Often parents fail to understand the changing personality dynamics of their child when he begins to show awareness of himself as a person, and as a result they interfere with legitimate modes of initiative and striving for autonomy. While adults have the duty of protecting the infant by stopping activities that may result in infection or injury, for sound personality development they must also respect the child as an individual with a mind and values of his own. In addition to having his infantile personality, he is now an autonomous child in at least some of his developments and activities. The distinction is this: the toddler is aware of only some aspects of his body and mind; a child is clearly a self-conscious person. As compared with the infant, general self-awareness is a distinctive mark of the child.

Joseph Church (1966, pp. 156–158) shows a child at twenty-seven months of age who has gained much in self-awareness. Benjy is able to express most of his wishes, and he is gradually realizing that yelling for something is less effective than asking for it and saying "Please." Sometimes he will spank himself and say "Benjy spank," or "Mommy spank," but this is usually when he seems to realize that he is cranky and may be punished. He is still confused about possessive pronouns—uses "my" to mean "your" and has not discovered "mine" at all. Recently, he discovered "no" and occasionally uses it to mean "yes," but mostly he appears to know that he means "no"—often very vehemently, as a matter of fact. He knows all the large parts of his anatomy and can find them on his animals, dolls, or parents. He has taken a renewed interest in looking at himself in the mirror; Benjy is becoming a child.

QUESTIONS FOR REVIEW

1 What makes the expansion of environmental exploration possible? Under what conditions may neighborhood investigation begin?

2 Indicate the characteristic changes in physiological growth from the first to the second year.
3 What are the major factors contributing to behavior organization?
4 Describe the toddler's exploration of the home environment. Explain the mother's role.
5 What characteristic language developments occur when the infant arrives at the speech level?
6 Name the criteria of speech and explain what processes and controls are necessary for speech.
7 Consider the infant's search for names and explain its significance.
8 What is the first question that the toddler discovers in the process of speech acquisition? What does he or she want to know?
9 List the developmental tasks of late infancy, and discuss one of them.
10 What are the outstanding signs of self-awareness? Describe infant behavior with this development in mind.
11 In what ways does the infant resemble the child? Identify some of the signs indicating the dawning of childhood.

REFERENCES

Selected Readings

Brackbill, Y. (Ed.) *Research in infant behavior: A cross-indexed bibliography.* Baltimore: Williams & Wilkins, 1964. A catalogue of predominantly empirical studies of normal infant behavior from 1876 to early 1964.

Chandler, C. A., Lurie, R. S., & Peters, A. D. In Laura L. Dittmann (Ed.), *Early child care: The new perspectives.* New York: Atherton, 1968. This book provides much information about the first three years of life; it was written by seventeen participants in a series of four conferences sponsored jointly by the National Institute of Mental Health and the Children's Hospital of Washington, D.C., 1964–1965.

Church, J. (Ed.) *Three babies: Biographies of cognitive development.* New York: Random House, 1966. A cross-referenced study of two girls and a boy recorded by their mothers, all following an identical observation schedule, devised by the editor, stressing intellectual growth and learning.

Illingworth, R. S. *The development of the infant and young child: Normal and abnormal.* Edinburgh and London: Livingstone, 1972. Prenatal and postnatal factors are presented at length, including assessment of the newborn baby and patterns of early child development.

Smart, M. S., & Smart, R. C. *Infants: Development and relationships.* New York: Macmillan, 1973. This book includes prenatal and early postnatal phases of life. All five parts are accompanied by selected readings.

Specific References

Bandura, A., & Barab, P. G. Conditions governing nonreinforced imitation. *Developm. Psychol.,* 1971, **5,** 244–255.

Bayley, N. *Bayley scales of infant development: Manual.* New York: Psychological Corporation, 1969.

Berlyne, D. E. Curiosity and exploration. *Science,* 1966, **153,** 23–25.

Church, J. (Ed.) *Three babies: Biographies of cognitive development.* New York: Random House, 1966.

Coopersmith, S. Studies in self-esteem. *Sci. Amer.,* 1968, **218** (2), 96–106.

Gesell, A. *Developmental schedules.* New York: Psychological Corporation, 1949.

Golden, M., & Birns, B. Social class and cognitive development in infancy. *Merrill-Palmer Quart.,* 1968, **14,** 139–149.

Golden, M., Bridger, W., & Montare, A. Social class differences in the use of language as a tool for learning in two-year-old children. *Proc. 80th Ann. Convention APA,* 1972, **7,** 107–108.

Katz, P. A. Effects of labels on children's perception and discrimination learning. *J. experim. Psychol.,* 1963, **66,** 423–428.

McCall, R. B., Hogarty, P. S., & Hurburt, N. Transitions in infant sensorimotor development and the prediction of childhood IQ. *Amer. Psychologist,* 1972, **26,** 728–748.

Murphy, L. B. Individualization of child care and its relation to environment. In L. L. Dittman (Ed.), *Early child care: The new perspectives.* New York: Atherton Press, 1968, pp. 68–106.

Pavenstedt, E. Development during the second year: The one-year-old. In L. L. Dittman (Ed.), *Early child care: The new perspectives.* New York: Atherton Press, 1969, pp. 40–56.

Piaget, J. *Play, dreams and imitation in childhood.* New York: Norton 1951. (Original French ed., 1945.)

Piaget, J. *The construction of reality in the child* (M. Cook, Trans.). New York: Basic Books, 1954. (Originally published, 1937.)

Užgiris, I. Č. Patterns of cognitive development in infancy. *Merrill-Palmer Quart.,* 1973, **19,** 181–204.

Chapter 9

Foundations of Personality during Infancy

HIGHLIGHTS

At birth infants have distinct physiological, emotional, and cognitive bases for personality development. Physical constitution, temperament, perceptivity, energy level, and other dispositions—all vary and set different styles of reaction as the infant begins to respond to physical and parental stimuli. Each infant develops his or her own way of behaving, thereby forming consistent modes of acting upon surrounding persons and conditions.

Individuation of responses permits consistencies and patterns that reflect an emergent personality. Behavioral consistencies give rise to traits and attitudes, which are molded by parents in terms of their conceptions of what the baby needs and what is good for him or her. In direct or subtle ways, parental responses encourage and press infants to modify their behavior.

Genetic and environmental influences are supplemented by the infant's growing awareness that he is a person with a mind of his own. Objections to parental management rise then as the toddler begins to struggle for autonomy yet is not able to free himself from the need of frequent support from a maternal person.

The early part of the third year is marked by the appearance of many childhood features, including distinctly human forms of language. Generalized awareness of himself turns the infant into a child—a role and position to be played over many years of lessening dependence on parental figures and an ever-expanding environment.

In the first two years of life an infant develops greatly and increasingly begins to react as a whole person rather then as a mere reflexive organism producing somewhat isolated stimulus-elicited responses. In order to understand the child in terms of an emerging personality pattern, he or she must be viewed in terms of total behavioral capacity. Appraising the human being as an evolving personality necessitates an integrative approach to the psychological study of development and behavior. *Infant personality* refers to a cluster of biopsychic tendencies generating contingent traits and attitudes which are expressed in behavioral consistencies. In early personality appraisal, attention must be focused on psychomotor and temperamental activity and later on the emerging behavioral style.

NEONATAL POTENTIAL

Each neonate enters the world with numerous *Anlagen*—dispositions toward a number of strengths and weaknesses. These behavioral roots are at first latent or dormant, some beginning to unfold during the early phases of life. A number of them will not receive sufficient stimulation to be actualized, while others will be suppressed by external pressures at their embryonic levels of development. Innumerable genetic complexities, as well as varied patterns of social stimulation, give rise to a distinct personality pattern for each infant.

No two individuals have identical genetic materials, even with respect to proteins or ribonucleic acids, although identical, or monozygotic, twins are often assumed to have the same heredity and early physical environment. Neither the temperamental or cognitive endowment associated with a capacity to learn nor the potential for individualization is the same for any two infants. Chapter 3 analyzed the internal and external factors, indicating why the sequences of experience and the environments of all infants, including identical twins, are different in many aspects. As a result of specific environmental events and uneven opportunities for the internal constituents to grow, the variation among infants is magnified with the passage of time. Most tendencies, traits, and features stemming from inborn capacities and organizational qualities are molded by the environment while emerging.

During infancy, changes in sensitivity and in adjustment to a host of environmental stimuli begin to take form as distinct patterns of reactivity emerge. By the third year, all the major personality qualities and traits have become more closely interrelated and the psychosocial equilibrium established—a process due particularly to the child's increased self-awareness, which is now developing at a considerable rate into a concept of self and a distinct pattern of relationships with others. Emergent traits and attitudes, as well as habits, are also contingent on the formation of the core of personality—the sense of self.

DEVELOPMENT OF TRAITS

There are many styles of reaction in young infants. Many infants show good adaptability, moderate sensitivity, and relaxed, easygoing behavior. They are seldom emotionally upset by normal internal and external changes such as getting wet, waiting for or taking in food, falling asleep, or experiencing natural temperature changes. Other infants are more reactive and excitable: each new influence and manipulation—for instance, being laid down, having a diaper or shirt changed, being put in a different position, or waiting to be fed—causes much emotional tension and crying. Such children are difficult to manage. Reports of the New York Longitudinal Study of Child Development (Chess, Thomas, & Birch, 1965, pp. 28–34; Thomas et al., 1963) show great variability within all modes of behavior defined: activity level, regularity, rhythmicity of approach or withdrawal, adaptability, sensory threshold, mood, intensity of response, distractibility or persistence, and attention span. The investigators found very quiet, moderately quiet, active, restless, and hyperactive children, for example, for the first mode of behavior. They found easy and difficult children, as well as various ways of parenting good for each type of child. Later modifications and refinements of these varied behavioral modes in children greatly influence the pattern of personality organization, especially in terms of adjustment and development of defense mechanisms. These basic modes of behavior are the beginnings of fundamental traits and attitudes of various intensities and shades.

Personality patterns and traits shown during the early years of life tend to develop further and attain a relatively persistent structural status (Bowlby, 1969, pp. 342–348; Coopersmith, 1967; Erikson, 1963; White, 1972). Madorah E. Smith (1952), in her study of six siblings of the same family as children and adults, found many traits of individual adults to be significantly consistent with their traits as children. Fifty years later, the traits appraised as most consistent were affection, ambition, attractiveness, brightness, carelessness, irritability, jealousy, nervousness, quarrelsomeness, and determination (p. 179).

Later developments of basic patterns and traits are more an expansion and supplement than a metamorphosis. They are evolutionary rather than revolutionary. Apparently one of the fundamental explanations of why the first two or three years are so crucial for later well-being and adjustment is that the developments of these years define the parameters for structural growth of personality. Strengths or defects become more glaring in later years, first at the entrance into school, then later during the pubertal years. The years of infancy are not only a very critical period for later personality development; they are the most formative for the growth of a child's personality because patterns are set at this time. In this period the structural foundation is laid for the complex psychological structures and functions which "will be built in a child's lifetime" (Bijou, 1975). Much later behavior can also be understood in the light of the achievement of trust and autonomy during the first three to four years of life (Erikson, 1963, pp. 247–253).

ROLE OF PARENTAL ATTITUDES

Parental influences upon the child are far-reaching; the hereditary, constitutional, social, and cultural qualities and attitudes of the parents are the most potent conditioning factors in the life of the child.

Parents' attitudes toward their children range from affectionate acceptance to hostile rejection, immoderate indulgence to carefree neglect, excessive pampering to extreme lack of nurturing, autocratic dictatorship to great permissiveness, intense pressure for achievement to indifference and abdication of regular parental responsibilities.

The parents' level of emotional acceptance of the child and their resulting attitudes toward him or her play a leading role in laying the foundations for the type of personality the child will develop. The value of the warm and affectionate mothering that produces an emotionally and cognitively maturing child is probably beyond scientific verification. Understanding the natural language of the baby—cry, smile, head and hand motions—and responding appropriately is an art that sets the pattern of genuine communication arising from early emotional symbiosis. This sensitivity to an infant's signals in a one-to-one relationship furthers dyadic love—a by-product of deep maternal involvement (Ainsworth, 1973).

The infant has intrinsic needs for close human association, for gentle and affectionate handling, and for direct protection against common dangers and undesirable influences. Ordinarily the mother and father are best suited to help the infant gratify these needs. With a favorable child-parent relationship, the parents are likely to be the two dominant human models when the process of identification starts during the third year of life. The extent and quality of the ability to identify with other persons and relate to them is closely and largely determined by the nature of the relationships within the family during these early years of life.

The myriad complexities of behavior in the parents and others who surround the child inevitably tend to elicit from the child a particular behavior pattern and direct him into it. Thus the infant acquires means and techniques for gratifying needs that those around approve. Within the family matrix a child also acquires tendencies to desire or fear certain objects and situations and learns what to do and what to avoid doing. Such learning, in turn, gives direction to sundry components of the emerging personality. Later the child learns reasons for the do's and don'ts that the parents have established early.

David M. Levy's study *Maternal Overprotection* (1943) provides classic illustrations of reinforcement of specific traits in children by their dominant or submissive mothers. Dominant mothers, by their extensive use of control and punishment, tend to create submissive and dependent traits in their children. An empirical study by Halla Beloff (1957) shows that anal traits exhibited by the mothers are substantially related to the child's acquisition of an "anal character," marked by parsimony, orderliness, pedantry, egoism, desire to dominate, and other related features. These and many other studies (Frank,

1966; Freedman, Loring, & Martin, 1967) illustrate how infantile experiences can have major effects and repercussions in later life. For example, reinforcement increases the strength of the oral drive, and coercive toilet training heightens separation anxiety and magnifies negativistic behavior (Bernstein, 1955). Each child has his or her own reinforcement history, determined by a particular ratio of gratifying and aversive contingencies used by parents (Bijou, 1975).

Jerome Kagan (1971, pp. 180–182) reports that mothers "have different conceptions of the ideal boy or girl, and engage in different practices in order to attain these idealized goals." Many mothers believe that boys should be "strong and proficient at gross motor talents," and they encourage the rough-and-tumble play activities of boys to attain this goal. By contrast, most mothers value vocal expression in their daughters and condition them to increased vocalization by reciprocating more often than they do with boys.

The effects of a satisfied need for affection are deep and pervasive. Arthur T. Jersild (1954, p. 894) stresses the fact that love enters into and greatly affects the quality of the total environment and conditions its relationship with the child: a central feature of parental love for a child is that "the child is liked for his own sake; he is viewed as something valuable per se; he is respected as a personality in his own right. The child who is loved for himself is free to be himself." He is then in a position to experiment for himself and learn to mold his own self in terms of his natural endowments and gifts.

INDIVIDUATION OF RESPONSES

The latter part of infancy is marked by individuation of responses, feelings, and attitudes. As the emotions of self-competence and self-esteem emerge, self-initiative increases and the child's resistance to parental control and suggestions is magnified. "No, no!" and "Johnnie won't!" are frequent expressions of the two- or three-year-old child. The developing child gradually realizes that he is an individual with a mind of his own. He can make choices and has desires of his own. From this point on, he plans many of his activities and uses strong emotional outbursts to demonstrate his opposition to any interference from parents, siblings, or other children. Dawdling, stubbornness, and contrariness seem to constitute the nucleus of postinfantile self-assertion. For most children, the peak of negativistic behavior occurs early in the third year.

Negativism in children represents a kind of fundamental conflict at the critical periods of self and personality organization. The third year is one of the first genuinely critical years in human development, because at this age the child's self emerges as a directing force. One of the major principles of development is that, when any new ability appears, it is practiced vigorously for some time until it is smoothly functional. Personal likes and dislikes, as well as tendencies toward specific foods, drinks, and activity or rest, now become self-related and are therefore powerfully defended against most threats and barriers. While at first this intensified negativism does not seem to make much sense, it soon becomes integrated with the needs and desires of the emerging personality. David P. Ausubel (1950) considers negativism not merely a developmental necessity in promoting individualization but also a distinct phase of self-development. Negativism seems to typify an awkward stage of transition from the helplessness, docility, and dependence of infancy to the relative autonomy and partial self-reliance of the preschool child who can feed, bathe, and dress himself with little assistance, can plan his own amusement and activities and utilize a variety of means for his own goals, and can go alone to play with children in the neighborhood. Thoughtful and partially permissive handling of this negativistic behavior by parents is a prerequisite for adequate self-esteem and personality integration in the child.

But the period of dawning childhood can bring problems. Probably the most serious is a failure to maintain warm emotional relations with the parents. High dependency feelings also militate against wholesome ego growth. The toddler must be given ample opportunity to initiate activities and learn from them. Erikson (1963) recognizes the core problem as the alternation of feelings between autonomy and shame or guilt. He feels that if the child is not allowed to experience autonomy for his own choice, he will become constricted and overinhibited. Additionally, he will turn the urge to manipulate against himself rather than against the environment.

When parents provide opportunity for the child to choose toys, food, and clothes, they contribute to his desire for self-regulation. They should guard themselves against perfectionism and strictness, particularly in inconsistent discipline. Parental and social influences that contribute to the child's experimentation, planning, and active participation in family, neighborhood, and community life promote the development of individuality and selfhood. Under adverse circumstances the suppressive effects of overrestriction will produce emotional strain and tend to unbalance behavioral controls. From the early years of life the human being operates as a self-organizing, self-directing, largely self-repairing and self-stabilizing open system, which becomes progressively patterned after and aligned with the culturally established dimensions of the environment (Freedman, Loring, & Martin, 1967).

Revision of Habits Another fundamental capacity closely allied to personality growth is the early appearance of the ability to form and revise habits. Even during the first phase of infancy, but especially in the second, habit formation extends to practically all aspects of the child's activity. Any orderly sequence introduced by parents readily molds natural tendencies into habitual patterns of responding. Nevertheless, the infant shows high malleability in structuring his or her behavior. Even when a habit is practiced for some time, only moderate or mild resistance is shown to most attempts to change it. At this age, behavior modification is an ongoing business. Normal infants show openness to attentional, social, affective, linguistic, and cognitive cultivation. They stay with most new tasks for a long time—they are absorbed by each of them and show pleasure in prediction. In several conditioning experiments, three- or four-month-old babies smiled when they had learned to predict, on the basis of a signal tone, the side on which a feeding nipple would appear—smiling before taking the nipple (Bruner, 1973, pp. 129–131; Kalnins, 1970; Papoušek, 1967).

Since an infant has a limited number of ready-made response systems for gratifying his needs, it is his task to acquire many new and socially acceptable behavior patterns for the expression and satisfaction of hunger, thirst, elimination, curiosity, sex, and other drives. Newly acquired strategies and skills are usually extensively used and therefore become habitual. Provision of a variety of similar and contrasting stimuli offers opportunities to play and to utilize emerging schemata or skills and develop them into new adaptive responses. The repertoire of habitual responses increases with each passing week, month, and year. Addition of novel patterns of behavior depends on the provision of subsequent social reinforcement (Gaines, 1973; White, 1972).

INFANT MANAGEMENT

Parental treatment of the infant either forms a basis for security and growth or handicaps both. A desirable and secure relationship between mother and infant is marked by genuine acceptance and love, under which a variety of social and affective communication occurs. Unconditional acceptance of the infant as he is helps the parent to avoid all extreme forms of interaction and assists much in the establishment of a harmonious balance: the infant is loved but not overprotected; the parent is firm but not domineering. Infant management is thus flexible but not too permissive. The parent makes efforts to satisfy the infant's needs adequately but refrains from indulgence.

The last three decades have been marked by a general decrease in the austerity of infant training and an encouraging trend toward emotional warmness and tolerance of the infant's setbacks in toilet training and expressions of autoerotic impulses. Parents are beginning to understand more fully than their forebears that young children are soft and pliable beings who profit much from fairly lenient rearing practices. Maternal attention, for example, when given following appropriate infant behavior, increases the frequency of its occurrence. It is one kind of reinforcer that begins to be effective quite early in infancy, perhaps by the end of the third month (Herbert & Baer, 1972; Sears, 1972).

The infant needs to form an attachment to at least one stable figure, and the mother is usually the attachment figure. "The amount of mother-infant interaction determines whether the infant becomes attached, but the kind of interaction shapes the quality of the attachment" (Ainsworth, 1973). Sears (1972, p. 14) hypothesizes that "the attachment capacity is part of some maturational sequence more precisely related to developmental status measured by other indices than by the crude one of chronological age." Attachment behavior is expressed by moving toward the mother, touching and holding her, staying near her, and seeking permission and reassurance in moves away from her.

Separation from the mother can be tolerated for increasingly longer periods as the child grows, yet continued separation "inevitably leads to a disorganization of the child's whole behavior system," since much of the behavior is organized around dyadic interaction with the mother (Sears, 1972, p. 22). When searching does not restore the mother's presence, separation usually becomes heavily charged with fear and its prolongation with agitation, then depression, and finally expressions of frustration and despair. Maximum damage from separation occurs during later infancy (Ainsworth, 1969, 1973; Yarrow, 1964).

There are many forms of social reinforcement in which the verbal and other forms of expressive behavior of one person have the effect of modifying the behavior of another person. The parent is usually the social reinforcer for a child. Appropriate or desired behavior increases when the mother pays attention at the time the behavior occurs, or shortly afterward. If operant behavior-modification principles are correct, behavior is a product of the prevailing contingencies that play upon it. When these contingencies are rearranged by design, systematic behavior changes occur (Risley & Baer, 1973). Deliberate development of behavior is thus possible. Herbert and Baer (1972) report that it takes little time for professionals to train mothers to become efficient behavior modifiers of their infants; gains in desirable behaviors showed durability over the following five-month period for two mothers of deviant young children who participated in the training. For making children more sensitive to behavior modification, limited deprivation of social stimulation may be used; it enhances the effectiveness of subsequent social reinforcement (Horowitz, 1969, p. 39).

The infant's inner dispositions leading to a variety of developments and to a gradual self-realization may be suppressed, inhibited, or distorted if his or her fundamental needs for acceptance and love, protection, and respect are not understood and gratified. The way the mother or her substitute approaches and reacts to the infant is one of the first lasting influences contributing to differential responsiveness (appropriate and particularized, not generalized, responses). The mother's attitude is therefore a major factor in the formation of not only the foundation but the later surface dimensions and traits of personality. White (1972, p. 100) declares that the basic quality of the entire life of a child is determined by the mother's actions during the first three years of life: the mother's actions at this early age are "the most powerful formative factors in the development of a preschool-age child."

During late infancy, the parental "yes" and "no" are the beginnings of the semantics of morality. The child also begins to look for "go" and "stop" signs. It is better for a child to feel the safety of limits then the pain of unbounded reality. Even if not fully understood by the toddler, verbal directions are a good start for child management. They allow for communication at a distance. The child learns to respond to voice modalities. In effective child-rearing practice a mother talks a great deal to her children; provides access to many persons, objects, and diverse situations; and explains things to the child, but mostly on his instigation rather than her own. She prohibits certain activities firmly, saying "no" to the child without fearing that he will not love her. In various ways she strengthens the toddler's motivation to explore and to learn. It is desirable for her to have an ability and willingness to take the child's point of view and to discover his concerns. She capitalizes on the moments when the child's attention is maximal to teach him basic lessons of life (White, 1972, pp. 100–102).

Adjusting the home to the older infant is a major task for parents. Toddlers need a great deal of sense stimulation; they take the initiative and "get into things" because of their endless exploration. Various provisions and precautions are therefore necessary (Murphy, 1968). Evelyn Duvall's *Family Development* (1971, pp. 236–239) presents a detailed treatment of such household arrangements. The child's safety and freedom of activity are, however, much more important than any damage to furniture or other objects. Valuable objects must be placed out of reach. Old or dispensable objects can often be substituted for the more fragile ones the child may ruin or damage. Toys, fixtures, and other items must be large enough not to be swallowed or easily broken. Sharp tools, knives, forks, insecticides, paints, medicines, and various cleaning compounds must be locked up or placed out of reach. Low electric outlets should be capped and, when possible, fenced off with heavy furniture so that the toddler cannot insert the tongue or a piece of wire into them. It is a wise precaution to have gates at the top (and bottom) of stairs, allowing the toddler to roam freely about each story of the house. By careful placement of objects, pain and injury can be prevented. However, there will always be minor bruises, cuts, and bangs to be taken in stride, and some more serious ones that may need medical attention. One child, for example, within less than one year, tipped a corner cabinet over on herself, drank from a gasoline can, and ran into a bicycle ridden by another child. In each instance she needed emergency treatment.

Various growth processes can be disturbed by any strong and lasting fear or hostile reaction, which may condition the child, for the sake of safety, to encapsulate himself in one particular phase of maturity or even press him to regress. Parents have a responsibility to exert their own resourcefulness in making the need to grow seem more attractive than the choice of fixation or regression (Maslow, 1956).

Since fear is a universal response of infants and children, intense or frequent displeasure or frustrating situations associated with parents will tend to stimulate fearfulness and insecurity. The affection- and attention-seeking of the fearful infant will be exaggerated and difficult to gratify; hence anger, temper tantrums, jealousy, and hostile reactions will become predominant as the child grows into the preschooler. John B. Watson (1928) feels that if this kind of negative relationship is imprinted on the mind of the infant it will block the responsiveness of the child to parental discipline and later to societal controls. Later attempts by parents to facilitate the formation of sound habits and desirable attitudes will be less effective. Poor training by parents during the late stage of infancy and early childhood will lead to various forms of aggressive behavior or undue dependence and a feeling of inferiority. These traits then become dominant qualities of individual reactivity and ego defense. Referring to the emotional patterns of life, Watson (1928, p. 43) declared:

> At three years of age, the child's whole emotional life plan has been laid down, his emotional disposition set. At that age the parents have already determined for him whether he is to be a whining, complaining neurotic, an anger-driven, vindictive, overbearing slave driver; or one whose every move in life is definitely controlled by fear.

All the conditions and situations that foster insecurity and promote undesirable intrafamily relationships interfere with the personality development of the child and in some instances predispose him or her to specific patterns of a maladaptive or deviant nature. Psychologists and psychoanalysts agree that once such a pattern is stabilized, long-term reeducative or reconstructive psychotherapy is necessary to "correct" the past, stimulate interest in more adaptive goals, and fortify the ego by more efficient strategies of coping.

Neglect of parental responsibility often has powerful repercussions on later developments. Some parents, for example, show little concern when a child lags in performing a developmental task. A mother may not care that her two-year-old infant does not make progress in eating new solids and may continue providing "baby foods." As a result, feeding difficulties arise and plague the family for years. This undesirable behavior of the infant, which could readily have been averted or corrected early in the second year, when the child is eager to put into his mouth practically anything, becomes more generalized and may expand into resistance to other developmental tasks. Difficulties in adjustment thus increase with the years. The child's dressing himself is another example. Because the mother can do it better and faster, she continues doing something that the infant is capable of doing; in this way she deprives the growing child of early initiative and acquisition of skill. The child may then have a problem on school entry because of having failed to learn something that contributes to adjustment.

Although the potential for personality development is present at birth, the core of personality structure seems to solidify somewhat when the neonate continues responding to external stimuli and to parental feelings and attitudes. Infants seem to be predisposed toward exhibiting and developing certain traits partly because of their awareness of parental reactions to them and partly because of the unique constitutional qualities and features that contribute to more or less distinct behavioral tendencies. Personality foundations laid during infancy are wholesome only when parents are aware of and satisfy their infant's somatic, psychological, and cognitive needs and when their care is affectionate and their guidance gentle. As time goes on, parents can promote the child's development by providing appropriate stimulation for emerging qualities and traits. They should play with the infant and provide age-related educational toys, specially designed to foster sensorimotor and cognitive growth as well as to give the infant frequent opportunities for self-expression. Such positive social encouragement enables the infant to express his or her individuality fully and to take the initiative in learning to interact effectively with other persons and with various external objects and situations. Under such circumstances the individual personality becomes organized in terms of its own potentialities and may function near its optimum level of performance.

EMERGENCE OF A SELF-CONCEPT

Early in the second year, the infant becomes aware of his own bodily organs, some of their functions, and his capacity to control them. At this age he differentiates clearly between his own organism and various persons and objects in the environment. Inner needs and organically aroused desires are distinguished from persons, objects, and conditions that can gratify such wants. If there is enough love and the infant is respected as an individual, he will easily progress in awareness of his abilities and will seek opportunities to use them.

The child's self-awareness plays a leading role in unifying drives, emotions, needs, motives, and incentives. As the infant gains control over his motor functions, he begins to manipulate various environmental objects to an increasing degree. For William Stern (1930, pp. 471–475), the rising ego consciousness is first of all a desire-filled, emotional experience of the organism, meeting with obstacles and boundaries in fulfilling its needs and desires. He recognizes the presence of self-willed behavior in the fifteen-month-old infant's attempt to accomplish little tasks. A toddler often demands things not even in sight in order to repeat activity engaged in days before (pp. 443–447).

Freudian psychoanalysts assume that during the early phase of infancy a considerable amount of the baby's activity consists of the taking in of nourishment and affectionate care, centered on the oral zone. The retentive and later active or expulsive orality—biting, spitting out, and masticating—may be partially inhibited through lack of objects of gratification. If conflict and anxiety become fixated at this level of psychosexual development, an orally motivated character ensues. It is further hypothesized that such a person will continually seek oral gratification in a variety of forms, for example, excessive interest in and compensative activities with food and drink, or potent verbal activity characterized by a sharp tongue, verbal aggression, and sarcasm.

The great majority of theories of personality development assign a leading role to the events and experiences of infancy (Freedman, Loring, & Martin, 1967; Freud, 1923; Levy, 1943/1966; Murphy, 1968; Ribble, 1955; Watson, 1928; White, 1972, pp. 94–98). As the infant develops his or her capacity for perception and retention, certain images of the parents appear and stabilize. If, in the mind of the child, the parents are a source of protection, affectionate care, and security, the subsequent positive experiences will help contingently to engrave favorable images and characteristics. Conversely, when early experiences of an unpleasant kind predominate, the perception of parents as threatening and punishing agents will be fixated, and many positive experiences will be needed to alter or transform these original impressions (Ainsworth, 1969; White, 1972, pp. 100–102).

The Freudian doctrine of personality types maintains that personality is largely determined in the first five or six years of life, during which social occurrences within the family triad (father, mother, child) are the decisive factors. Parents have practically unlimited opportunities to "duplicate themselves" in their children and to develop either desirable or deviant personality traits and features. Many cognitive and emotional disorders result from infant mismanagement, which often permanently damages the child's ego (Freud, 1923). Kagan and Klein (1973) question this strongly supported continuity assumption and provide some cross-cultural data suggesting polarized changes in many Indian children in Guatemala from passivity and quietness at three years of age to activity and competence at eleven. They observe that "it is misleading to talk about continuity of any psychological characteristic—be it cognitive, motivational, or behavioral—without specifying simultaneously the context of development." Both change and continuity mark human development, and the emphasis on one of them is probably of theoretical making.

At the beginning of late infancy, the baby progresses greatly in ability to exercise control over the basic skeletal muscle groups and to respond directly to the simple requests of parents and siblings. Because of this positive attitude, this is the proper time to begin toilet training. The age of seventeen to twenty months seems to be the most favorable period for undertaking this task with most children. At about the one-and-a-half-year level, the innervation of the sphincters has sufficiently matured and the function of the lower intestine has become perceptible to the child. The child's cooperation, however, persists only when the emotional ties to the mother are appreciated. Madsen et al. (1969) reported moderate success with buzzer pants at sixteen to eighteen months and high success at eighteen to twenty months of age, especially if rewards were given for dryness.

Bowel control is frequently achieved before the infant's second birthday; complete bladder control is established a year or two later. Two basic conditions underlie this significant accomplishment: neuromuscular readiness and no undue strain in the emotional ties with the parents. Within several months after bowel control is achieved the infant usually learns to use the toilet facilities with only occasional assistance from the mother.

Many American parents are eager to establish control at a much earlier age than that indicated above, and earlier than parents of most other cultures or nations. Psychoanalytic and psychiatric literature points out that in doing so they expose their children to unnecessary strain and conflict, which foster excessive resistance and excessive negativism at this time and stinginess, cleanliness, orderliness, obstinacy, and compulsiveness later in life (Beloff, 1957; Bernstein, 1955). Psychoanalytic theory emphasizes the close relationship between various types of feeding and toilet control and contingent libidinal cathexes (object attachment and means of need satisfaction as well as the subsequent patterns of personality organization). In infancy, any intensified frustration leads to a disturbed coping pattern and to difficulty in outgrowing early patterns of adjustment. If severe, the disturbance often persists for many years, causing a person to act as if situations that have long since disappeared are still in existence.

The Freudian matrix of human development, based on the child-mother-father triangle, in which the child's self-concept and adjustment depend chiefly upon the ways his or her instinctual needs are met by parents, is seen by Robert R. Sears as a *dyadic* model. The child relates to a single person at a time. His development occurs within the dyadic relationship; most stimulus-response (S-R) sequences are based on mother-child and father-child dyads. For Sears, behavior is the product of S-R learning, reinforced by gratification of organismic needs. During the early years of life, parents are the chief agents of reinforcement. A child's inherent desire to learn and a mother's urge to do right create a dyadic situation in which the proper interactions occur. Personality is a product of a "lifetime of dyadic action which modifies the individual's potential for further action" (Maier, 1969, pp. 144–154).

Any theoretical interpretation answers as well as raises questions. Is the toddler blindly acquiring attitudes, motives, and habits in concordance with reinforcing agents and environmental contingencies, as Watson, Sears, White, and Skinner assume? Yes, he is a recipient of what is provided for him, but he is more than that—he actively molds his environment and disregards many stimuli as he searches for what is internally gratifying to him—to his budding sense of himself. He is making a significant contribution to his own development. "The mind, like the nucleus of a cell, has a plan for growth and can transduce a new flower, an odd pain, or a stranger's unexpected smile into a form that is comprehensible" (Kagan & Klein, 1973).

Throughout the latter part of infancy, the self continues to establish the order of priority among the infant's activities, feelings, goals, and desires. Unless some kind of powerful interference or pressure reverses the trend, motives and actions are determined by this evolving hierarchy of wishes and desires. The power to draw from one's own resources and environmental givens for self-directed experiences expands with the following months and years. More and more frequently as the infant grows into childhood and later into adolescence, he makes decisions in terms of himself rather than the persons and ideas he identifies with. Anderson (1948) and Eisenberg (1972) go one step further by assuming that any human being is the chief architect of his or her personality. Anderson (p. 416) concludes that the child "is very much a creature in his own right, moving through his own experiences and creating his own world."

Self-preservation, achievement of physiological equilibrium, and adjustment to the demands of external reality are the tasks of the neonatal phase. Control over large muscle groups and overcoming helplessness are among the main tasks of middle infancy. Learning verbal self-expression and fundamental concepts of the physical and social environment and acquiring toilet control are tasks of late infancy. The satisfactory completion of the developmental tasks of infancy lays a wholesome foundation for growth and maturation in childhood and later periods. Generally, infancy establishes a basic pattern of relating and coping for all the later periods of life, during which modifications of behavior are frequent but revisions are rare.

QUESTIONS FOR REVIEW

1. What major internal and external factors lay the foundations for human individuality?
2. Name and illustrate several general reaction patterns of neonates toward environmental stimuli.
3. Describe several traits of a model infant and indicate the fundamental relationships among traits, attitudes, and habits.
4. What is meant by individualization of the infant's responses? Relate it to negativism.
5. Why are parental attitudes toward and treatment of an infant important for his or her development and personality growth?
6. Describe the role of learning in personality development.
7. Explain and illustrate how parents can modify an infant's behavior.
8. Explain how to adjust the home environment for a toddler.
9. Describe attachment behavior of an infant and its relationship to the mother's care.
10. Explain the emergence of the self-concept and its effects on the infant.
11. What does Watson urge parents to do to further children's independence?
12. Describe Sears' theory of dyadic relationships and compare it with the Freudian triangle of child-mother-father.

REFERENCES

Selected Readings

Baldwin, A. L. *Theories of child development.* New York: Wiley, 1967. Discusses theories of developmental psychology from early naive propositions to the currently accepted theories (Piaget, Freud, Levin, and Werner), including the sociological systems of Parsons and Bales.

Erikson, E. H. *Childhood and society* (2d ed.). New York: Norton, 1963. A psychoanalytically oriented work that deals with the period of childhood in a genetic frame of reference. Core crises and social conditioning processes are treated in detail. Many sociocultural problems are discussed, including reports of Hitler's childhood and Maxim Gorky's youth.

Maier, H. W. *Three theories of child development* (2d ed). New York: Harper & Row, 1969. A fairly comprehensive study of the theoretical and research contributions of Erikson, Piaget, and Sears, with their applications.

Specific References

Ainsworth, M. D. S. Object relations, dependency, and attachment: A theoretical review of the infant-mother relationship. *Child Developm.,* 1969, **40,** 909–1023.

Ainsworth, M. D. S. The development of infant-mother attachment. In B. M. Caldwell &

H. N. Ricciuti (Eds.), *Review of child development research* (Vol. 3, pp. 1–94). Chicago: University of Chicago Press, 1973.

Anderson, J. E. Personality organization in children. *Amer. Psychologist,* 1948, **3,** 409–416.

Ausubel, D. P. Negativism as a phase of ego development. *Amer. J. Orthopsychiat.,* 1950, **20,** 796–805.

Beloff, H. The structure and origin of the anal character. *Genet. Psychol. Monogr.,* 1957, **55,** 141–172.

Bernstein, A. Some relations between techniques of feeding and training during infancy and certain behavior in childhood. *Genet. Psychol. Monogr.,* 1955, **51,** 3–44.

Bijou, S. W. Development in the preschool years: A functional analysis. *Amer. Psychologist,* 1975, **30,** 829–837.

Bowlby, J. *Attachment and loss.* Vol. 1, *Attachment.* New York: Basic Books, 1969.

Bruner, J. S. *The relevance of education.* New York: Norton, 1973.

Chess, S., Thomas, A., & Birch, H. G. *Your child is a person: A psychological approach to parenthood without guilt.* New York: Viking Press, 1965.

Coopersmith, S. *The antecedents of self-esteem.* San Francisco: Freeman, 1967.

Duvall, E. M. *Family development* (4th ed.). Philadelphia: Lippincott, 1971.

Eisenberg, L. The human nature of human nature. *Science,* 1972, **176,** 123–128.

Erikson, E. H. *Childhood and society* (2d ed.). New York: Norton, 1963.

Frank, L. K. *On the importance of infancy.* New York: Random House, 1966.

Freedman, D. G., Loring, C. B., & Martin, R. M. Emotional behavior and personality development. In I. Brackbill (Ed.), *Infancy and early childhood,* pp. 427–502. New York: Free Press, 1967.

Freud, S. *Das Ich und das Es.* Leipzig: Psychoanalytischer Verlag, 1923.

Gaines, R. Matrices and pattern detection by young children. *Developm. Psychol.,* 1973, **9,** 143–150.

Herbert, E. W., & Baer, D. Training parents as behavior modifiers: Self-recording of contingent attention. *J. appl. beh. Anal.,* 1972, **5,** 139–149.

Horowitz, F. D. Social reinforcement effects on child behavior. In *The young child: Reviews of research,* pp. 27–41. Washington, D. C.: National Association for the Education of Young Children, 1969.

Jersild, A. T. Emotional development. In L. Charmichael (Ed.), *Manual of child psychology,* pp. 833–917. New York: Wiley, 1954.

Kagan, J. *Change and continuity in infancy.* New York: Wiley, 1971.

Kagan, J. & Klein, R. E. Cross-cultural perspectives on early development. *Amer. Psychologist,* 1973, **28,** 947–961.

Kalnins, I. *The use of sucking in instrumental learning.* Unpublished doctoral thesis, University of Toronto, 1970.

Levy, D. M. *Maternal overprotection.* New York: Columbia University Press, 1943. (Paperback reprint, New York: Norton, 1966.)

Madsen, C. H., Jr., et al. Comparisons of toilet training techniques. In Donna M. Gelfand (Ed.), *Social learning in childhood: Readings in theory and application,* pp. 124–132. Belmont, Calif.: Brooks/Cole, 1969.

Maier, H. W. *Three theories of child development* (2d ed.). New York: Harper & Row, 1969.

Maslow, A. H. Defense and growth. *Merrill-Palmer Quart.,* 1956, **3,** 36–47.

Murphy, L. B. Individuation of child care and its relation to environment. In L. L. Dittmann (Ed.), *Early child care: The new perspectives,* pp. 68–104. New York: Atherton Press, 1968.

Papoušek, H. Experimental studies of appetitional behavior in human newborns and infants. In H. W. Stevenson, H. L. Rheingold, & E. Hess (Eds.), *Early behavior: Comparative and developmental approaches,* pp. 249–277. New York: Wiley, 1967.

Ribble, M. A. *The personality of the young child.* New York: Columbia University Press, 1955.

Risley, T. R., & Baer, D. M. Operant behavior modification: The deliberate development of behavior. In B. M. Caldwell & H. N. Ricciuti (Eds.), *Review of child development research.* Vol. 3, *Child development and social policy.* Chicago: University of Chicago Press, 1973, pp. 283–329.

Sears, R. R. Attachment, dependency, and frustration. In J. L. Gerwitz (Ed.), *Attachment and dependency,* pp. 1–27. New York: Winston & Sons, 1972.

Smith, M. E. A comparison of certain personality traits as rated in the same individuals in childhood and fifty years later. *Child Developm.,* 1952, **23,** 159–180.

Stern, W. *Psychology of early childhood.* New York: Holt, 1930.

Thomas, A., Chess, S., Birch, H. G., Hertzig, M. E., & Korn, S. *Behavioral individuality in early childhood.* New York: New York University Press, 1963.

Watson, J. B. *Psychological care of infant and child.* New York: Norton, 1928.

White, B. L. *Preschool project: Child rearing practices and the development of competence. Final report.* Washington, D. C.: Office of Economic Opportunity, 1972.

Yarrow, L. J. Separation from parents during early childhood. In M. L. Hoffman & L. W. Hoffman (Eds.), *Review of child development research* (Vol. 1, pp. 89–136). New York: Russell Sage Foundation, 1964.

Part Five

Childhood

The span of childhood covers three maturational levels: early childhood, middle childhood, and late childhood or preadolescence. Significant social and cognitive developments, as well as modifications in behavior and personality, occur in the early phases of each of these related levels. During the preschool years amazing progress is made in the child's verbal self-expression and fantasy, in cognitive differentiation, and in environmental expansion. Because of multiplying social relationships and increasing value judgments, the child reacts in many new ways to various environmental factors and situations. Individual differences become magnified as young persons unfold their endowments and selectively utilize social, educational, and other opportunities given to them.

As the child gains autonomy from parents and efficiency in concrete mental operations, he or she is ready to enter primary school. There the child meets and interacts with a teacher and many other children of the same age. What the teacher affirms to be right is right because "the teacher says so." After one or two years of increased association with peers, the child becomes eager to join various peer groups. He modifies his behavior and eagerly learns the skills necessary to be accepted into varieties of group activities. The advancing years of childhood are usually marked by increasingly numerous and intimate peer identifications. The child develops many insights into parental and peer traits and attitudes and into his own motives for acting his way as he continues to grow in knowledge of his physical, social, and cultural environment. The child values highly what he has not yet become and often models his behavior on children one or two years older.

Chapter 10

Early Childhood

HIGHLIGHTS

The young child becomes graceful as his physiological growth decelerates and he gains practice in applying his abilities and sensorimotor skills. As the child's play activities diversify, he uses language not only to identify objects and activities but also to engage in various "make believe" transformations. His fantasy carries him into many intangibles as he vicariously creates and solves many problems.

Further emotional differentiation elicits self-centered emotions such as shame, awe, remorse, guilt, and hostility. Anger and temper outbursts and experiences of fear, jealousy, and envy are frequent during the early years of childhood. Up to about ten years of age, the number of fears increases. Situations associated with noise and possible lack of safety, animals, and ghosts call for unreasonable fright and tears in many young children.

Progress in speech development centers on pronunciation, sentence formation, and conversational skill. Gains in comprehension and vocabulary are often spectacular. There is no end to the child's curiosity and questioning if there is someone who provides educational stimulation and answers in a way he can understand.

Self-awareness rises throughout the early years of life. It is marked by frequent resistance to parents—the child eagerly uses his own mind and forms his own expressive style. Sensitivity to incentives and needs sharpens the child's self-

awareness and presses him to act in accordance with what his environment can offer him. Much ambivalance is experienced as the child now wants to be himself but is also moved to gain affection, attention, and approval from parents, peers, and other significant persons constituting his "intrapsychic family."

The period of early childhood begins at about two-and-a-half, when the toddler gains a general awareness of self as a male or female person, makes progress in toilet control, and begins to recognize common dangers. It is a time when symbolic function appears in many play activities, and the child's complex action patterns become internalized by advances in semiotic representation. The generative use of language and self-awareness are the landmarks that distinguish the child from an infant. Early childhood extends to the age of six or seven, when decentration occurs and when concrete operations empower the child to learn major school subjects—fast and well. Because of the dearth of analytic depth studies of transitional phases, any adequate division of childhood is necessarily arbitrary. In studies of large groups of children, wide individual differences mask rather than clarify transitional phases—many curves appear deceptively smooth and progressive rather than jagged and direction-changing. Chronological age per se is a poor predictor of important developmental changes leading to higher levels of operational maturity.

Consequently, the division of childhood into three operational levels raises many questions. Well-planned longitudinal research studies are necessary in order to answer the many questions related to phases in childhood and throughout life.

PSYCHOMOTOR AND PLAY ACTIVITIES

Throughout early childhood, the rate of physiological growth decelerates rapidly. The child now has "more time" to gain and refine neuromuscular controls. He or she can also engage in more differentiated physical and play activities than before. Increased facility in a growing repertoire of motor skills helps the child to gain competence in daily activities. From the beginning of this period, the child can leaf through a book or magazine, build a tower of nine or ten large cubes or blocks, take care of some daytime toilet needs, and express some desires verbally. At times the child readily responds to simple requests of the mother or father and cares about their presence. The child can climb, stand on one foot, go unaided up and down stairs alternating the feet, practice dancing steps, run and stop wherever necessary, imitate various sounds, sing, and follow and reproduce the behavioral patterns of others.

In paper-and-pencil expression, he draws circles and vertical and horizontal lines, and by age four, he draws objects and identifies them, even when adults cannot recognize them. Complicated psychomotor patterns, such as swimming and skating, writing the ABCs, and piano playing can also be acquired in a step-by-step procedure, if they are properly introduced and if the child's attention and effort are sustained by an internal or social reinforcement.

Whatever the child learns to do he performs faster and with increasing competence and ease. Ball playing and group singing appeal to children at four and five years of age, and they often eagerly practice many of their advanced activities to increase mastery. Facility in performing basic and specific motor skills leads to rhythmic patterns and gracefulness, which begin to mark the child at about the midpoint of early childhood.

The preschool child appreciates being noticed by everyone and shows much delight in whatever he accomplishes. He is proud to show what he can do in order to receive attention and recognition. Although his "products" are often crude, the value of his experience is great. The child therefore deserves attention and encouragement for any competence attained in day-by-day engagements. A five-year-old may bring a picture he has just drawn, show it to his mother and say: "Mommy, that's you!" Even if she does not recognize any features of her own in the picture, she should acknowledge the accomplishment by showing attention to both the picture and her child. He deserves it because this is the best picture of his mother he can make. It is a pity for a child to carry a production from person to person without receiving any attention or praise. Lack of reinforcement will reduce the child's productivity of this kind. At this age, his search for avenues of success is yet limited, and he needs success to maintain his adjustment.

During the early childhood years, play activities become increasingly creative and dramatic as the child's imagination flourishes. In imitating adults, other children, and television characters, various make-believe situations are enacted using miniature human figures and animals, household tools, dolls, balls, clay, and whatever else is available. A child's equipment must be not only age-related but varied enough not to restrict "let's pretend" and related activities. It is good for the child to be able to imitate play activities according to his concerns and interests of the moment. He must be free to explore his close and distant environment, as well as to seek out parts and details in order to gain understanding of his own surroundings and of himself. This is why the child rehearses many times his less usual and exceptional experiences until they begin to make sense to him (Almy, 1966; Piaget, 1936/1952).

From the beginning of this period, the child reacts with much excitement to hide-and-seek and run-and-catch games. Double pair and ring games appeal greatly from about four years of age. Tag and hopscotch attract five-year-olds. These and other social games add much to the child's success in relating to others.

Ordinarily, preschool children balance strenuous physical activities like running, swimming, and tag with passive and sedentary play. Some young children, however, need frequent encouragement for more strenuous exercise, while others need parental restraint in order to avoid exhaustion. When other children are present, activities often gain in speed to the point of actual danger or injury to younger children. Parents then ought to redirect children to other gratifying activities of lesser speed. Many four- and five-year-olds are great show-offs, and their need for attention is difficult to satisfy. In fulfilling their children's need for attention, affection, and acceptance, parents must allot a reasonable amount of time for each child but should not allow children to dominate their time completely. Of course, the quality of time spent with a child matters much more than the mere quantity.

At nursery or kindergarten or at home, the early years of childhood are very active years of life. The child possesses a natural urge to play, i.e., to apply his budding powers and abilities in a rising variety of ways to explore himself and his environment. In interpreting play, Karl Buhler and Schenk-Danziger (1968, pp. 145–149) emphasize the "function pleasure" that results from play activity and acts as a stimulus for it. As the child unfolds language and fantasy, function pleasure becomes expanded into *Schaffensfreude*—the joy and pleasure of projecting and creating. This tenet is elaborated by Charlotte Bühler, who speaks of play as a precursor to creativity. Her implication is that children who play much enhance their future creativity—a salient cultural trait of human beings (Bühler & Marschak, 1968, pp. 93–97).

In one of his rare references to play, Freud (1959, pp. 174–176) made a wish-fulfillment interpretation of play:

> The play of children is determined by their wish—really by the child's one wish, which is to be grown-up, the wish that helps to "bring him up." He always plays at being grown-up; in play he imitates what is known to him of the lives of adults.

For Freud, the child "takes his play very seriously and expends a great deal of emotion on it. The opposite of play is not a serious occupation but—reality."

According to Erik Erikson, play is one of the major *ego* functions by which the child organizes his inner world in relation to his outer world. Play enhances self-teaching as the child often attempts to organize and master, to think and plan, through the medium of "playing it out." By means of play the child attempts to make up for his defeats and frustrations. By learning new modes of play, the child, according to Erikson, advances his or her readiness to enter higher developmental stages (Maier, 1969, pp. 25–26). Paraphrasing Freud, Erikson (1963, p. 209) calls play the royal road to the understanding of the infantile ego's efforts at synthesis. The industriousness of the child in various play activities enhances his achievement and adds much to his sense of both identity and mastery. By making children partners in adult activities, parents further the child's efforts in finding identity (Erikson, 1963, pp. 237–239).

In its fundamental aspects, Piaget's theory of play is that of *assimilation* of reality to the self. Play expresses the child's representation of reality with two existential modes: adaptation to what he already knows and response to what is novel to him. In play, means often become ends, and many responses are emitted for their own sake. Play schemata are often ritualized, and accommodation becomes subordinate to assimilation. A single object like a box may be treated as a table, a carriage, or a bed; it depends on the play activity the child is engaged in and on the means available to him to portray the objects significant to the continuation of the symbolic schema of play (Furth, 1969, pp. 95–96).

For a young child, play is a medium of converting dormant powers into various abilities and skills. The child's playful approach toward persons, objects, and situations permits him to test them without assuming responsibility for consequences. "I was only playing" is the child's frequent excuse when something goes wrong. Play is a key mode of learning about the laws of nature as well as about the relationships inherent within and among persons and objects. Play also serves the purpose of adaptation to frustrating situations of unfulfilled longings and desires. When a child plays and enjoys playing, he is completely a child—his alpha brain waves gain in amplitude and "eureka experiences" (feeling of joy upon discovering something new) occur frequently to the point of domination over long periods of time. Theoretical aspects of play are analyzed in detail by Herron and Sutton-Smith (1971), who also list extensive bibliographic resources. Equipment needed for play by various age groups is presented by the Davises and Hansens (1974) and by Hartley and Goldenson (1957).

EMOTIONAL GROWTH

The transitional phase of the toddler's entrance into childhood is marked by a major reorganization and further differentiation of emotionality. Self-centered emotions and attitudes of competence and incompetence are vividly shown. Personality-core-centered emotions, attitudes, and sentiments of a social, esthetic, moral, and religious nature are readily evoked if the necessary social and environmental stimuli appear in the life of the child. The interaction of emergent emotions within the child magnifies his or her temperamental reactions—immature at this time but a source for humanized reactions in later childhood and to a degree throughout life.

Early childhood is a very temperamental age, marked by diffuse and distinct expressions of vital biological reactions and refined psychosocial coexperiential sentiments. *Fear,* which is evoked by anything threatening or unusual, is a frequent affective reaction of the preschool child. As self-awareness increases, so does personal vulnerability to fear objects. Many young children fail to realize that animals and events which they fear are rarely harmful. At four or five years of age most children learn to fear remote and imaginary dangers, coming from tales and stories about giants and ghosts, kidnappers and strange places, accidents, and death. If the mother is afraid of thunder, the child is likely to experience it even more intensely. Television adds its share to the increase in fear objects at this age. Fears of getting hurt, of being alone, and of the dark disturb many children, since their imagination tends to increase their awareness of the terrifying (Pikunas & Clary, 1962; Scarr & Salapatek, 1970).

Anger and temper outbursts are frequent during the years of early childhood. Any deprivation of the child's needs or desires, or any conflicts or frustrations resulting from unsatisfactory management procedures multiply these reactions. Some children exhibit anger without any provocation—their own ineptitude is sufficient reason for it. Many of the "good-natured" babies now show little capacity to tolerate any amount of thwarting or interference.

Affective outbursts often help children gain the attention they need and seek when other means fail them. By temper fits children can gratify their whims if parents yield to them for the sake of peace. Internal conditions and external situations that provoke anger and temper outbursts in young children include (1) need and stimulus deprivation, (2) interruption of habitual activity, and (3) restraint and inconsistent punishment that the child cannot predict. Excessive tiredness, a strange visitor, an overprotective parent, or the onset of an illness—all dispose children to anger and tears. The following case may shed some light on temper and its handling.

> Beth was a spoiled little girl of three years of age. If she could not have something exactly as she wished, she would lose her temper, scream, kick, and throw herself on the furniture or floor. Often she would hold her breath until she lost consciousness. Her mother would become terribly frightened and give Beth what she wanted. Much fuss would be made over her, and she would be put into bed with her doting mother at her side promising her what she wanted.
>
> The temper outbursts became worse and worse until the girl was a nuisance not only to her parents but to everyone else in the family. Finally they consulted a clinical psychologist, who made them see that they must ignore the excessive demands of their young daughter. But they became more aware of Beth's emotional and social needs and adept in finding ways to gratify them. Beth's behavior improved dramatically in a matter of days.

Envy and jealousy are other frequent emotional experiences of the preschool child. If a child is interested in an object, he wants it until something else gains his attention. A tendency to collect objects of interest often begins at four or five. Thus, the child becomes envious if he cannot take and explore the "treasures" his peers have and play with.

All children crave attention and affection from parents and other persons who stimulate them greatly. Their relative helplessness and comparative smallness complicate the situation. As a result, every child keenly notices any attention or favor one or both parents show to somebody else. His jealousy is readily aroused and he tries hard to regain the central position, often by a display of immature behavior—helplessness, anger, or self-punishment of some kind. The peak of jealousy in childhood comes between three and four years of age.

Emotional Needs The need for *affection* is more than a feeling or expectation; it is a fundamental human need. Most experts in family and human relationships stress the paramount importance of receiving maternal affection in terms of sensory stimulation, attention, and gentle care during the early years of life (Ainsworth, 1973; Spitz, 1951). Spitz found that the child's very survival apparently depends on the quality and frequency of human contact with the same adult. As a medical expert, he came to the conclusion that where affectionate contact is inadequate, many infants suffer permanent damage and some infants die without sufficient medical cause. Emotional and cognitive well-being are furthered by the affectionate care and loving attention of parents and their surrogates. Maternal feelings and attitudes toward a child seem to influence greatly his course of development and adjustment. Issues related to maternal deprivation and its long-term consequences have recently been surveyed and critically evaluated by Bronfenbrenner (1972).

The need for emotional *security* is another basic need in childhood. Generally the child experiences a sense of security when his parents and siblings gratify his needs and treat him as a desired member of the family and when his past is not marred by any permanent ego damage. Blatz (1966, pp. 41–43) defines the desirable midpoints of parental fields of influence greatly affecting the child's security. In administration of discipline parents must be patient, impersonal, and nonviolent; in feeling, affectionate and tender; in bestowal of status, accepting; in family organization, considerate and tolerant; and in protection, resourceful and concerned.

Parental expectations must be in harmony with the child's ability and readiness to respond. Usually the child tries hard when success is in sight. At the right time, he or she is ready for the next step. In most situations, parents should use the positive approach of encouragement and reinforcement rather than prohibition or correction. The child's self-respect is furthered by attention and affection. A question may often surface in his mind: "Is there enough love to go happily around?" The child wants to be loved and respected as a person with a right to do things his own way. From parents the young child needs unconditional acceptance as he is, plenty of affection, and large amounts of age-related stimulation.

Coexperiential emotions are important in fostering desirable interpersonal relationships at this age. Sympathy, for example, denotes identification with another's sorrow or pain and a manifest desire to assist the other. Empathy refers to identification with another's emotional or cognitive state. Compassion is a more developed response to another's sufferings, conflicts, or problems. How much progress will occur in developing these emotions depends on the parents' own ability to experience and express them in ways observable to the child. Coexperiential emotions help the young person to understand others and to enter more fully into the dynamics of family life. Later these emotions, if fully developed, permit sensitive interaction with other persons and with peer groups.

TASKS IN LANGUAGE DEVELOPMENT

The efficient use of speech at the preschool age gives the child a valuable means of self-expression and interpersonal communication. With continued rapid progress during the third year, the child moves toward complete sentences and increases the sentence length to an average of four or five words. Then, before he is five years old, there is a gradual disappearance of infantile forms of speech such as breaks in rhythm, slurring, and lisping. For Chomsky (1972, pp. 26–27, 115), human language is based on a specific type of mental organization rather than simply on a higher degree of intelligence, since even the normal use of language is a creative activity. Generative grammar underlies the observed use of language, incorporating the creative aspect of normal language use and distinguishing it from any known system of animal communication. Language competence means that a child "has mastered a system of rules that assigns sound and meaning in a definite way for an infinite class of possible sentences" (pp. 100–103).

The internal mechanisms that operate on the data of sense and produce linguistic competence for one or more languages are not known. Postulation of universal, genetically programmed biological structures and perceptual mechanisms leads to the plausible theoretical viewpoint assumed by Chomsky (1972), Lenneberg (1969), Piaget (1945/1951), and many other researchers in language development. At preschool age children are never taught the rule system of any language directly—they are merely exposed to individual situations; yet as Chomsky (1965, pp. 58–59) and Slobin (1971, pp. 55–57) show, they do not fail to acquire the underlying linguistic system with striking uniformity.

The linguistic performance of schoolchildren changes in terms of the models they are exposed to. If the models use long and complex sentences, children respond with longer and more complex sentences than when models use short, simple sentences. Harris and Hassemer's empirical study (1972) shows that adaptation to the model occurs even in the absence of reinforcement or instruction to imitate.

Cognitive organization of a certain set empowers the young child to acquire speech with no direct tutoring at all; the mere presence of language stimuli is enough to enable him or her to arrive at something like a transformational grammar. From the beginning of childhood at about two-and-a-half, the child models his expression on adult ways of interpersonal communication. Parents can do much to facilitate speech acquisition by using single words, phrases, and simple sentences during the second and third years of life. Lenneberg (1969) acknowledges that there is a "critical period for primary language" acquisition, related to the maturation of special structures in the human brain performing various language functions.

Speech progress at the preschool age centers on seven interrelated tasks: (1) improving pronunciation and diction, (2) comprehending the speech of others, (3) combining words into sentences to express thoughts and feelings, (4) expressing needs and emotional experiences, (5) increasing conversational skill to secure attention and acceptance, (6) using all the parts of speech, and (7) building a vocabulary of concrete and abstract words. Linguistic competence is based on progress in all these tasks. Most children, including those emotionally or socially disturbed, continue to improve in these speech tasks at their own individualized rates.

Many preschool children have difficulty in enunciating *th, j, r, s, z, h, g,* and *ch,* usually in this order of frequency. When the child begins to use all the parts of speech in sentences, he enters the level of *adult* speech. In the majority of cases this occurs before the age of four. The average active vocabulary at this time includes about three hundred words that the child uses vocally, and the passive vocabulary comprises about fifteen hundred other words that are readily comprehended. The ability to use words by combining them into sentences thus lags considerably behind the ability to understand and respond to them. The child responds with comprehension to a verbal request long before he can repeat the words spoken to him. In many ways he demonstrates an understanding of specific words long before they become part of his active vocabulary. Figure 10-1, based on a sample of 480 children in Minneapolis, illustrates the progress in articulation during preschool and early school years. A sharp rise in accomplishment takes place between three and four years of age, then progress continues at a slower rate during the later years of childhood.

Comprehension is one of the easier skills to develop; however, the preschool child takes literally what is said to him. Because at first he accepts

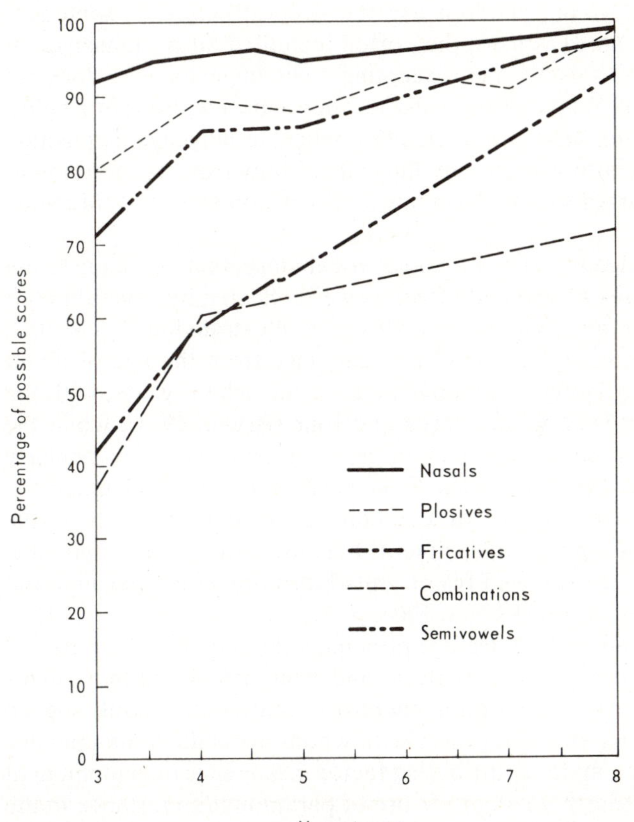

Figure 10-1 Progress in articulation: total possible scores on nasals, plosives, fricatives, combinations, and semivowels. [*Source:* M. C. Templin. *Certain language skills in children, their development and interrelationships.* Minneapolis: University of Minnesota Press, 1957, p. 39. By permission.]

only one meaning per word, he can misinterpret some parental and adult communications, including some affectionate gestures toward him. Occasionally these can anger or frighten him as he comprehends them word by word. The ability to recognize several meanings, to abstract, to see analogies or metaphors, and to understand humor are difficult tasks for the early school years. Facial expressions, gestures, and development of emotional rapport help refine the child's interpersonal communication.

The most rapid increase in vocabulary occurs in the early part of the preschool years and then gradually levels off. In addition to nouns and verbs, which dominate the two-year-old's speech, the child also begins to use other parts of speech one by one until all parts are used. The article and certain pronouns are the last to come into use. Increase in sentence length is a major index of progress in language development. Most three-year-olds use incomplete sentences of two to four words. Many four-year-olds can use complete sentences of three to seven words. The increase in the length of the sentence spoken is about one word per year, nearly doubling between two and three but increasingly less after four years of age (McCarthy, 1954, pp. 546–549). What is really surprising is that young children use the deep structure of sentences before they learn grammar. Even so, they often include a bit of nonsense, or blunder, or start or finish falsely. Yet, excepting some linguists, who does not make errors in speaking? Who can state all the rules and exceptions of English grammar? Young children have a universally applicable language-acquisition system that enables them to acquire any language. Many aspects of grammar are universal and are part of the child's innate ability (Chomsky, 1965; McNeill, 1970, pp. 1087–1088).

The amount of language structure and vocabulary that is acquired by children between the ages of one and fifteen is well reflected by several major studies (Smith, 1926; Bryan, 1954; Loban, 1966) and illustrated in Figure 10-2. The Smith study indicates a rapid rise in vocabulary from the age of about one-and-a-half; a similar increase continues during the school years, with the exception of a year between grades three and four (Bryan, 1953), while the number of words used in communication units rises rather spectacularly between the second and the sixth grades. From kindergarten through grade six, children also increase the number of communication units as well as the average number of words spoken in each unit. In grades seven, eight, and nine, *S*s of the Loban study (*N*=72) used fewer units but more words per unit and more complexity of expression (Loban, 1966, p. 21).

Since the young child learns language principally by modeling parents and others, he readily accepts the speech patterns and pronunciation of those in his immediate environment. It is important, therefore, that correct adult speech and diction be used by parents and others with whom the child communicates often. Television is now a major contributing factor in language development at all ages. The third and fourth years of life are of paramount importance in the development of language, for it is during these years that most speech deviations begin: articulatory errors or stuttering, as well as voice characteristics, some of which may last throughout life (Van Riper, 1950, p. 92).

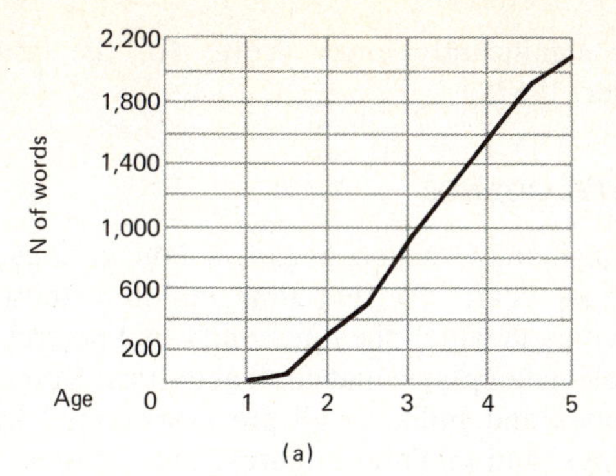

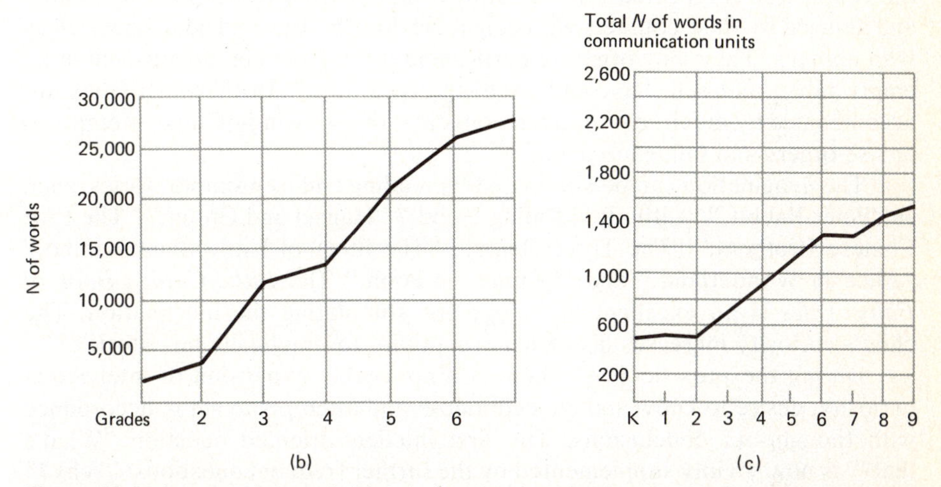

Figure 10-2 Increase in vocabulary during the preschool and school years assessed by different methods and expressed by means. [*Source:* (a) M. E. Smith. An investigation of the development of the sentence and the extent of vocabulary in young children. *Univ. Iowa Stud. Child Welf.*, 1926, **3**(5). (b) Fred E. Bryan. How large are children's vocabularies? *Elem. School J.*, 1954, **54**, 210–216. (c) Walter Loban. Language ability: Grades seven, eight, and nine. U.S. Department of Health, Education and Welfare (Cooperative Research Monograph No. 18). Washington, D.C., 1966 (fig. 2, p. 22).]

Bilingual families seem to have no clear-cut detrimental effect on a child's speech development provided his cognitive capacity is about average or better, each language is taught correctly, and undesirable emotional attitudes that would make the child resistant to one of the languages are not involved. In vocabulary and other verbal tests, the child must be tested in both languages to secure valid results. When a child is mentally retarded, learning a single language is a practically insurmountable task and a second one becomes an additional burden. Many mentally deficient children never reach the full-sentence stage. Socioeconomic status is a recognized factor that usually creates a highly stimulating environment with many educational opportunities for upper- and middle-class children and conversely a dull setting with few educational advantages for children of the lower class. Language and intelligence tests show significantly lower scores for the latter (Rohwer, 1970; Stodolsky & Lesser, 1967).

FANTASY AND INTELLIGENCE

The growth of imagination, which began in late infancy, increases sharply during the preschool years. At that time children show much interest in make-believe activities in which they personify and portray fragments of their past adventures, television plays, and adult activities. Home and family, doctor and patient, cowboys and Indians—all are represented by the use of dolls, household tools, toys, and miniature figures. The incident of Washington and the cherry tree is repeated in a variety of contexts. Children are also stimulated and amused by their contacts with neighborhood children and adults, as well as with animals. They may organize parties and drink from empty cups, eat at an empty table, and sell, buy, and exchange their toys. All these activities are accompanied by much self-centered conversation, showing-off, and attempts to amuse others and influence them.

The imagination can be stimulated by reading children simple stories, such as "Peter Rabbit," "Little Red Riding Hood," "Hansel and Gretel," "The Five Chinese Brothers," "The Three Bears," "The Story of Little Black Sambo," "Alice in Wonderland," and "Winnie-the-Pooh." The *Little Golden Book of Fairy Tales* is an excellent anthology for stimulating the imagination. The child's curiosity increases his or her receptivity to practically any stories.

During the preschool age there is a noticeable expansion of intellectual curiosity, desire to know and conceptualize, and attempts to act in accordance with thought-out conclusions. The first intellect-oriented question "What's that?" is now vividly supplemented by the further frequent questions: "Why?" "How?" "What for?" The child begins to understand and appreciate the purposes various objects serve, what makes things work, and where they come from. It is good for the child if parents and other adults answer such questions adequately and in this way form correct concepts, attitudes, and expectations, because the child needs an accumulation of workable knowledge before beginning formal education. Impatience and irritability with the child's frequent questioning is interpreted by many children as rejection. Several carefully designed studies show that hostile mothers have sons who score high in intelligence in the first year but have comparatively low IQs from four to eighteen years of age (Bayley & Schafer, 1964; Hurly, 1965). Question-and-answer learning is intellectually stimulating and eliminates much trial and error. Fahey's study (1942) estimates that questioning accounts for 10 to 15 percent of the preschool child's conversation. He suggests further investigations to determine the meaning and implications of the child's questioning activity. Berlyne and Frommer (1966) reported two experiments in which 180 Canadian children from kindergarten through grade six were exposed to a series of stimuli consisting of stories and pictures and invited to ask questions after each item. Novel, surprising, and incongruous items elicited more questions than others, supporting Berlyne's theory of *epistemic* (cognitively critical) curiousity. Generally, questioning increased with age, indicating a higher sensitivity to informational gaps and a need to relieve the uncertainty of any kind.

An accumulation of various patterns and skills is largely a result of opportunity. Most of the concepts acquired by the child result from his communication with others. Native intelligence cannot be evaluated, since test results depend in large part on the amount of knowledge acquired through interaction with the environment or with persons and objects in it. Piaget (1952) rightly observes that "the more a child has seen and heard, the more he wants to see and hear." The Montessori teaching method, which uses visual material appealing to preschool children and makes provision for self-corrective matching, is helpful for most children and is greatly needed by culturally disadvantaged children of preschool age. Only an educationally rich environment offers the child sufficient opportunity for obtaining stimulation of all modalities. Such an environment satisfies the nearly inexhaustible curiosity to look, listen, poke, sniff, and feel.

DEMANDS OF SOCIALIZATION

At the preschool age the child shows much more interest in others than before. Group activities become more and more appealing. At four years of age, he begins to need individuals of his own age or slightly older. At five years of age, any long isolation from adults or other children becomes unbearable. The child is eager to join and to learn to assume roles in group situations. Socialization is not something children can do for themselves, nor is television a good substitute for it. Exposure and interaction with adults and other children of different ages are necessary for a child to acquire proper attitudes and social skills. Parents have a challenging responsibility to provide opportunities for early social experience with other children and adults. It is their task to stimulate adjustive and cooperative behavior through verbal instruction and direct control. They may have to moderate the child's desire to dominate, to secure the limelight, and to be excessively possessive, all of which interfere with progress in relating to others. When parents do their part, social progress is satisfactory, since the child of four and beyond has a strong desire to please the adult and to find a peer companion. In Bronfenbrenner's words (1972, pp. 663–664):

> *Children need people in order to become human. . . .* It is primarily through observing, playing, and working with others older and younger than himself that a child discovers both what he can do and who he can become—that he develops both his ability and his identity.

When behavior modification is necessary, parents ought to learn how to play the role of a positive reinforcer rather than the traditional role of a disciplinarian. Often just a proper shift of parental attention increases the frequency of desirable behavior. When a mother is clearly instructed and succeeds in becoming inattentive whenever her child engages in any deviant behavior and very attentive and pleasant following desirable behavior, the child's behavior usually shifts for the better. The child's desirable behavior is now the most effective means he has to obtain her attention and her smile. These are powerful rewards for most children. The mother's consistency often evokes dramatic overall reorganization of the child's activity. In some cases, however, certain punishment contingencies (isolation, time out, etc.) are necessary to help the child inhibit certain oppositional forms of behavior. At present it is felt that many parents, and even some teachers, are too attentive to deviant actions of children. With some children, however, who rarely if ever show any improvement and in whom there is a dearth of desirable behavior, it may not be easy to find opportunities for attention (Cohen, 1972; Wahler, 1972).

VALUATION

Most experiences involve valuation—a personal judgment of worth, merit, desirability, or significance. Cognitive and emotional powers seem to be inclined toward valuation of one kind or another. What the child likes or dislikes is not likely to agree fully with what his mother or father seems to favor. From the very early years of life, parents are engaged in behavioral and verbal conditioning; they see some of their child's activities as desirable and others as regrettable. Consciously or not, they introduce the morality of what is good in general and possibly what is right for the child. Moral behavior is simply "doing what is good and what is right." At the preschool age "what is right" is determined largely by parents and older siblings. Other persons and sources like television are also involved to a considerable degree. The child models himself on his parents and to a lesser degree on various television characters (Stein, 1972).

Jean Piaget (1932) distinguishes restraint and cooperation as two different types of moral experience and behavior evoked by interpersonal situations. Situations of restraint force conformity to parental demands until these responses become habituated, while cooperation fosters agreement, motivating the child to want to conform. The young child tends to change rules according to his momentary dislike for them, while the older child accepts established rules. When the rule of reciprocal cooperation replaces the rule of adult restraint in motivating an action, it can be considered a moral act (Piaget, 1932, p. 62).

In more obvious situations, the moral sense of right and wrong gradually deepens unless parents are unwilling to use proper instruction and example. Moral conduct is fostered mainly by a deepening awareness of fundamental moral concepts. The understanding of moral concepts is closely related to intellectual maturation and emotional identification with the values and ideals suggested. In stimulating the child's awareness of moral values and virtues, parents must guard against a preaching attitude, a "Do as I say, not as I do" philosophy. At the preschool age, the child can be taught principles of honesty, justice, and fair play in many concrete situations of his activity.

In Western culture, religious experience often begins through observation of situations and actions such as the presence of a cross, a short prayer before meals, or any religious ceremony. The presence of religious articles often evokes questions about them, providing another opportunity for religious experience. Taking the child to church or to religious shrines can help him sense the awe-inspiring atmosphere of a religious environment.

By six or seven the child is capable of understanding by analogy most fundamental concepts, such as the idea of God as creator and heavenly father or the purpose of prayer. Illustrated stories of the Bible or other religious classics can provide material for further growth in religious knowledge. The child's curiosity and natural openness to many religious experiences aids him or her in acquiring moral guides and understanding the purpose of right behavior in human life. Early childhood is the right time to lay the base for religious and humanistic value orientation, since "nothing is more important than the person that child becomes; he alone holds the key to future life" (Rheingold, 1973).

PERSONALITY DEVELOPMENT

Since the young child is constantly exposed to the social stimulation of parents, other children, and other adults and since he or she usually establishes strong interactional ties with them, the natural tendencies are molded by them. Through verbal reinforcement, contingent responses grow stronger and engender trait, attitude, and habit patterns. Moreover, parents promote maturity by showing pleasure with and verbally reinforcing relatively mature responses to everyday demands. Conversely, if they pay attention to or otherwise reinforce the immature responses of the child, they lessen the incentive to grow and mature. With some additional professional training, parents could greatly raise their ability to modify children's behavior (Herbert & Baer, 1972).

Social reinforcement rates high with children. In Warren and Cairns's experimental study (1972), the primary treatment involved two levels of frequency of prior presentation of the social reinforcer ("right"), and two levels of signal reliability (discriminative and ambiguous). The results with 100 second-grade children indicate that the higher level of signal satiation had results opposite to those expected: they enhanced the effectiveness of the social reinforcer ($p < .001$), while extended use of the ambiguous signal reduced its effectiveness.

During the preschool years, the child's beliefs, attitudes, and traits are often significantly altered by association with neighborhood children and other persons. If any major difference from other children is observed regarding, for example, limits of free movement, language, clothes, or hairstyles, the child will try to imitate what is new. This, in turn, often creates conflict with parents and occasionally forms the basis for an undesirable attitude.

Preschool training often neglects one or another aspect of the total personality necessary for the growth of humanness in the child. For example, many parents disregard cultivation of the positive emotions, such as affection or empathy. Others neglect to impart lessons in what is right and what is wrong. Foundations for esthetic, ethic, and moral values may lack support. Many parents fail to provide their child with peer companions until he goes to school, where he is suddenly overpowered by peer demands and expectations. Any such omission handicaps progress in wholesome personality development and may produce weaknesses, giving rise to feelings of inadequacy and distorted perception of some aspects of reality. Compensation by continued dependence on adults or by inferiority symptoms may result and plague the child for many years.

Beginning with the preschool age, the child's concept of himself guides much of his behavior. Psychologically, then, the child becomes somewhat removed from the environmental context and from other persons. Sometimes he stops to think or to delve into his feelings or relationships with others. By disregarding many objects and persons, he restricts his psychological frame of reference. Some children when punished or angered run to their rooms or closets and sob quietly or chew their fingernails for as long as three or four hours. Such preoccupation with self is frequently overcome by a rising desire for social interaction.

The child recognizes his abilities and limitations through the appraisal of others. He perceives himself and those close to him and draws his own conclusions. Loss or change, as from a haircut or tooth extraction or anything producing a major change in a parent's physical appearance, often stimulates deep concern about self-identity and the identity of others close to him.

The child of five or six consolidates most of his new developmental gains and usually integrates them according to his particular personality pattern. If this occurs, he becomes a more secure and self-reliant person, able to adjust to problems and to tolerate some privation and anxiety. Indeed, many studies support the view that personality in its basic structure is largely set, if not determined, during the years of early childhood (Freud, 1959; Kagan & Moss, 1962; Mischel, 1969; Scott, 1962). Perhaps pubertal, adolescent, and early adult changes will show that personality is not fixed in childhood but continues to grow and change in many significant ways over the total life span of the person.

QUESTIONS FOR REVIEW

1 What qualities or abilities distinguish a child from an infant?
2 List some of the fine motor controls attained during the preschool years and discuss basic implications of these achievements.
3 What are the leading functions of play activity? What provisons are needed to encourage constructive and creative kinds of play?
4 Discuss and relate two major theories of play, e.g., Bühler's and Piaget's formulations.
5 Describe the emotional developments occurring during the preschool years. Explain the need for and role of parental affection.

6 Enumerate and relate the major tasks in speech development. What constitutes adult speech?
7 What are the chief questions frequently asked by four- and five-year-old children? Why should parents be careful to answer the child's questions correctly?
8 What social activities have to be encouraged at the preschool age? Give some reasons.
9 Discuss the child's capacity for valuing. How does he understand what is right?
10 Of what value is it to teach parents to condition the young child?
11 How does the child's self-awareness affect his relationship with parents and others?

REFERENCES

Selected Reading

Mussen, P. H., Conger, J. J., & Kagan, J. *Child development and personality* (4th ed.). New York: Harper & Row, 1974. A major work on various aspects of child growth and behavior. Part I presents the prenatal period; Part II covers the first two years of life; Part III deals with language and cognitive development; and Parts IV through VI take up the preschool years, middle childhood, and adolescence, with emphasis on personality development and adjustment.

Smart, M. S., & Smart, R. C. *Children: Development and relationships.* (2d ed.) New York: Macmillan, 1972. A major text on the stages of child development. Part II, in its four chapters, analyzes the preschool child, emphasizing play and socialization.

Zaporozhets, A. V., & Elkonin, D. B. (Eds.) *The Psychology of preschool children* (J. Shybut & S. Simon, Trans.) Cambridge, Mass.: MIT Press, 1971. (Originally published, 1964.) A book by eight authors dealing with the development of perception, attention, memory, speech, thinking, imagination, and motor habits.

Specific References

Ainsworth, D. S. The development of infant-mother attachment. In B. M. Caldwell & H. N. Ricciuti (Eds.), *Review of child development research* (Vol. 3, pp. 1–94). Chicago: University of Chicago Press, 1973.

Almy, M. Spontaneous play: An avenue for intellectual development. *Child Study,* 1966, **28,** 2–15.

Bayley, N., & Schafer, E. S. Correlations of maternal and child behavior with the development of mental abilities: Data from the Berkeley growth study. *Monogr. Soc. Res. Child Developm.,* 1964, **29** (Ser. No. 97).

Berlyne, D. E., & Frommer, F. D. Some determinants of the incidence and content of children's questions. *Child Developm.,* 1966, **37,** 177–189.

Blatz, W. *Human security.* University of Toronto Press, 1966.

Bronfenbrenner, U. Early deprivation in monkey and man. In U. Bronfenbrenner (Ed.), *Influences on human development,* pp. 256–301. Hinsdale, Ill.: Dryden, 1972.

Bryan, F. E. How large are children's vocabularies? *Elem. School J.,* 1954, **54,** 210–216.

Bühler, C. & Marschak, M. Basic tendencies of human life. In C. Bühler & F. Massarik (Eds.), *The course of human life,* pp. 92–102. New York: Springer, 1968.

Buhler, K., & Schenk-Danziger, L. *Abriss dec geistigen Entvicklung des Kleinkindes* (9th ed.). Heidelberg: Quelle & Meyer, 1968.

Chomsky, N. A. *Aspects of the theory of syntax.* Cambridge: MIT, 1965.

Chomsky, N. A. *Language and mind* (enlarged ed.). New York: Harcourt, Brace, Jovanovich, 1972.

Cohen, H. L. Programming alternatives to punishment: The design of competence through consequences. In S. W. Bijou & E. Ribes-Inesta, *Behavior modification: Issues and extensions*, pp. 63–84. New York: Academic Press, 1972.

Davis, D., Davis, M., Hansen, H., & Hansen, R. *Playway: Education for reality.* Minneapolis: Winston Press, 1974. (*Playway handbook* and boxed kit available from publisher.)

Erikson, E. H. *Childhood and society* (2d ed.). New York: Norton, 1963.

Fahey, G. L. The questioning activity of children. *J. genet. Psychol.*, 1942, **60**, 337–357.

Freud, S. *Collected papers* (Vol. 4). New York: Basic Books, 1959.

Furth, H. G. *Piaget and knowledge: Theoretical foundations.* Englewood Cliffs, N.J.: Prentice-Hall, 1969.

Harris, M. B., & Hassemer, W. G. Some factors affecting the complexity of children's sentences: The effects of modeling, age, sex and bilingualism. *J. exp. Child Psychol.*, 1972, **13**, 447–455.

Hartley, R. E., & Goldenson, R. M. *The complete book of children's play.* New York: Crowell, 1957.

Herbert. E. W., & Baer, D. M. Training parents as behavior modifiers: Self-recording of contingent attention. *J. appl. beh. Anal.*, 1972, **5**, 139–149.

Herron, R. E., & Sutton-Smith, B. *Child's play.* New York: Wiley, 1971.

Hurley, J. R. Parental acceptance-rejection and children's intelligence. *Merrill-Palmer Quart.*, 1965, **11**, 19–31.

Kagan, J., & Moss, J. A. *Birth to maturity: A study in psychological development.* New York: Wiley, 1962.

Lenneberg, E. On explaining language. *Science*, 1969, **164**, 635–643.

Loban, W. Language ability: Grades seven, eight, and nine. U.S. Dept. of Health, Education, and Welfare. Cooperative research monograph No. 18. Washington, D.C., 1966, Fig. 3, p. 22.

Maier, H. W. *Three theories of child development* (2d ed.). New York: Harper & Row, 1969.

McCarthy, D. Language development in children. In L. Carmichael (Ed.), *Manual of child psychology* (2d ed.). New York: Wiley, 1954.

McNeill, D. The development of language. In P. H. Mussen (Ed.), *Carmichael's manual of child psychology.* Vol. 1, pp. 1061–1161. New York: Wiley, 1970.

Mischel, W. Continuity and change in personality. *Amer. Psychologist,* 1969, **24**, 1012–1018.

Piaget, J. *The moral judgment of the child.* New York: Harcourt, Brace, 1932.

Piaget, J. *The origins of intelligence in children* (Margaret Cook, Trans.). New York: International Universities Press, 1952. (Originally published, 1936.)

Piaget, J. *Play, dreams, and imitation in childhood* (C. Gattegno & F. M. Hodgson, Trans.). New York: Norton, 1951. (Original French ed., 1945.)

Pikunas, J., & Clary, J. Fears in normal and emotionally disturbed children. *J. psychol. Studies*, 1962, **13**, 157–164.

Rheingold, H. L. To rear a child. *Amer. Psychologist*, 1973, **48**, 42–52.

Rohwer, W. D., Jr. Implications of cognitive development and education. In P. H. Mussen (Ed.), *Carmichael's manual of child psychology*, pp. 1379–1454. New York: Wiley, 1970.

Scarr, S., & Salapatek, P. Patterns of fear development during infancy. *Merrill-Palmer Quart.*, 1970, **16**, 53–90.

Scott, L. H. Personality at four. *Child Developm.*, 1962, **33**, 387–411.

Slobin, D. I. *Psycholinguistics.* Greenview, Ill.: Scott, Foresman, 1971.

Smith, M. E. An investigation of the development of the sentence and the extent of vocabulary in young children. *Univ. Iowa Stud. Child Welf.,* 1926, **3,** (5).

Spitz, R. A. The psychogenic diseases in infancy: An attempt at their etiologic classification. *Psychoanal. Stud. Child,* 1951, **6,** 255–278.

Stein, A. H. Mass media and young children's development. In Ira J. Gordon (Ed.), *Early childhood education,* pp. 181–202. (Seventy-first Yearbook of the National Society for the Study of Education, Part II.) Chicago: University of Chicago Press, 1972.

Stodolsky, S. S., & Lesser, G. Learning patterns in the disadvantaged. *Harvard educ. Rev.,* 1967, **37,** 546–593.

Van Riper, C. *Teaching the young child to talk.* New York: Harper & Row, 1950.

Wahler, R. G. Some ecological problems in child behavior modification. In S. W. Bijou & E. Ribes-Inesta, *Behavior modification: Issues and extensions,* pp. 7–18. New York: Academic Press, 1972.

Warren. V. L., & Cairns, R. B. Social reinforcement satiation: An outcome of frequency or ambiguity? *J. exp. child Psychol.,* 1972, **13,** 249–260.

Chapter 11

Middle Childhood

HIGHLIGHTS

Middle childhood is an age of varied play and refined perceptual-motor activity. The child is ready to try practically anything he or she sees others doing.

It is a period of intense learning at home and in school. Most aspects of the child's behavior are molded through reinforcement and verbal conditioning as well as through modeling and identification.

The concrete operational mode of thinking begins at about seven, when the child frees himself from perceptual dependence and egocentric language. Interest in classification, seriation, and other systematic groupings and transformations makes the child a good subject for school learning.

The child's developmental tasks center on gains in emotional control, recognition of his own social role, and learning to get along with age-mates. Recognition of his own limits and sensitivity to adult demands help to expand his gains in self-regulation of behavior.

The period of middle childhood extends from the age of about six or seven to nine or ten. The milestones in this phase of development are (1) readiness for school and actual school entry, (2) the broadening of intellectual horizons by concrete operations, (3) a keener interest in the peer group, (4) a growing independence of parents, (5) gains in resilience, and (6) enhanced self-identification.

Entrance into middle childhood is marked by a moderate physical imbalance caused by the periodic loss of baby teeth, emergence of the first permanent molars, and a greater susceptibility to colds and infectious childhood diseases. Passive withdrawal, dawdling and impulsiveness, increased excitability and inconsistency, and oscillation and conflicts characterize the motivational tendencies of the typical six-year-old. At about seven years, these tendencies lessen and the child becomes less demanding and more sociable.

PLAY ACTIVITIES

Since height and weight increase slowly, the child is able to gain vigor and balance in sensorimotor operations. Control over the large muscles is perfected, and control over the small muscles is moderately advanced. Since the child has much energy, he is in nearly constant motion, but he also becomes somewhat cautious. Active play, such as running, playing ball, riding a bicycle, jumping rope, and swimming, can be so intensely absorbing that the child continues to near-exhaustion unless adults direct his activities. Water, sand, and mud play can involve both boys and girls of this age for hours.

Most children enjoy constructive play in which they can use simple devices—hammer, knife, scissors, and keys. Bricks and blocks, boxes and beads, pegboards, and picture puzzles also are stimulating materials for construction. Both boys and girls enjoy drawing, one of their finest means of unconscious self-projection. As the child matures, the drawings take on definite form, color, and accuracy. Coloring and finger painting also appeal to children at this age, especially when they are emotionally upset. Many children begin saving objects they like; they enjoy collecting coins, coupons, marbles, stamps, and comic books. Accomplishments bring great satisfaction, since the child now longs for the appreciation of adults and children alike. Parents' or teachers' interest in the child's achievements easily stimulates further activities and manifold improvements.

SCHOOL ENTRY

As indicated in Chapter 3, many factors continually shape the personality of the preschool child; in individual cases they vary both in kind and in degree. Some exert a considerable influence at a certain age and then decline as new influences appear. By the time the child enters school, he or she has been exposed to many family influences, to mass media of communication, and to physical and social experiences in the neighborhood. From the parents he has received his first training and education, which outside contacts enriched and modified. He has met many other adults and children who have influenced his emotions and motivation. Through these contacts with parents, other adults, and neighborhood children and through his experiences in kindergarten, the child builds up response patterns and attitudes in regard to school activity.

Liking kindergarten and wanting to go to school are probably the most fundamental indications of the child's readiness for school. Verbal communication of needs and the ability to care for himself at toilet are other important ones. "Yes" answers to the following questions confirm the child's readiness: Can the child get along with other children of similar age with only occasional difficulties? Does the child like new adults and adjust to them quickly? Does the child carry out any simple task he is asked to do more often than not? Does the child indicate his toilet needs and go to the toilet unassisted? Can the child recognize his coat, hat, and rubbers or boots and put them on properly without assistance?

It is an important day for the child when he or she leaves home for the first time to begin kindergarten or first grade. The separation from the confines of the home and surrounding neighborhood is exciting and often frightening. Once the child has left his little world of loved ones, he may not return until the school day has ended. Yet if he is ready for school, the child eagerly anticipates entry into school and exhibits a certain maturity of behavior when he enters school, often as a result of having attended kindergarten.

A study by Valletta, Brantley, and Pryzwansky (1972) suggests multiphasic screening of first-grade children in an attempt to formulate early intervention strategies for the correction and prevention of learning difficulties. In addition to ratings by teachers, they used the following validating instruments: Keystone telebinocular survey, Beltone audiometer, developmental test of visual-motor integration, Arizona articulation proficiency scale, survey form, verbal expression subtest of the Illinois Test of Psycholinguistic Abilities (ITPA), and Primary Mental Abilities (PMA), Form K-1. The teacher could make two types of errors: failure to identify a problem case and false identification. They found that in vision and hearing, respectively, 67 percent and 63 percent of selections by teachers were false identifications. Teachers were least efficient in identifying vision problems and most efficient in identifying visual-motor integration problems. Instrumental screening results point to proper intervention strategies for correcting difficulties and for preventing further difficulties. The researchers assumed that many children referred from middle and later elementary grades showed characteristics that might have been detected and corrected during the primary years. Some "emotionally disturbed malingerers" are actually children who have undetected learning disabilities. This is one of the reasons why comprehensive psychological services should be extended to all children who may need them. At present, most states restrict psychological testing and counseling to only certain exceptional groups of children (Cordon, 1972).

In recent years there has been increasing interest in children who show motor hyperactivity, impulsiveness, emotional lability, high distractibility, shortness of attention span, and poor academic performance. Some of them have a specific learning disability, others do not. Wender (1971) proposed to broaden the concept of "minimal brain damage" to include neurotic, prepsychotic, and other variants. Some of these children change greatly and lose many of their symptoms, while others take on more fully neurotic and impulsive personality disorders, during adolescence or early adulthood. Such a broadly conceived syndrome simply encourages the use of the same medication—amphetamines, Benzedrine, or Ritalin, for example—to depress overactivity. Some of these children are hyperkinetic and need this medication, but many others probably do not. For the majority, increased physical exercise might reduce restlessness; they need little else but good ability-related teaching. Whatever the proper corrective measures may be, these children need exposure to differential diagnostic psychological and medical testing before any corrective or chemical treatment measures are applied. Minimal brain dysfunction is a vague label that probably obscures as much as it enlightens. Excessive application of psychoactive drugs may make many children susceptible to drug misuse for life. While some children must suffer long-term drug addiction to save them from even more damaging effects of their hyperkinetic condition, for many others drug measures are simply repressive means—chemical straitjackets to induce tranquility in classrooms for the advantage of teachers and other children. Hyperactive children need special education of their own.

While drugs make children's symptoms less intense and thus more manageable, behavior modification techniques often bring out their best qualities. Behavior modification often allows children to discover their impulses and to apply their own powers of self-management in dealing with them. Behavioral patterns acquired through the use of such techniques can generalize to many other situations at school and elsewhere (Doubros & Daniels, 1966; Grinspoon & Singer, 1973).

Because the child's education has already developed considerably, the public school usually acts as a major supplement to the family in promoting the child's intellectual and social development. Beyond that, the school situation is unique in many ways. Here, for the first time, the child is entrusted daily to another adult and relates with a large group of peers, most of whom are unfamiliar at first. He will spend a large part of his day in this group situation, and his linguistic and cognitive abilities and social adjustments will be challenged. From this point of view, the child entering school is a typical beginner. Going to school is successful only when the child is ready to make satisfactory adjustments to the novel aspects of school life. His success or failure depends on his level of maturity, which implies mastery of preschool tasks, including separation from parents and readiness to assume reality orientation in tackling school responsibilities.

The child's cognitive ability and social skills influence his or her school adjustment. School performance depends on cognitive development and on progress in doing things for oneself. Intellectual maturation can be moderately advanced by educational activities at home, and abilities and skills are enriched by opportunity and encouragement. Praise for attempts to get dressed without help, for buttoning clothes, and for putting on shoes works virtual miracles in the child's mastery over clothing. Success in these simple daily activities fosters feelings of adequacy and a desire to do things independently.

The child's preparation for school also includes frequent exposure to social play situations, in which he or she is taught to assume roles and to cooperate with others. Ability to interact well with other children is usually indicative of proper emotional development and control over fear and anger, envy and jealousy.

Language skills, too, are directly related to a child's education. A schoolchild must be able to communicate his needs, thoughts, and experiences. His intellectual curiosity is verbalized by "How?" and "Why?" These and similar questions indicate the child's level of understanding, as well as the subject matter in which he is most teachable.

Before the child enters school, his interests have been self-centered and parent-centered. Not much in the way of sharing has been experienced, unless the child has learned from other members of the family to give and take. Emotional impulsiveness and other forms of self-assertion have typified many of his responses, while cooperation has been restricted to dyadic relationships with the parents. The child likes to relate to a single person at a time, but in school he has to contend with many children.

Let us examine some of the fundamental situations and tasks that determine whether a child has a sufficient reservoir of experience to be able to make appropriate responses at school.

In the school situation the child must function without family support and must learn to accept not only authority outside the family group but also competition. Teacher-child interaction and group dynamics are more complex than the interpersonal relationships experienced at home and in the neighborhood. At school, for example, there is no escape from the regular demands of the teacher. Moreover, the class often acts as a corrective factor. The child must fuse his behavior with the group pattern or face the rejection and aggression of the class. If a child's appearance or behavior differs greatly from that of other children in the group, he is likely to become a scapegoat for the more aggressive children. If he is smaller than others, some will take advantage of him. A child's acceptance by the group is conditional, while at home he may have been unconditionally accepted without making a contribution of his own. A schoolchild has indeed new adjustments to make and new material to learn.

A good primary school corrects many misconceptions acquired at home, promotes openness to the world, introduces the complexities of human nature and interpersonal relationships, and prepares the child for a lifelong career of learning. The changing functions of the school and the teacher, the need for higher education, and the forces influencing curricula at all levels of education are well presented by various authors in the Seventy-second Yearbook of the National Society for the Study of Education (Goodlad & Shane, 1973). Behavior modification related to school is presented in detail in Volume 3 of the *Review of Child Development Research* (Caldwell & Ricciuti, 1973).

In modern culture, school experience is an indispensable expansion of home training and education. A good school sparks the child's curiosity. It creates situations for learning subjects of vital importance in his life. To a great extent, school relieves parents of educational tasks and provides them with rest periods and a chance to objectify their relation to their children. As discussed in Chapter 3, many prominent psychologists and educators question whether the typical first-grade environment is compatible with the inevitably stormy beginning of a new developmental phase. For one thing, many children now at school because of compulsory school attendance are not prepared for it either in their maturational level or in their psychosocial development. In school, therefore, many six-year-olds may appear to be regressing rather than progressing in knowledge and behavior. One day a child writes the letters of the alphabet without apparent difficulty, yet a week later he may make some letters backward. One day he reads a few lines in a story and on the next fails to recognize any of the words. Because of his inexperience in group relations, he will often make mistakes, since he is confronted with more problems than he can solve. His errors often alternate from one extreme to the other. He becomes either insistent and aggressive or shy and hesitant, or he may attempt tasks beyond his ability. Many boys and girls in school and on the playground have the deep-seated urge to win, to conquer, and to subdue. Is it then surprising that verbal and physical violence flares up from time to time?

Physiological instability and the new demands of school produce moderate to severe strain for many six-year-olds. First-grade teachers know that each year at least one beginner will cry, try to run home, refuse to participate in group activities, or cling to his mother when she brings him to school. The reasons for this type of behavior vary; one child may not be accustomed to a group, while another has a fear implanted by an older child or adult through stories of cruel punishment a teacher may inflict.

For the child, the teacher is a mother substitute. Daisy Franco's study (1955) shows that children tend to see their teacher in terms of qualities and traits they attribute to their mothers. During middle childhood, this transference shows considerable stability. For Piaget, what the teacher needs most is to develop effective modes of communication with children. He or she must learn to comprehend genuinely what each child is saying and to respond in the same mode of discourse. This achievement alone makes a teacher-child dialogue possible (Elkind, 1974, pp. 108–110).

The schoolteacher is, however, more than a mother substitute; he or she creates and maintains an atmosphere conducive to learning and helps children meet their new academic and social demands. Studies indicate that pupils prefer teachers who can present subject matter clearly and elicit enthusiasm for learning, who are well-balanced and even-tempered, fair and consistent, democratic, and helpful. Because of the teacher's importance in the socialization of children and the development of their intellect, it is a matter of growing concern that well-adjusted adults be selected as educators. Since teaching is a legally regulated occupation, requiring academic competence and having its own professional organization that sets standards of competence, it is a profession comparable to law or the ministry.

A teacher's guiding influence is vast. He must use his imagination, a friendly enthusiasm, and a playful spirit to live partially in a child's world and to enter into the feelings, attitudes, and emotions he wants to develop in the class. He must make personal efforts to be alert, compassionate, and well-balanced. Self-confidence and poise, high moral standards, a sense of humor, and leadership are other personal qualities the effective teacher needs to show. Generally, the teacher is responsible for every situation that emerges in the classroom. He must encourage projects and activities that will reap satisfying results and discourage those in which success cannot be anticipated. He guides children to form their goals and to plan, execute, and evaluate their performance. He needs a genuine interest in and love of children, knowledge of child development, and good training in educational methods and behavior-modification techniques.

The child has the capacity to evaluate—he likes and appreciates some objects, types of behavior, and events and dislikes others. Evaluation is an ever-increasing feature of his total behavior, often a very decisive feature. Robert L. Brackenbury (1966) assumes that no teaching and learning occur without some transmission of values, since education is teaching children to behave as they do not now behave. The real question is: What values are children taught? In school, attitudes toward universal and contemporary issues are manifested both overtly and covertly. "If a teacher makes every effort to hide his beliefs, will his students learn to think for themselves or will they learn that their teacher apparently thinks it does not matter what one thinks or whether one thinks?" (Brackenbury, 1966). Indifference, even apathy, is the logical consequence of instruction that is sterile because of its neutrality.

Since childhood is a process of "coming out of enclosures and taking new risks and meeting new and exciting challenges" (Piaget, 1919/1951, p. 36), for most children school is neither all drudgery nor all play. During the first year in school, the child's abilities are magnified and expanded in many directions, because new stimuli and new subjects are introduced and reinforced by the group reaction. The child is given tasks and projects that demand planning and persistence. For superior achievements he is praised by the teacher and admired by the other children. This fosters a sense of self-esteem, of pride in his own ability, and thus promotes self-reliance. The child is then ready to master new and more difficult subjects and problems.

Through success in group interaction and in performance of routine duties and tasks, the child develops a certain amount of freedom and self-reliance. Yet he frequently experiences the need for companionship and the desire for assistance, which promote the process of socialization. As the child's experience broadens, he learns to work for more remote goals and in this way contributes to his own education. It is readily seen, from what has been said, that going to school is one of the great milestones in the life of any child.

COGNITIVE DEVELOPMENT

Piaget studied intensively the early and middle years of childhood. In a series of books and articles, he reported on many facets of a major theory concerning the child's cognitive development, already outlined in Chapters 1 and 4. By means of systematic, firsthand observations and ad hoc experiments (experiments not planned in advance), he gained factual data about language development, the process of thought, and concepts of various aspects of reality. At the Maison des Petits in Geneva, Piaget and his associates recorded most of the speech of a number of children for a month and supplemented their free talk with questions to test the validity of the hypotheses suggested by earlier observations. The findings show a very high frequency of egocentric speech for the group of children aged three to five, and significantly lower frequency for the group aged five to seven. From about seven on, the speech of children becomes sociocentric (Piaget, 1952, p. 257). Socialized language is a necessary condition for the development of mental operations (Piaget, 1973, pp. 110–119). Perceptual centration and egocentrism prevent children from seeing points of view they are not experiencing. For Piaget, children are similar to adults in their feelings but very different in their thoughts—they are intellectual aliens in the adult world (Elkind, 1974, p. 108).

Many apparently obvious relationships are not clearly recognized by children. This is illustrated by Piaget's observation (1952, p. 122) that in Geneva children aged eight said they were Genevan yet denied being Swiss, although they stated correctly that Geneva is in Switzerland. Another example given by Piaget may be added: "Small boats float because they are light." "Big boats float because they can carry themselves." In such answers, younger children fail to recognize a contradiction. The images and ideas children acquire and rely on differ significantly from the ideas acquired in school and later in life.

Before *concrete* operations (classification, seriation, correspondence, etc.) of manipulable objects become functionally systematized, the child depends largely on direct perceptual experience. This perceptual centration may be illustrated by a conservation experiment. Two different glass containers are used, and water is poured from the first, which is low and wide, to the second, which is tall and narrow, with the result that the water level is higher in the second container. A five-year-old child focuses on the height of the water in the second container, ignoring the fact that the amount of water is unchanged: he will say that there is more water in the second container than in the first. An eight-year-old child can reverse the operation mentally (that is, he can imagine what would happen if the water was poured back into the first glass); he knows, therefore, that the amount of water is the same as it was in the original container. Up to about seven years of age most children are usually unable to give reasons for their solutions of questions or problems, because they rely heavily on intuition and on the situation at hand rather than on any sequence of direct or mental operations which preceded it. Children making intuitive solutions depend highly on expectation and the perceptual elements of the moment, and their cognitive processes are irreversible. The notion of *reversibility* of operations (returns and detours in the process of exploration and judgment) is crucial to the development of concrete operational thinking (Piaget, 1973, pp. 56–59, 61). During the middle years of childhood, boys and girls must refer to actual objects in order to solve problems. By this time they coordinate two or more dimensions (e.g., height and width of a container or the fact that a friend can be a boy and a child at the same time) and increasingly depend on logicomathematical experience involving induction and deduction rather than on mere association of objects or events (Elkind, 1974, pp. 111–115).

The amount of knowledge the child acquires outside the home rises sharply during middle childhood as the teacher's influence is felt. As the child grows and learns to adapt himself to an ever-enlarging environment, he acquires increasingly abstract concepts and becomes increasingly objective and, consequently, less self-centered. The middle phase of childhood reaches its end as the child grows into peer society and embraces its values and attitudes.

DEVELOPMENTAL TASKS

A boy or girl in middle childhood has at command most of the human qualities and abilities, and many are already specialized into skills. As new areas for application of these skills are presented, the child must be ready to meet the challenges that arise. Most children are sufficiently confident and aggressive to use their early years of education at home and in school in an effective way.

1 *Recognizing one's social role.* To develop and maintain effective relationships with parents, peers, and others, individual differences and idiosyncrasies must be recognized. The child is curious about others and their social roles. He is attracted to those who support him and develops loyalty to them; he also tries to come to terms with others who are seemingly antagonistic toward him. He is eager to acquire behavior and manners appropriate to his roles and sex and learns to relate himself to the roles of others.

2 *Emotional control.* The five- to ten-year-old has greater control of his feelings, emotions, and drives than the toddler, perhaps because he begins to see the necessity for control. He finds acceptable ways to release the energies of negative emotions and thus makes his temperament more acceptable to others. A balance between readiness to help himself and to be helped is often established before the completion of middle childhood.

3 *Knowledge of school subjects.* Learning the fundamentals of school subjects, such as reading, spelling, writing, and arithmetic, is crucial in middle childhood. The child's ability to utilize expanding intellectual powers reaches new heights and becomes science-oriented. Because he is interested in practical applications of the knowledge acquired, the child tends to ask many questions of his parents and teachers.

4 *Physical fitness.* The schoolchild needs about as much vigorous physical activity as before. Although the kind and amount of physical activity needed varies, all children must have some vigorous exercise during which most muscles, especially those of the torso and limbs, are strenuously used. For physical fitness, children "must have from four and a half to six hours daily vigorous muscular exercise" (LaSalle, 1957, p. 56). Unfortunately, many schoolchildren today are driven to school by parents or take the bus. After they return home the same way, there are additional sedentary activities, such as homwork and television. In winter, early darkness and the cold keep them inside. These children are disadvantaged through no fault of their own and are deprived of their right to acquire sound bodily growth and to improve bodily endurance and cardiac reserve, all of which are necessary for good health. Physical fitness also helps a child improve his or her appearance and gain self-esteem.

Havighurst (1972) defines the following developmental tasks for the ages of six to twelve years:

Learning the physical skills necessary for ordinary games
Building wholesome attitudes toward oneself as a growing organism
Learning to get along with age-mates
Learning the appropriate male or female sex role
Developing fundamental skills in reading, writing, and calculating
Developing the concepts necessary for everyday living
Developing conscience, morality, and a scale of values
Achieving personal independence
Developing attitudes toward social groups and institutions

GROWING INTO CHILD SOCIETY

During middle childhood advanced socialization among peers begins. The child's desire to participate in peer-group activities is usually strong. Whenever groups of children assemble in the neighborhood, playground, or classroom, they soon form lines of association and interaction. Socially cooperative activity often comes into prominence. Occasionally quarreling, rivalry, and fighting break out, but generally the child attempts to adjust to others by doing what they expect. As the social interaction progresses, the child assumes various roles, often parallel with his interests and abilities. He realizes that in order to be accepted he must act in a prescribed manner. Many children show great appreciation for their playmates and are pleased and secure in an admiring group, but resist the efforts of others to join them, especially if the would-be joiners are younger children. Some of the decisive factors in the matrix of attraction and rejection are linked with security in the home environment, common goals and interests, common values and attitudes, friendliness, and resourcefulness.

From the earliest stages of life, an infant's behavior is in many ways influenced by siblings, first by older brothers and sisters and later by the newcomers to the family circle. Among many studies of the family, Bossard and Boll's *Large Family System* (1956, pp. 205–221) gives much information concerning the effects of siblings on one another. This empirical study of sixty-four large families identifies eight roles assumed by siblings: the *responsible* one, the *popular* one, the socially *ambitious* one, the *studious* one, the family *isolate,* the *irresponsible* one, the *sickly* one, and the *spoiled* one. The *responsible* one is often the firstborn child, who learns early in life to assume responsibilities for himself and for those younger. Parents often take advantage of him to help younger children and, in this way, further stimulate his growth in responsibility.

The process of assuming roles is influenced by the family constellation and the child's emerging personality type. Role acceptance is usually related to the child's self-concept. In a small family, first identifications with a role are influenced by parents when they assess, interpret, and weigh the child's behavior: "Let's keep little Zeke inside; he catches flu every time he goes out." In a large family, siblings attribute roles to each other; they help parents set distinct roles for each child. Children seize upon differences in appearance, abilities, traits, and idiosyncracies to distinguish one sibling from or contrast him with another. With various degrees of gratification, many children accept the roles attributed to them. Sometimes such distinguishing features help a child to stand out as an individual. Some children reject certain roles and show deep resentment toward those who emphasize them. Any form of adjustment to one's specialized role within the family is, in some ways at least, the key to one's status orientation.

At home children develop their basic orientation to life; there they learn to adhere (or not to adhere) to moral standards and religious ideals. In many good homes the children develop a sense of values and a morality that, in modified form, endure throughout their lives. It is good if school and personal interactions with neighbors reinforce this value fiber; otherwise, slight or intense conflicts occur. The family is thus a workshop for developing a lasting style of life, while the community exerts a major modifying influence.

As the child matures, the need for companions grows stronger; and toward the end of middle childhood, the child approaches the "gang age." Group identification and sentiments of pride, loyalty, and solidarity become powerful drives toward social intimacy in late childhood. The next chapter will deal extensively with group life in later childhood.

The expansion of intellectual horizons and parent and peer interaction are factors that provoke questions concerning what is right or wrong. The ability to distinguish between right and wrong gradually deepens. It can be applied to many situations if the child's inquiry is supported by instruction in moral concepts and principles; otherwise it remains precarious and at times confusing. The self-initiated practice of moral virtues such as honesty, justice, and fortitude is reinforced if moral education is given. Doing good for the sake of others also begins to be appealing. The child first shows a desire to please his parents, siblings, and a few close friends, then expands his interest to include persons and groups with whom he is only slightly associated. Children recognize some norm, however vague, that dictates aiding others. Mutual experience of pleasure reinforces the child in following that norm. Altruism is also often reinforced by parents who like and encourage such behavior (Bryan & London, 1970).

The child's striving for approval and praise grows with age. Fairness often becomes a leading virtue as, about the age of nine or ten, the child recognizes more fully than before the needs of others. At this age also, playing games and doing things according to established rules and regulations become important. When disciplinary action is necessary, the child will tend to accept certain forms of punishment as long as he realizes he deserves them. He is now sensitive to public criticism and reproof, because this is a threat to his social standing. "Losing face" pulls a child down and often creates feelings of inferiority or hostility.

GROWTH OF SELFHOOD

With increased independence from parental supervision and daily contacts with a large peer group and a teacher, a child has ample opportunity to develop a realistic self-concept. A schoolchild's self-appraisal is largely based on the appraisals given by parents and teachers, siblings and peers. Parental estimation, whether favorable or not, is often somewhat one-sided. If it is negative, the child's self-concept will be distorted by a lack of self-acceptance or by emotionalized self-assertion; later, aggressive tendencies will begin to be generated, resulting in internal conflict. This situation is somewhat relieved when relationships with others increase and gains in objectivity are made.

Growth in selfhood is only in part autogenous (self-generated). Much of it is elicited by others with whom the child identifies. The child's efforts to change his behavior and conceal disapproved traits also influences his concept of himself. Any improvement or expansion of self-regulation is a sign of increasing ego strength, unless it involves repression of strong drives or emotions, in which case internal conflict develops and increases tension.

A child's self-control often begins outside himself. He starts to curb his impulses to please his parents. He will inhibit one of his drives or demands when he sees that he will be punished. A child is ingenious in his efforts to win recognition or praise. His parent's or teacher's sensitivity in responding to his efforts encourages formation of desirable traits and attitudes, including acceptance by himself.

In the middle and late phases of childhood a child is well-adjusted if an affirmative answer can be given to the following questions (Gesell & Ilg, 1974, p. 10): Does the child have reasonable control over his emotions? Does he play well with other children for long periods? Is he helpful to siblings and classmates most of the time? Is he achieving near his capacity? Can he be depended on for simple chores and homework?

Parental direction of the activities of an eight- or nine-year-old often helps the child further his internal control. Through external control he acquires new habits and skills. Without some environmental reinforcement the child's endowments are often neglected, and undesirable and maladjustive reactions may take root. Extensive conditioning, however, severely lessens the child's flexibility and initiative, increases his dependence, and stunts his curiosity. If restricted too much, the child cannot fully utilize his native endowments, and later adjustment and development of self-direction become too difficult for him. The dependent child has little confidence in his drives for autonomy and self-reliance; more often than not he merely transfers his dependence from one person to another. From parental dependence to peer dependence is just one step. Mutually satisfying transactions and reciprocal relationships with parents and peers are what children need to further their socialization and self-esteem.

QUESTIONS FOR REVIEW

1. What changes mark the entrance of a child into middle childhood?
2. What are the typical play activities and interests at this stage of development?
3. Indicate some abilities and skills that are prerequisites for school adjustment in the first grade.
4. What are the key functions of the teacher in the primary grades?
5. What symptoms are associated with "minimal brain damage"?
6. Identify and describe the major findings of Piaget's studies on the reasoning and language of children.
7. What personality factors operate in group acceptance and peer companionship?
8. According to Bossard and Boll, what roles do siblings in large families tend to assume?
9. What often handicaps the child in his peer relationships?
10. Under what conditions are social and moral standards best assimilated by children?
11. Identify several developmental tasks of middle childhood and explain two or three of them.
12. How does appraisal by parents and peers affect the child's self-acceptance?

REFERENCES

Selected Readings

Elkind, D. *Children and adolescents: Interpretive essays on Jean Piaget* (2d ed.). New York: Oxford University Press, 1974. This paperback presents Piaget's ideas on cognitive development, egocentrism, and children's educational needs.

Goodlad, J. I. (Ed.) *The changing American school.* (The sixty-fifth yearbook of the National Society for the Study of Education, Part II.) Chicago: University of Chicago Press, 1966. The second part of the sixty-fifth yearbook of the Society, this work includes studies by twelve contributors. It analyzes the current school situation and the various forces molding it.

Havighurst, R. J. *Developmental tasks and education* (3d ed.). New York: McKay, 1972. Presentation of tasks to be accomplished during childhood and how their accomplishment is related to school adjustment.

Specific References

Bossard, J. H. S., & Boll, E. S. *The large family system.* Philadelphia: University of Pennsylvania Press, 1956.

Brackenbury, R. L. Values: Developing through education. *Child Fam.,* 1966, **5,** 51–61.

Bryan, J. H., & London, P. Altruistic behavior by children. *Psychol. Bull.,* 1970, **73,** 200–211.

Caldwell, T. M., & Ricciuti, H. N. (Eds.) *Review of child development research.* Vol. 3, *Child development and social policy.* Chicago: University of Chicago Press, 1973.

Cordon, B. W. School psychology for the total school. *Profess. Psychol.,* 1972, **3** (1), 53–56.

Doubros, S. G., & Daniels, G. J. An experimental approach to the reduction of overactive behavior. *Behav. Res. Therapy,* 1966, **4,** 251–258.

Elkind, D. *Children and adolescents: Interpretative essays on Jean Piaget* (2d ed.). New York: Oxford University Press, 1974.

Franco, D. The child's perception of "the teacher" as compared to his perception of "the mother." *J. genet. Psychol.,* 1955, **107,** 133–141.

Gesell, A., & Ilg, F. L. *Infant and child in the culture of today* (2d ed.). New York: Harper & Row, 1974.

Goodlad, J. I., & Shane, H. G. (Eds.) *The elementary school in the United States.* (72d Yearbook of the National Society for the Study of Education, Part II.) Chicago: University of Chicago Press, 1973.

Grinspoon, L., & Singer, S. B. Amphetamines in the treatment of hyperkinetic children. *Harvard educ. Rev.,* 1973, **43,** (4), 515–555.

Havighurst, R. J. *Developmental tasks and education* (3d ed.). New York: McKay, 1972.

La Salle, D. *Guidance of children through physical education.* (2d ed.). New York: Ronald, 1957.

Piaget, J. *The child's conception of the world* (Joan and Andrew Tomlinson, Trans.). New York: Humanities Press, 1951. (Originally published, 1929.)

Piaget, J. *Judgment and reasoning in the child.* New York: Humanities Press, 1952.

Piaget, J. *The child and reality.* New York: Grossman, 1973.

Von Valletta, J., Brantley, J. C., & Pryzwansky, W. B. Multiphasic screening of first-grade children. *Proc. 80th Ann. Convention APA, 1972,* **7,** 571–572.

Wender, P. H. *Minimal brain dysfunction in children.* New York: Wiley-Interscience, 1971.

Chapter 12

Late Childhood

HIGHLIGHTS

Late childhood marks the completion of childhood development and vividly anticipates adolescent changes to come. What the child learns at this period of life depends greatly on what he or she is eager to know and understand.

The preadolescent years are a time of intimate peer companionship and esprit de corps—the child feels most comfortable among others of the same age, maturity, and status.

Sexual typing is advanced greatly by a homogeneous peer group and by the companionship of the parent of the same sex. Close association with members of the same sex strengthens the child's sex identity and related features of the personality.

Late childhood is a period of preparation to face the tasks of pubertal and adolescent growth. Age-related sex information and moral instruction, when properly combined, work wonders for the child's adjustment and facilitate the transition to sound adolescent styles of life.

Most children reach late childhood—or preadolescence, as it is also called—at approximately nine or ten. The years of late childhood are marked by an increase in critical thinking, by theoretical questioning about causes and effects, by resistance to adult opinions, and by emotional identification with peers of the same sex. Interests and activities begin to reflect the child's sex more closely than before. Late childhood is further characterized by substantial gains in emotional self-control and greater readiness on the child's part to assume responsibility for his or her actions. The peak of childhood development is now reached. This is accompanied by a strong drive for self-expansion and adventure, as the child strongly feels his competence and skill. Late childhood ends as major pubertal changes begin and childhood merges into adolescence.

The preadolescent is strongly influenced by his peer group, while the control of parents and other adults weakens. Time and again parental controls are weakened by the child's insistence on his social group life. While a younger child does not notice parents' moods and foresee their probable responses, most older children know how and when to get privileges from them. They cleverly play upon the feelings and sentiments of either parent. When younger, they saw their parents as all-seeing and all-knowing, but now the dethroning process is about complete. This can be very disturbing to a parent who does not understand the change taking place in the child's social standards and expectations. The preadolescent often rejects what grown-ups consider to be good manners; at times he criticizes almost everything and everyone, lacks consideration for parents, and often behaves boisterously. Adults find such conduct difficult to overlook. Nevertheless, the child needs the warmth of a harmonious home, where he can heal the wounds inflicted on his ego by some of his classmates and peers. In times of trouble, whether he or someone else has caused it, he needs the emotional support of his parents. He longs to be fully understood by the parents when the fruits of growing up in peer society turn a little bitter.

In relation to his siblings, the preadolescent is impressed by older brothers or sisters, while he sees younger children as inferior and tries to make servants of them or avoid them. Squabbling and rivalry among siblings are unavoidable. The preadolescent may be friendly one minute and as scrappy as an alley cat the next. Occasionally he seems to derive sheer delight from embarrassing, bullying, or tormenting others, yet these persecutions are seldom carried to an extreme. Sibling support and companionship are often sought, as preadolescents readily gang up against parents as well as teachers. Individual differences among children, including siblings, are intensified, and interests and activities reach a peak of diversity.

In any family with school-age children, sufficient and effective communication is a problem. Although the family is dispersed much of the day, a good start in the morning is important. The mother and father must "forget" the quarrels and unpleasant remarks of yesterday and turn bright and encouraging faces toward their children each day, for it must be recognized that family life serves as an "emotional reconditioning center" for all its members (Duvall, 1971, pp. 277–280). A child who comes "home from his rigorous day in the classroom and playground full of . . . pent-up emotions" is likely to take them out "on the first available member of the family." From time to time this is to be expected. Sometimes feelings will explode as children annoy each other and begin quarreling, even fighting. Preadolescents as well as other people need occasional opportunities to express their rising emotions, even to the point of breaking ties with opponents. A resourceful mother can do much to restore the disturbed lines of family communication, especially if she takes time to evaluate the needs of the whole family. It is important that family members share their feelings, thoughts, and activities. Only when the communication system within the family is restored can parental love flow through and remove the "waste products" of everyday living with children. Love is vital at this age, as it was before.

NEW HORIZONS OF UNDERSTANDING

By the end of the fourth grade, the child has acquired much competence in the fundamental skills and abilities that are necessary for further, more advanced learning. Most children are now ready for a more complex curriculum, and their interest in extracurricular activities also increases noticeably. The fourth grade is often the first in which the child must use abstraction and judgment, in addition to retention. Arithmetic, science, and social studies begin to demand something more than memory work. Some children therefore progress rapidly, while others begin to have serious difficulties. When this occurs, it is important that the less intellectually able children should not be left with a frustrating sense of defeat. A teacher should therefore encourage such children to improve their own accomplishments rather than compete against more gifted students.

After the fourth grade the child is ready to read independently, to deal with fractions, to refine his sense of history, to abstract and generalize, and to notice individual idiosyncrasies. He has little difficulty understanding explanations, whether these concern moral, social, or cultural matters. Now the child often develops interests and ideas of his own. He is eager to learn more about his immediate environment, his country, other nations, and the universe as well. Interest in and knowledge of world history, geography, and the secrets of nature gradually gain in depth and understanding. The motivation to master new skills and techniques is dynamic and consistent. Often a ten- or eleven-year-old will spend the entire afternoon with a chemistry kit, an interesting book, or a hobby. The child works hard to increase his knowledge and his feeling of accomplishment. Much time is spent in group games and projects, athletics, and other social activities, which now take precedence over schoolwork and time spent with parents.

The progressive-education policy of passing a slowly developing child to the next grade level and not detaining him in a grade he has failed creates serious obstacles to the development of academic and social self-assertion. Children who have not acquired the knowledge and skills of previous grades have little foundation upon which to be eager and interested in more advanced and complicated subjects. In some children of a sensitive nature, fear, worry, insecurity, and feelings of inadequacy are stirred up and generate anxiety, which in turn disturbs their physiological organs and systems. How much anxiety will be generated depends on the child's ego strength, flexibility in handling the stresses of school and home life, and ability to face stress without resorting to defenses that lead to neurotic and other symptoms (Kaplan, 1970).

The policy of promotion is beneficial, however, for the child's social integration, since group identity, it is believed, plays a key role in the child's security, promotes feelings of adequacy, and contributes substantially to emotional growth and adjustment. The question is whether the value of promotion outweighs the undesirable consequences of academic failure and failure in other school activities. Such social promotion often leads to more serious academic deficiencies later, from which further emotional problems can germinate. In general, repeating a grade at the early primary level can fortify the child's academic foundation if remedial teaching is included. The social maladjustment, however, that results from failing a grade often hinders scholastic progress. Ungraded primaries, in which each child progresses at his own rate and may take two or three years for one grade, eliminate the concern and stigma attached to not being promoted.

PEER-GROUP LIFE

During preadolescence boys and girls are eager to join others of their own age, sex, and status. They readily develop emotional attachments and are proud of their friends and the groups to which they belong. Group play, team games, and seasonal athletic activities are very appealing at this age. Everyone feels obliged to assume a role assigned by others and to contribute to the preferred group activities. Obedience to a leader and conformity to basic group standards are generally necessary for complete acceptance. Children are attracted to groups by common interests, standards, and expectations.

Peer companionship is so urgently needed and so constantly sought that many children come to school mainly to play with their companions. Often two or more children with similar needs or interests form an attachment, for they understand each other's desires and derive satisfaction from their friendship. Preadolescent groups offer opportunities for the development of deep interpersonal friendships and for identification with selected members of the same sex. Generally in preadolescence, if not before, the child finds a particular friend to confide in and share his or her conflicts and problems with. Such a friend becomes one of the major influences in the child's life.

Groups at this age are frequently homogeneous; members of the opposite sex are rarely included in the more compact and emotionally toned groupings. Generally girls engage in fewer group activities than boys. Their groups are small, consisting of three to five girls. Usually they meet at the home of one of the girls, since parents generally grant less freedom to girls than to boys, even though for more than a decade there has been a trend toward treating boys and girls equally (Sutton-Smith, 1973, p. 401).

Engaging in exciting adventures is often one of the objectives of the boys' groups. Much depends on the leader's initiative. The boy who surpasses other group members in strength or achievement in preferred activities usually assumes leadership of the group. Awareness of the likes and dislikes, interests and social ideals of other children adds much to this leadership potential. Competition and cooperation run high during preadolescence and must be satisfied. A child often develops an intense drive to surpass others, including his friends. The motivation to make a showing for himself in order to gain approval or prestige is intense. When rivalry between groups is involved, he is likely to exert himself as much for "his side" as for personal recognition. Competition has both advantages and disadvantages for older children. In competing, a child may discover capacities within himself that he had not otherwise realized. It also helps him discover the limits of his abilities and efforts. On the other hand, competition can become harmful when it gives rise to feelings of inferiority or when it makes other children unhappy. In addition to leadership, friendliness, enthusiasm, daring, and originality are other qualities highly valued by children. It is not unusual for a child tending to delinquency to become a leader and to persuade an entire group to follow his ideas and engage in his preferred type of misdemeanor.

The strength of group identification increases as the child grows older. He or she begins to transfer some emotional identification from parents to companions. Since peer ties are marked by loyalty and solidarity, group life places its imprint on the personality of the child, and parental influence gradually declines as the group influence becomes stronger. Naturally, these two influences operate simultaneously in molding children's attitudes and interests. Under desirable circumstances parental and group influences reinforce each other. More often, however, they clash, at least in some ways, and cause conflicts and anxieties within the child. As participation in a peer society increases, resistance to adult standards and guidance seems to be reinforced.

In a group situation the child is less inhibited than in the home. Some undesirable tendencies are therefore expressed when the support of others is secured. Group activities also often become so time-consuming that a child finds it difficult to complete his schoolwork or neglects home responsibilities.

Gangs result from spontaneous efforts by children to form groups able to meet their needs for self-expansion and security. Some gang activities are cloaked in secrets, such as special codes. Acceptability depends to some extent on the members' social class, ethnic origin, and residential district. Gangs in the lower socioeconomic neighborhoods may have fights and also encounters with law enforcement agencies. Nevertheless, through gang activities a child receives important training in group dynamics and social relationships that he could not obtain from adults. Peer cooperation is fostered, as well as communication skills. The peer group is second only to parents in socializing the child. It is indispensable in role rehearsal (cooperative and competitive behavior, sex-role adoption, expression of aggression, and so on), but whether it is more important than parental models in these overt behavior areas is an open question. The peer group confirms (reinforces) the child's self-judgments of competence and self-esteem, although the foundation of these is probably more influenced by the family. A high degree of acceptance by peers (the sociometric star) on a combination of friendship and status variables is perhaps the best indicator of a child's personal-social adjustment (McCandless, 1969, pp. 808–810).

It has been suggested that "gang spirits" can be channeled into supervised clubs for children sponsored by schools and adult organizations. Admittedly Boy Scouts, Girl Scouts, Camp Fire Girls, and Four-H and C.Y.O. clubs exert a powerful influence on the social and personality traits of the older child, yet it is questionable whether the adult-sponsored clubs can completely fill the need of the preadolescent to "go it alone." It is important that children's clubs and camps ensure (1) sufficient guidance to initiate wholesome activities and (2) enough freedom to satisfy the child's need for independence in individual and group situations. When the child learns ethical conduct in an adult-supervised organization, he will probably be guided by these standards when he is on his own.

SEXUAL TYPING

Sexual typing may be understood as a process of intrapsychic identification with those personality qualities and traits that pertain to one's own gender. It begins with a closer identification with the parent of the same sex in early childhood. From about three years of age, a male child labels himself a boy and likes associating with his father, and a female child labels herself a girl and likes associating with her mother. Already at this age a girl is more interested in the mother's activities than a boy is. Most four-year-olds dichotomize people into male and female. In our culture, then, the gender role is apparently imprinted during the first two-and-a-half to three years of life in most children. Its strengthening (or weakening, in some cases) occurs during the later years of childhood (Rosenberg & Sutton-Smith, 1972, pp. 49–52).

For lack of any significant sexual development during the early phases of childhood, the young boy's sex awareness increases only moderately. But many older boys are interested in reading about human origins and about interpersonal sex relations, with some interest directed toward the father's role in sex. They are also interested in—if not fascinated by—female nudity. Affective attachments, however, are usually directed toward other boys who have similar interests and needs; playmates are almost exclusively boys. Similarly, girls' social and emotional ties are directed to other girls. Sex roles probably become well interiorized during the early years of school, if identification with the parent of the same sex is sufficiently rooted in the mind of the child. Parents then effectively reinforce sex-role behavior by urging the boy to control his expression of feelings and affects while girls are being taught to control aggression and to restrain self-assertion. Many studies suggest that the father is a more crucial agent in directing and channeling the sex typing of the child, both male and female, since his attitudes appear to be more strongly determined by the child's sex than are the mother's attitudes (Block, 1973; Block, von der Lippe, & Block, 1973).

At the nine- to ten-year-old level, the segregation of boys and girls is almost complete. Social contacts with members of the other sex are marked by aloofness, lack of response, mockery, annoyance, and apparent contempt, as well as shy withdrawal. The cleavage is pronounced in the later part of preadolescence. Most preadolescents find it difficult to accept members of the opposite sex as play companions. Disparaging and deriding of members of the opposite sex and ganging up on them are frequent in late childhood. Play activities and interests also diverge sharply. Boys tend to prefer vigorous and competitive activities—sports, bicycling, hiking, and mechanics. Girls' interests often center on clothes, handicrafts, art appreciation, household work, and other quiet or sedentary activities. Jumping rope, dancing, swimming, and skating are their more active interests.

It is natural for a child to associate with members of his own sex, for through these relationships he learns to identify with his own sex and adjust to it. Since a girl is expected to show feminine qualities and engage in typically feminine activities, close relationships with other girls help her meet these demands. Conflict or confusion from contrasting pubertal urges is thus reduced or avoided. Kagan (1964) distinguishes core attributes of masculinity and femininity. Boys want height and large muscles. They show more self-assertiveness, independence, athletic activity, and aggression than girls. As compared with boys, girls are conforming, dependent, and nurturant. They want to be attractive and to have social poise. The impressions girls feel they make on others influence significantly their own self-concepts.

Occasionally some preadolescents hesitate to identify themselves with a sex-linked role. When a boy relates excessively to his mother and other females, various aspects of his behavior begin to reflect female characteristics. Such an effeminate boy is usually rejected by other boys, and fixation of a feminine style of life becomes a possibility. During the preadolescent years and at the time of pubertal changes, some girls strive to be masculine, but these efforts usually lessen as adolescence advances. Masculine traits frequently mark a girl who has only brothers and naturally is forced to compete with them. It is advantageous for girls to have understanding mothers and for boys to have fathers who show interest and encouragement. This enhances proper sex-role identification—a basic necessity for the healthy development of the child's personality.

The child needs two mature sex models in order to fashion his or her own sex identity and later relationships to members of the opposite sex. A dangerously close and too deep association with members of the same sex is infrequently encountered but can occur, especially when cross-parent relationships are poor. If the cross-parent relationship is missing because of death, divorce, or separation, the child is deprived of a model on whom to base the qualities and traits of the sex role as it pertains to members of the opposite sex.

Probably because of her close association with her mother, a girl's perception of her sex-typed role is more advanced than that of a boy; she is more embarrassed by conversations concerning sex, and she thinks often about her future feminine role. Somewhat paradoxically, though her interest in boys is aroused even earlier, she usually continues to show aloofness or scorn during preadolescence. Beyond the facade of indifference, however, she is eager to learn to dance and to develop her social manners and conversational skills. In fact, in many instances she is ready to interact with boys and uses some means to attract their attention.

Boys, however, find it more difficult to achieve sex-role identification, and their early relationships with girls are awkward and uncomfortable if not filled with anxiety. The desired sex-role behavior for a boy is often defined as something he should not do: to gain in masculinity, he must learn to avoid "girl-like" activities. Feminine interests and activities are a sign that the boy is not progressing satisfactorily toward the genuine pattern of masculine behavior. As a result, boys are more anxious than girls about their sex-role identification and hold stronger feelings of hostility toward girls than girls do toward boys (Lynn, 1964).

For a man to be masculine and a woman to be feminine is what makes each an *authentic* human person (Bieliauskas, 1965). A good start is created by warm, accepting mothers and fathers, whose children usually have appropriate sex identity and high self-esteem (Coopersmith, 1968; Sears, 1970). With increase of age, as Table 12-1 shows, the number of children who apparently rely on

Table 12-1 Age and Relative Confidence in Self and in External Source

Questions	Percent agreeing, by age			
	8–10	11–13	14–15	16–18
"How smart you are"				
Mother right	76	61	58	31
Self right	24	39	42	69
	(*N* = 549)	(*N* = 629)	(*N* = 315)	(*N* = 318)
"How good you are"				
Mother right	77	70	53	47
Self right	23	30	47	53
	(*N* = 559)	(*N* = 634)	(*N* = 317)	(*N* = 321)
"How good-looking you are"				
Father right	70	55	49	45
Self right	30	45	51	55
	(*N* = 506)	(*N* = 581)	(*N* = 291)	(*N* = 294)
"Knows best what you are like deep down inside"				
Mother	50	33	19	20
Father	6	5	4	2
Yourself	40	57	61	62
Best friend	3	5	16	15
	(*N* = 542)	(*N* = 655)	(*N* = 339)	(*N* = 355)

Source: Morris Rosenberg. Which significant others? *Amer. behav. Scientist,* 1973, **16,** 829–860 (Table 3, p. 843). By permission.

themselves in judging "how good they are" and similar matters increases, yet mothers and fathers still play a leading part for many older children and teenagers in the age brackets fourteen to fifteen and sixteen to eighteen (Rosenberg, 1973).

PERSONALITY AND SELF-CONCEPT

During preadolescence, the child's self-concept undergoes new developments as his or her identity becomes increasingly related to the peer society. The child attains a new level of self-expression through advanced schoolwork, extracurricular projects, and complex group activities. As he reaches the peak of childhood development, he has greater harmony within himself, often accompanied by a superior ability to apply himself. Feelings of self-respect and optimism are mingled with buoyant cheerfulness and audacity.

The child's experiential background has outlined the contours of his self-image. Except where considerable damage has been done to his concept of self-worth, he sees himself as being good and capable of accomplishing his tasks. His own abilities and talents are usually evaluated in terms of school standing, athletics, peer acceptance, and popularity. He is ready to use his powers and prefers activity that tests his ability. He establishes fair standards and wants to perform well. The preadolescent's attitude toward self is thus based on an expanded frame of social relationships and comparative performances.

The child has an increased sensitivity to the approval and disapproval of people he considers important. He is particularly concerned about winning the approval of his peers, and this desire increases throughout late childhood. Because the less skilled child often has difficulty in asserting himself when confronted with new peers, the emotional support of the parents is very important for his self-acceptance and adjustment to others. The development of some special interest or ability is of much help. Encouragement of skill in athletics, crafts, or art or ability to play a musical instrument usually assists in socialization and personal maturation. Relating oneself to others is very important, expecially for girls. Learning the skills of relating successfully to others, as in modes of conversation and etiquette, is helpful in personal adjustment.

Occasional problems in relating to others, parents and peers alike, may give a child the impression that he is changing for the worse. Many parents strongly reinforce this self-devaluation, especially if the child's efforts at self-improvement are not given careful consideration. If encouraged and given opportunities, the preadolescent can capitalize on his own strengths and assets. If, however, the family atmosphere does not allow for shortcomings, he is forced to turn to the defenses of denial or rationalization of his own responsibility to continue trying until improvement occurs.

The preadolescent's activities and accomplishments are in some way expressions of himself. Various facets of his personality come to light in projective evaluation of his motion patterns in play, art, and social situations. Figure 12-1 reproduces drawings of a house, a tree, and a person by two eleven-year-old boys from the same middle-class environment, attending the same school. The differences between the boys are vividly reflected in all three subjects. When these illustrations are compared with the same boys' drawings of the same subjects at eight years of age (Pikunas, 1965, p. 265), two impressions arise: continuity and change, with the second overshadowing the first.

The stabilization of psychological gains that takes place during preadolescence strengthens identity and furthers character development. *Identity* refers to the sense of being the same despite internal and external changes. Identity results from identification with significant persons in one's life, as well as from assimilation of their values and attitudes. Consistent parental behavior in regard to the child furthers his sense of identity. While character is primarily an inner system of traits, outside factors delay or accelerate its development. Among these factors are consistent parental behavior in regard to values and principles, consistent parental discipline, and a sense of parental moral responsibility. Emotional identification with parents promotes children's assimilation of parental values.

Most parental behavior affects children either directly or indirectly. Parents who are guided by a definite philosophy of life accept a certain hierarchy of values and act in accordance with it. Often children are taught these values and the value-based ethical principles, such as honesty, truthfulness, fair play, loyalty, and personal responsibility for their acts. Adoption of such values and the resultant behavior standards is reflected in many aspects of a child's thinking and acting. A preadolescent usually understands practical necessities and is ready to modify principles or rules to fit circumstances. When social reinforcements complete a child's reliance on values, codification of principles and ideals begins.

By the age of twelve, most children are ready to assume greater responsibilities toward others, such as baby-sitting, safety patrol, or a part-time job. Their identification with a duty or task assigned to them improves as they become older. Their sense of right and wrong becomes more refined; in fact, many children will not cheat when they are sure of being trusted. As control increases over the impulsive and emotional tendencies, parental and school controls help to develop self-regulation based on ethical and moral principles. Thus the character foundation for later life gains in structure. Ideally, ethical motivation, when developed by example and religious instruction, reinforces moral sensitivity. The observance of religious practices can encourage the desire to act in accordance with religious precepts. As moral and religious values gain power, they direct the child's behavior toward personally and socially desired goals, fostering (as part of adjustment to reality) a philosophy of life based on religion. Development of identity and acquisition of a salient character are, of course, cumulative but slow processes, and they continue through adolescence and adulthood.

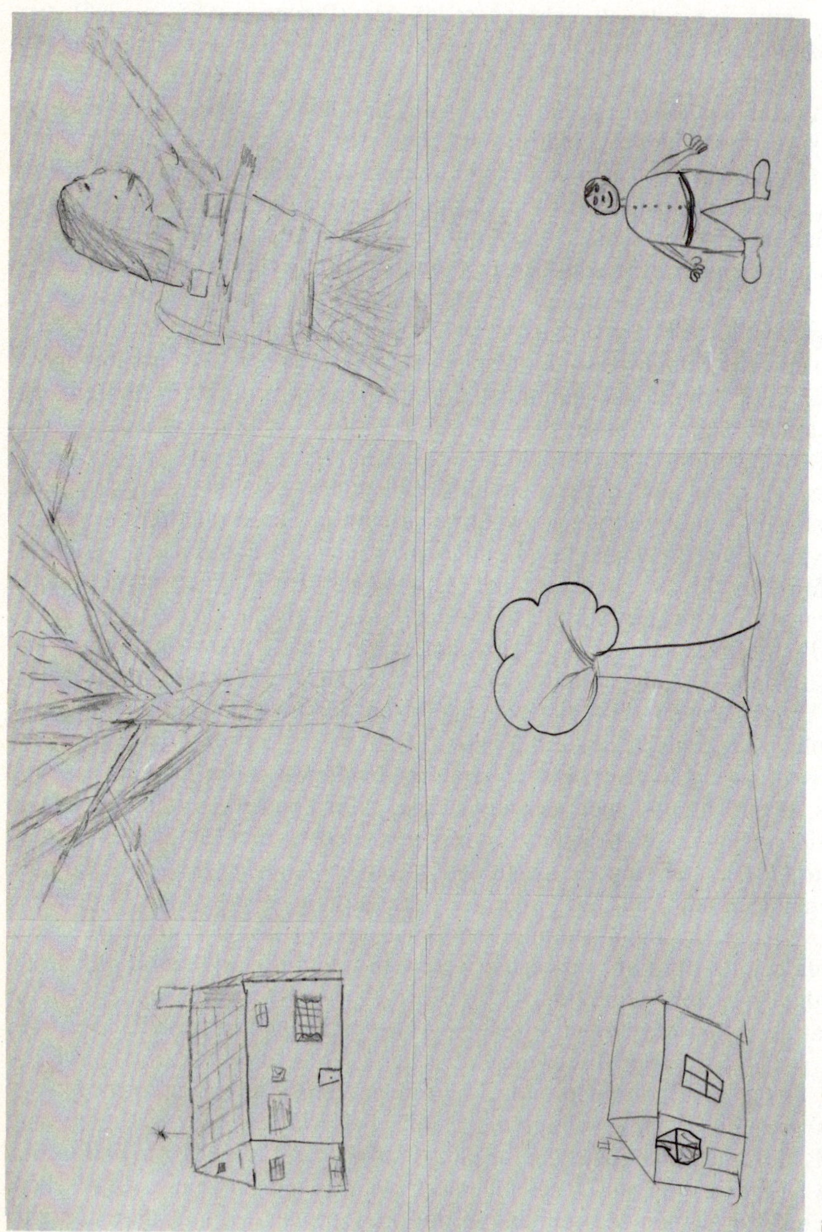

Figure 12-1 The same objects drawn by children of the same age.

The Case of Daniel The following case illustrates severe deficits of value and identity formation in a child.

> Daniel had spent most of the previous year in the County Juvenile Home. This year he returned to school to "try" a new teacher; the trial lasted only six weeks. His last act was to attack his teacher with a gnarled branch from a nearby tree. Once more he was ushered out by two policemen.
>
> Daniel was an "angry boy," and his background points to some of the reasons why. At home he was rejected by his mother, unsupported and occasionally abused by his father. They lived in a poor neighborhood. At the age of ten he was a master of deceit, quick in provoking his classmates and teachers, threatening and cruel to those physically inferior to himself. His delinquency ranged from truancy to acts of vandalism, from foul and lewd language to extortion and sexual attacks on small girls.
>
> There was virtually no day at school without some disciplinary problem, in spite of a great deal of patience and kindness shown him by school superiors. His particular talent was turning potentially bright days into sad ones. Understanding and leniency were not enough, and there were always other children to teach and to guide. Daniel made himself difficult to teach by hurting others and at times striking out blindly. Usually he hurt himself more than he hurt others.

PREPARATION FOR ADOLESCENT TASKS

Characteristics of each stage of development reach maximum expression some time before the period ends. Most children arrive at their preadolescent maximum at ten or eleven and appear to stop for a while. This, then, is a time for efficient preparation for the developmental tasks of puberty and adolescence.

Academic accomplishments are important for most preadolescents, because they need to be proud of their achievements. The child's desire to excel usually increases his efforts. He looks for an area of success, and when he finds one, either in academic or nonacademic fields, he is benefited. If his sense of accomplishment is strengthened, the feelings of inadequacy that readily afflict the nonachieving preadolescent are counteracted. The adolescent defeatist attitude often begins to take root during the years of late childhood. Procrastination at work and aggressiveness in social situations often accompany this attitude. Parents, through encouragement and control, can often correct the situation if they are aware of the importance of achievement in preadolescence.

Peer identification is one of the major antecedents of late adolescent and adult identification with persons of one's own age. In late childhood, as stated earlier, close peer associations are usually limited to members of the same sex. The emotional intimacy, however, is a precursor of later identifications. Frequent peer activities in large and small groups lay the foundations for personally gratifying relationships with cliques and crowds during postpuberty.

As a child becomes readier to respond to the information and suggestion of peer groups, as the leader or majority rule of the group is followed, the child recognizes alternatives and expands his perspective. Then, in some home or school situations, he insists on acting on his own choice. Peer identification thus represents a major step toward self-reliance. It is noteworthy that a child of twelve or thirteen is occasionally concerned with and projects himself into a new cycle of development that will continue until the twenties. He admires boys who are bold and daring. She admires girls who are beautiful and popular. He wants to be sixteen or eighteen, anticipating the privileges he will enjoy then, especially driving a car or a motorcycle. She wants to be sixteen or eighteen, anticipating more freedom from parental demands.

With adolescent developments and adjustments in mind, specific preparations can be undertaken to form later attitudes, especially toward sex and morality. Fundamentally, this preparation consists of parental and school instruction and the preadolescent's own learning about adjusting to sexual maturation. Because the school is primarily concerned with intellectual development, the home bears a large share of responsibility for sexual, moral, and related instruction at this level.

Since in the various phases of childhood, incoherent and incomplete information on sexual matters is bound to be encountered, advanced personalized instruction is needed to correct and complete sex information as it concerns the approaching adolescent phase of life. When to give this personalized instruction depends on many factors, especially the need and interest of the child concerned. The beginnings of pubertal growth in height and weight are a definite indication of such a need, because these changes are followed by sexual maturation. Generally, the parent of the same sex is expected to give this instruction. If he or she feels the task is too difficult, a satisfactory substitute must be found. Impersonal class instruction is not enough. Most physicians, psychologists, and social workers, as well as many teachers and clergymen, are well prepared to convey this information. The instruction centers on the oncoming sexual developments and their basic implications. Certain books and pamphlets (American School Health Association, 1967; Child Study Association, 1968; Fraser, 1972; Grams, 1970; Rosenberg & Sutton-Smith, 1972) may be of considerable assistance to the parent, and some are prepared for the prepubescent himself (Gruenberg, 1973). Sex instruction should present not only basic information about sexual changes during puberty but also healthy attitudes. Along with factual instruction, the major moral implications of sex should be tackled. Lyman's *Let's Tell the Whole Story about Sex* (Lyman, n.d.) consists of four recorded conversations between parents and their children: how babies are born, menstruation, problems of growing boys, and the marriage union. These stories are for parents and adult organizations, not for children. Frank and fairly complete, they consider some of the moral and religious implications of the subject.

Beyond this, most children need explanations of moral considerations in various peer relationships and activities, a clarification of the concept of what constitutes harm to others, and an analysis of the relation between moral conduct and religious precepts. Such information fosters the development of conscience, which in turn will make it easier to act in accordance with moral principles and humanistic values. Thus the foundation of a moral code will be strengthened before adolescent turmoil sets in. In childhood, much depended on what happened in infancy; so, too, in adolescence, much depends on the developments and adjustments of the preadolescent years of life.

QUESTIONS FOR REVIEW

1. What characteristics distinguish a preadolescent from a child?
2. Describe the child's advances in cognitive development and the resultant interests.
3. How does the preadolescent grow into child society? Describe the emergence of the gang.
4. How do peer influences affect the child's personality during the late years of childhood?
5. What are some of the outstanding goals in children's group life? How are they realized?
6. Explain the changes in competition and cooperation that occur in preadolescence.
7. In what ways do preadolescent boys and girls differ? How do boys express their dislike for girls and vice versa?
8. Why are children's clubs and organizations necessary?
9. How are self-identity and self-appraisal modified by peers?
10. Why does the preadolescent need parental emotional warmth and support?
11. Describe sexual typing and identify factors affecting it.
12. What are the factors promoting identity and character development? How does parental behavior affect the child's character development?
13. What does a child need in order to be prepared for adolescent developmental tasks?

REFERENCES

Selected Reading

Gesell, A., Ilg, F. L., & Ames, L. B. *Youth: The years from ten to sixteen.* New York: Harper & Row, 1956. A discussion of maturity profiles, gradients, and age trends on a year-to-year basis.

Kohen-Raz, R. *The child from nine to thirteen: The psychology of preadolescence and early puberty.* Chicago: Aldine, Atherton, 1971. Discusses late childhood and prepuberty as stages of human development; cognitive, emotional, social, and educational aspects are emphasized. Includes psychopathology and psychotherapy.

Maccoby, E. E. (Ed.) *The development of sex differences.* Stanford, Calif.: Stanford, 1966. A major work on various aspects of psychosexual development. It consists of five essays: three psychological, one anthropological, and one physiological.

Thornburg, H. D. (Ed.) *Preadolescent development.* Tucson: University of Arizona Press, 1975. A comprehensive book of readings by twenty-seven contributors on years of late childhood (nine to thirteen). Includes physical growth, sexuality, intelligence, discipline, drugs, and value and identity changes.

Specific References

American School Health Association, Committee on Health Guidance in Sex Education. *Growth patterns and sex education: A suggested program, kindergarten through grade twelve.* Columbus, Ohio: The Association, 1967.

Bieliauskas, V. J. Recent advances in the psychology of masculinity and femininity. *J. Psychol.,* 1965, **60,** 55–263.

Block, J., von der Lippe, A., & Block, J. H. Sex role and socialization patterns: Some personality concomitants and environmental antecedents. *J. consult. clin. Psychol.,* 1973, **41,** 321–341.

Block, J. H. Conceptions of sex role: Some cross-cultural and longitudinal perspectives. *Amer. Psychologist,* 1973, **28,** 512–526.

Child Study Association of America. *What to tell your children about sex.* New York: Meredith, 1968.

Coopersmith, S. Studies in self-esteem. *Scientific American,* 1968, **218** (2), 96–106.

Duvall, E. M. *Family development* (4th ed.). Philadelphia: Lippincott, 1971.

Fraser, S. E. (Comp.) *Sex, schools & society: International perspectives.* Nashville, Tenn.: George Peabody College for Teachers, 1972.

Grams, A. *Sex education: A guide for teachers and parents* (2d ed.). Danville, Ill.: Interstate, 1970.

Gruenberg, S. M. *The wonderful story of how you were born* (Rev. ed.). New York: Doubleday, 1970. (For reading by children from grade 3 up, paper ed., 1973.)

Kagan, J. Acquisition and significance of sex role identity. In M. L. Hoffman & L. W. Hoffman (Eds.), *Review of child development research* (Vol. 1, pp. 137–167). New York: Russell Sage Foundation, 1964.

Kaplan, B. L. Anxiety: A classroom close-up. *Elem. School J.,* 1970, **71** (2), 70–77.

Lyman, E. B. *Let's tell the whole story about sex.* New York: American Social Health Organization. (Album of four 78-rpm records or one 33⅓-rpm record.)

Lynn, D. B. Divergent feedback and sex-role identification in boys and men. *Merrill-Palmer Quart.,* 1964, **10,** 17–23.

McCandless, B. R. Childhood socialization. In D. A. Goslin (Ed.), *Handbook of socialization theory and research,* pp. 791–819. Chicago: Rand McNally, 1969.

Pikunas, J. *Fundamental child psychology* (2d ed.). New York: Bruce, 1965.

Rosenberg, B. G., & Sutton-Smith, B. *Sex and identity.* New York: Holt, 1972.

Rosenberg, M. Which significant others? In *Amer. behav. Scientist,* 1973, **16,** 829–860.

Sears, R. R. Relation of early socialization experiences to self-concepts and gender role in middle childhood. *Child Developm.,* 1970, **41,** 267–289.

Sutton-Smith, B. *Child psychology.* New York: Appleton-Century-Crofts, 1973.

Part Six

Puberty and Adolescence

Many long-term qualities and features have been observed in infancy and childhood, but not in the form in which they later present themselves in the adult's appearance and personality organization. Puberty is the phase of life when adult body features and personality traits begin to be formed. During the adolescent years these features and traits reach their fullness and often attain a high degree of functional integration.

Major pubertal changes elicit psychosexual maturation leading to major gains in masculinity or femininity. Preoccupation with self intensifies concurrently with these changes, and questions about one's appearance, social status, and identity are often raised. Adolescence is a time for further inquiry, and questions such as "Who am I?" "What do I want to be?" and "What is the real purpose of my life?" frequently emerge in the mind of the person moving toward adult maturity. Adolescent maturation is a highly personal task, and the adolescent must deal with himself or herself to find a place in peer and later adult society and to achieve a self-gratifying identity. Most of the growth-pattern changes of adolescence occur within the years of puberty, but behavior patterning continues for many years before the adult style of life takes hold. Parents and other adults can aid the adolescent by showing patience with and understanding of his behavioral oscillations and extremes in emotional and verbal behavior, as well as by communicating their values and expectations and thus assisting him with the leading task of defining himself. Chapter 13 deals chiefly with pubertal changes; Chapters 14 and 15 will concentrate on the developmental changes marking the adolescent span of life.

Chapter 13

Pubertal Developments

HIGHLIGHTS

Release of appropriate genetic information activates the thalamic-pituitary axis. The resultant growth acceleration furthers adult body proportions as well as sexual maturation in its physiological aspects. Rapid and extensive growth of primary and especially secondary sex characteristics is frequently troublesome and disturbing to pubescents.

Magnified social, emotional, and sexual self-awareness elicits a rather spectacular change in feelings and attitudes, marked by many reverses along with diffused uncertainty and ambiguity. How the adolescent feels about himself or herself and whether or not these changes are welcome depend on social feedback, especially from his peers.

Emotional turmoil and various psychosomatic upsets frequently occur. When intense, personal conflicts and difficulties sometimes result in an escape from the real situation. Compensation by fantasy and sex, alcohol and cigarettes, and marijuana and hard drugs can occur.

Developmental tasks arising at puberty often continue into the adolescent years. This is particularly true of gains in self-regulation of behavior and internal reorganization of goals and aspirations.

Puberty is usually considered a transitional stage between childhood and adolescence proper. Physical gains and sexual maturation prompt the young person to discard puerile forms of behavior and to embrace adolescent traits and features with their specific masculine or feminine characteristics. Puberty is an early phase of adolescence, when the process of intensified growth and maturation sets in. Besides sexual maturation and physical growth, pervasive social, emotional, cognitive, and personality developments also occur within the pubertal years and continue into middle and late adolescence.

CONCEPTS OF ADOLESCENCE

In American civilization the adolescent period of life has been variously looked upon as a time of storm and stress, an age of frustration and suffering, a span of intensified conflict and crises of adjustment, a phase of dreams and reveries, of romance and love, and an era of alienation from adult society and culture. Viewed from another standpoint, adolescence may be characterized as a stage of search for one's self marked by intimate peer affiliation and clique formation, by discovery of high values and ideals, by development of personality and identity formation, and by attainment of adult status with its challenging tasks and responsibilities.

Konopka (1973) sees adolescence as an important segment of continuing human development. She distinguishes early adolescence (the years from twelve to fifteen), middle adolescence (fifteen to eighteen), and late adolescence (nineteen to twenty-two). More exactly, it is "that span of a young person's life between the obvious onset of puberty and the completion of bone growth." This biological definition may be complemented by a psychological reference to "a 'marginal situation' in which new adjustments have to be made, namely, those that distinguish child behavior from adult behavior in a given society" (Muuss, 1968, p. 4).

Salzman (1973) assumes that whether adolescence will be popularly viewed as a sound developmental epoch or a disturbed state depends on "the attitudes of the adults toward blooming independence, burgeoning sexual interests, the tendency to become self-preoccupied and concerned with aesthetic values and moral issues in ways that disturb the existing culture." He concludes that "adolescence is a developmental epoch, fascinating in its varied possibilities of growth for each individual." Furthermore, it is a period that carries "marked loads and heavy burdens and is thereby characterized by considerable distress and mental anguish."

Generally the adolescent is deeply sensitized toward his social environment, and his motivational structure often appears to be a bundle of contradictions. He often vascillates between childish narcissism and youthful altruism, as he does between feverish activity and idleness. For several years his behavior is marked to some extent by instability and lack of coherence. He is at times confused with reference to his roles, tasks, and obligations. Pleasant surprises and shocking disappointments, stresses and exhilarations follow an unpredictable sequence. Some of adolescent life is like a never-ending dream in the dark night, where the occasional powerful flashes of light are blinding rather than illuminating.

Usually an adolescent becomes deeply aware of many life issues and questions relating to himself and others, to community and nation. He eagerly takes opportunities for a good time, for pleasure and excitement, but is also concerned over the expectations of others—especially members of the opposite sex—and his liberty to press for new experiences and adventures. He is often ready to go to any length to win peer acceptance, but he also gropes for the "right" way to behave. It is not easy to achieve a balance between the pursuit of pleasure and the rendering of service to family and humanity. In the years of late adolescence, his mind becomes somewhat attuned to both the challenges and opportunities before him. For many, some of the developmental tasks of this period remain unfinished and are carried over into the years of early adulthood.

The heightened experiences that occur in the course of adolescence vary greatly in intensity and duration but are probably quite similar in various culturally advanced societies. Norman Kiell (1964) gives an impressive amount of biographical data in support of his thesis that "the great internal turmoil and external disorder of adolescence are universal and only moderately affected by cultural determinants" (p. 9).

FACTORS PRECIPITATING PUBERTAL CHANGES

Pubertal, or early adolescent, growth consists primarily in rapid somatic maturation leading to an adult body structure, sexual developments resulting in the active functioning of the gonads, and the appearance of the secondary sex characters that clearly distinguish male and female structures. Heredity is again at work—the genetic code releases new information in the organizing cells, causing initiation of the pubertal growth spurt. The genetic information produces functional changes in the thalamic-pituitary axis. For the total growth cycle to occur, a balanced interaction among the cerebral cortex, hypothalamus, autonomic nervous system, pituitary, thyroid, adrenal cortex, and gonads is required.

On the average, the pubertal spurt occurs between eleven and fourteen for girls and between twelve-and-a-half and nearly sixteen for boys. The pattern of accelerated growth depends chiefly on change in the functioning of the endocrine gland system as modified by nutrition, health, and other environmental influences. A brief analysis of glandular activity will clarify major variations of structural growth.

The pituitary gland, located at the base of the brain, is a master gland that produces hormones for stimulating growth and activity in all the other endocrine glands. The hormonal activity of the pituitary depends largely on stimulation by the hypothalamus and the hormones of certain glands (DHEW, 1973; Tanner, 1964). The anterior lobe of the pituitary gland produces several hormones that directly and indirectly function as growth regulators. Somatotrophin (STH), one of the pituitary's growth-promoting hormones, controls the size of a person and especially the length of the limbs. Hyposecretion of the anterior portion of the pituitary causes a person to remain childlike in size, while hypersecretion makes growth spurt to giant proportions. Normally the increased activity of the pituitary's anterior lobe in terms of somatotrophic hormone production stimulates increased structural growth of the nose and the extremities of the body—arms and hands, legs and feet—which assume practically adult proportions. At the same time, the heart enlarges disproportionately to the slower increase in size of the arteries, resulting in temporarily increased blood pressure. The disproportion in capacity between heart and arteries and the elevated blood pressure cause many young teenagers to experience moments of dizziness and general weakness.

There is a close relation between the secretion of the pituitary's gonadotrophic hormone and the changes in sex glands: increased secretion of this hormone activates the growth and functional maturation of the sex glands. When the gonads—testes in the male and ovaries in the female—reach maturity, they begin to produce hormones of their own, which affect the pituitary's functions by slowing down its growth-producing secretions and which stimulate the development of secondary sex characteristics.

Other pituitary hormones stimulate the thyroid (TSH), the adrenal cortex (ACTH), certain metabolic processes, and to a certain extent the blood pressure. In cases of strain or infection, ACTH is produced rapidly and causes the adrenal cortex to speed up its corticoid production. The pro-inflammatory (P) and anti-inflammatory (A) hormones of the adrenals create an organized defense against any invasion by foreign matter. Too much P hormone, however, may produce more damage than bacilli or a splinter or other source of stress (Selye, 1966, pp. 178–181).

Another factor influencing structural growth is the thyroid gland, located in front of the larynx within the throat. Its secretion, thyroxin, consists chiefly of iodine (65 percent) and greatly influences metabolism—the nutrition and energy exchange within the organism and related cellular activity—especially oxygen consumption. Excessive thyroid functioning (hyperthyroidism) increases the metabolic rate, raises the blood pressure, and makes a person excitable, especially when the condition is accompanied by hypofunctioning of the parathyroids. Insufficient thyroid functioning (hypothyroidism) produces a low metabolic rate; in extreme cases, symptoms of myxedema (physical and cognitive retardation and sallow, wrinkled, thick skin) appear.

The adrenals, attached to the kidneys, increase androgen or estrogen secretion at the onset of major pubertal changes, thereby facilitating the acquisition of mature male or female features. The thymus gland, located in the chest anterior to the mediastinum and behind the sternum, grows smaller and deteriorates at about the same time. Its functions have a delaying effect on sexual maturation.

Ossification and muscle development play a considerable role in adolescent behavior, because physical strength depends largely on the bone and muscle systems. The number of muscle cells in the body correlates closely with age and sex. At ten years of age, a girl will have undergone a fivefold increase in the number of her muscle cells. Little further increase occurs in either size or number of muscles during the pubertal years. In contrast, a boy's muscle cells continue to divide until, at eighteen years of age, he has about fourteen times as many muscle cells as he did at birth, and his muscle cells will continue to enlarge for another five years (DHEW, 1973, p. 20). Despite lesser muscular strength, girls show about the same physical durability as boys. At the age of eight, the weight of their muscles approximates 27 percent of the gross body weight; at sixteen years it reaches about 44 percent. The greatest increase occurs with puberty. Strength usually doubles between the ages of twelve and sixteen. By 1975 the average height of young American women was nearly 66 inches (167 cm) and their weight was about 134 pounds, while the average young man measured 71 inches (180.5 cm) and weighed approximately 162 pounds.

While bone development is consistent with the general rate of physical maturation, the speed of muscular growth is influenced by the amount of physical exercise. Muscle growth that precedes bone development is likely to contribute heavily to psychomotor incoordination and clumsiness. Intricate relationships between various enlarged muscle systems and neural extensions, stimulating their activity, add much to the pubertal boy's difficulties in muscle coordination. Muscles become flabby and weak and shrink in size if they are not used for strenuous activities. Several surveys suggest that the average high school student spends only two hours a week in group play or exercise. Insufficient strength, flexibility, and stamina may be one reason why one-third of the 200,000 pupils tested for the President's Council on Physical Fitness failed the Kraus-Weber muscular fitness tests (La Salle & Geer, 1963, p. 5).

All the various biochemical and physical changes taking place expose the pubertal child to loss of symmetry and grace. This is especially true of boys in social situations. The adolescent's changing body and uncertain movements increase his or her self-consciousness. The emergence of new impulses and desires, especially those related to sexual maturation, bring about motivational and behavioral indecision and even temporary disorientation. Yet adolescents begin to search for the meanings and implications of these changes and make attempts to control and integrate them within the expanded self-concept. It is good for the young teenager to be well informed about the major effects of pubertal changes as they pertain to his or her own sex.

SEXUAL DEVELOPMENT

With the exception of the primary sex characteristics, boys and girls are quite similar in their physical appearance during childhood. The increasing difference in appearance during puberty is due to the growth of skeletal and muscular systems and is especially related to the appearance of secondary sex characteristics. Broadening shoulders, maturation of the testes, distribution and pigmentation of pubic hair, and changes in voice are the significant secondary sex traits of boys. First ejaculation is often recognized by boys as a sign of oncoming manhood. The young male's high-pitched voice drops approximately one octave. Broadening hips, breast formation from budding papillae to mature breasts, the appearance of pubic hair, and menstruation are the signs of a girl's progress in pubertal growth. Menarche is usually a very intense and demanding experience for the girl.

The physical changes of adolescence reach their peak just before sex glands achieve structural maturity and begin to function. The amount of gonadotrophic hormone in the urine, determined by laboratory analysis, furnishes a satisfactory but infrequently used criterion for measuring the level of pubertal development. Another criterion is provided by appraising x-rays of wrist-bone structure, for it has been discovered that gonad maturation occurs simultaneously with a certain level of ossification. On the basis of skeletal age, developmental norms have been formulated for both girls and boys (Bayley, 1946; Frisch, 1974). The entire process of sexual maturation requires about three years, but individual differences are great. Sexual maturity implies the capacity to generate offspring.

The onset of sexual maturation in girls in frequently an abrupt phenomenon requiring immediate emotional and social readadjustments. The process is less sudden in boys; their needs for readjustment come more gradually. The transition from immature to mature physiological sexuality requires approximately one year. This transition constitutes the peak of pubertal changes. Growth in the reproductive system greatly influences the experiences, attitudes, and behavior of the growing person. The effects of sexual growth and maturation on the self-concept of the individual are pervasive and cannot be overemphasized.

Slightly precocious sexual maturation usually has advantages for boys and probably has disadvantages for girls. Since the precocious boy looks grown-up, he often secures more popularity and social prestige among his peers than prepubertal boys. However, very early puberty often makes the individual unsuited for his own age group, while still unaccepted in older groups. The grown-up boy is often admired by his peers for his physical strength and athletic prowess. Early onset of puberty for girls gives them a growth spurt that makes them much taller and larger than boys. Suitable tall boyfriends are difficult to find among peers. Very often this is a decided disadvantage to girls. Many early-maturing girls and late-maturing boys are afflicted with moderate and severe social and emotional problems, particularly insecurity and anxiety. Their problems in identity formation have long-term repercussions (Frisk et al., 1966).

Delayed puberty for a boy means a small body with less strength than his faster-maturing peers. The late-maturing boy is likely to experience pressure to withdraw from the group with which he has associated and to compensate with a younger group—a situation that often causes the boy to entertain feelings of inadequacy or an attitude of failure. Considering teenage problems as a whole, it may be concluded that boys who mature late are beset with more problems and troubles than those who mature early.

When pubertal changes appear comparatively early, structural growth is more gradual and regular than in most cases of late growth. If it starts late, growth is often turbulent and less integrated. Rapid acceleration in growth often brings with it restlessness, fatigue, and disturbances in emotional experiences.

CHANGES IN EMOTIONS AND ATTITUDES

Puberty is a stage of heightened emotionality. The pubertal child begins to experience heightened feelings and undertakes to revise his or her own attitudes. The increase in emotional differentiation is apparent in the various moods experienced. The pubescent becomes increasingly sensitive and reacts strongly to events and social situations. While the advanced adolescent is able to control the emotions to a considerable extent, the young adolescent is swept along by the vivid currents of feelings and sentiments. When aroused, emotions are frequently out of proportion to the initial stimulus. Attempts to control emotional expression are frequent but not always successful. The emotionally charged actions resemble the heightened negativism and temper seen during the late second and third years of life at the beginning of early childhood.

Ambivalent feelings are the rule rather than the exception at puberty. The adolescent often experiences contradictory feelings—that is, love and hate, concern and apathy—toward persons and events. Trivial disappointments often arouse his antagonism without destroying his original feeling of cordiality or enthusiasm. The capacity for keeping affective experiences in harmony seems to be lost for some time.

The effort to progress in social, emotional, and sexual maturation is probably the most difficult task of the pubertal period. Transference of affection and love from parents to peers, including members of the opposite sex, represents a major change in emotional cathexis. Ambivalence and discouragement are common while this key shift of feelings is occurring. New adjustments at any age are accompanied by emotional tensions and heightened affective reactions. This transference is barely comparable to any earlier emotional changes. Development of new attitudes and integration of divergent peer values are parts of a painful process. Lack of preparation for the adolescent role, parental insensitivity and objections, and rising financial necessities all contribute to a state of emotional uneasiness. This leads to occasional tensions near the boiling point, the energies of which, when not sufficiently discharged physically, accumulate and interrupt the functioning of weaker or more sensitive organismic systems. Headaches, stomach aches, and heightened blood pressure are examples of these disturbances. As a result, organismic balance and personal health are disturbed, especially if such states arise often. New defense tactics are needed to restore and preserve some kind of balance. Dropping carefully developed plans and reverting to daydreaming, and aggressive and hostile reactions are among the defensive reactions often used by adolescents to protect themselves from the threats of the adult or peer world. The young adolescent experiences many obstacles in finding his or her place in peer society and culture.

Emotional instability is a counterpart of the physiological and social changes occurring during puberty. Strong likes and dislikes expressed in most teenage groups press the pubertal child to change his attitudes toward parents, teachers, and other groups he comes in contact with. In group situations more than anywhere else, the young teenager feels that many of his childhood beliefs and attitudes are not tenable and feels an urgency to construct a new set. The young teenager encounters many peers who have very different attitudes toward religious practice, academic achievement, sexuality, and drugs (Kagan, 1972). Ambivalence grows, as he cannot easily discredit the position taken by others. Since the integrity of the value and belief system is central to the conceptualization of self, it is necessary to keep it intact by defining the meaning of his existence and his role in adult and peer society and by selecting personal goals worthy of much effort. The resetting of his motivational hierarchy is often a result of peer influence that competes with and often overrules what parents have worked hard to establish. Kagan (1972) attributes great significance to the cognitive quality of motivation and treats motives as cognitive representations of goals. Nonetheless, emotional and sexual factors rank high at this age. Kagan suggests that mastery and resolution of uncertainty and hostility are primary motives; if sensory pleasure is added, this comes close to the goals and strivings of the young adolescent.

HEALTH AND ENERGY

Late childhood is usually a healthy age, marked by good adjustment to home, school, and peers, but the situation changes rather drastically after major pubertal changes erupt. Often the young adolescent feels ill and suffers headaches, bodily discomforts, and stomach pains. At times he or she has little energy for work or even for play, feels tired if not exhausted, or is annoyed by other minor disturbances of a psychosomatic nature. The adolescent may experience a generally run-down condition or be bothered frequently by influenza, sore throat, or tonsilitis; but sometimes "feeling ill" is used as an escape from disagreeable duties and responsibilities. There are no severe illnesses specific to this stage, and, excluding accidents and narcotics, very few adolescents die from illnesses or their complications.

There are some physiological explanations for the young adolescent's lack of energy and frequent colds and aches. First of all, the slower growth of blood vessels in relation to the heart raises the blood pressure and creates both strain on the heart and feelings of tension and tiredness. At this phase of accelerated growth, a young person should not be pressed into robust activities or excessive competition in athletics. Usually, though, the pubertal child feels inclined to expend a great amount of energy, even beyond his or her capacity. By overexpending energy in sports and late hours and endless unnecessary activity, often with an irregular intake of food, the adolescent occasionally develops an enlarged heart or respiratory disturbances.

At puberty the oil-producing glands increase their productivity, resulting in skin eruptions and acne. Rapid growth, emotional turmoil, and unpredictable eating and activity patterns also contribute to skin disturbances. Difficulties in social relationships and the resulting conflicts and frustrations are additional contributing causes. But following the completion of major pubertal changes, physical health usually reaches its highest level in late adolescence and early adulthood.

The smoking of tobacco is seen as a symbol of sophistication by many teenagers. Peer pressure is often a factor in starting to smoke cigarettes. Since cigarette smoking is fairly common among teenagers today, its detrimental effects on growth and health should be considered. After each cigarette the heart beats faster for about thirty to sixty minutes. This superfluous activity lowers the cardiac reserve. The use of tobacco mildly constricts blood vessels, slowing down oxygenation and producing quicker loss of breath in strenuous activities. Another frequent result of smoking is loss of appetite because of slower than normal stomach movement, and eating less may prevent the full nutrition required for growth. Medical research studies indicate that non-smokers have the best records for health and length of life; moderate smokers have poorer records and heavy smokers the worst records. Athletes who smoke experience greater difficulty in strenuous competition (LaSalle & Geer, 1963, pp. 212–213; Vermes, 1973). A recent study of 1,007 high school students in central Indiana (Husni-Palacios & Scheuer, 1972) reports that about half the sample did not smoke at all; 16.8 percent smoked either very heavily or one pack a day; and 17.3 percent smoked only a few cigarettes per day. About 13 percent had stopped or wanted to stop smoking. About 23 percent of this sample had experimented with barbiturates, marijuana, or a combination of drugs. The use of drugs will be further explored in Chapter 15.

DEVELOPMENTAL TASKS

The developmental tasks of puberty are not particular to this one period, since most of them extend into later adolescence. Some pubertal developments, however, including body control, peer identification, social sensitivity, reorganization of personality and self, rise of external interests and activities, and progress in controlling impulses—especially sexual urges and emotional moods—are prerequisites for later adolescent adjustment and achievement. At any age, gains in growth must be accompanied by increases of control.

1 *Bodily control.* Pubertal growth indirectly contributes to awkwardness, poor posture, and physical discomfort. Regaining control over the body becomes a continual task for the teenager. Frequently effort and exercise are necessary to achieve sufficient strength and the control required for mature gracefulness or athletic prowess. Poor back posture from sedentary activities can be corrected by swimming, running, and similar exercise.

2 *Peer identification.* At puberty the adolescent seems to be confronted with two sets of motives: the prepubertal egoistical ones, which aim to satisfy pleasure-seeking drives, and the social ones by which he strives for affiliation, approval, and interpersonal intimacy. The desire to make a good impression on others, especially on peers of both sexes, becomes intense. The pubescent gradually learns to adjust his actions to fit the peer-group pattern and strives for deep identification with its goals and activities. Achieving success in this developmental task is a major step in identifying socially and emotionally with persons and groups other than parents and family. This ability is crucial for social adjustment during later years. Identification with more mature models is a further step in this direction.

3 *Social sensitivity.* Acquiring knowledge of others' needs and expectations helps the teenager improve his sensitivity to the wants, likes, and preferences of his peers. He is deeply concerned about the impression he makes on them. Through his manner of speech, dress, and specific interests, the pubertal child attempts to gain popularity and become a member of a prestigious group. He alters his thoughts and opinions to fit group structure and dynamics as well as social conventions in order to be in good standing with his peers.

4 *Self-reorganization.* A general breakup of the childhood trait and attitude structure occurs in the early phase of adolescence. Many new attitudes and interests, the changing relation to parents and peers, awakened sexuality and a grown-up body, and new moods and fantasies lead to a reorganization of the total personality. The pubescent makes frequent attempts to prove to himself that he is not a child any longer, and he objects to being treated like a child when parents or teachers do not recognize his newly developing self-concept.

5 *External interests and activity.* The pubertal child is usually preoccupied with self and frequently engages in brooding fantasy and prolonged daydreaming. Withdrawal into oneself at puberty is a threat to learning and to advance in socialization. At this stage the pubescent can learn much by activity and participation and by experimenting with his own endowments in order to utilize and assess them for his present and future needs. It is a time for expanding interests in social life and peer culture.

6 *Growth of self-regulation.* Self-control must be exerted for school, athletic, art, or any other achievements and also to curb sexual impulses and base emotions, especially those which disturb interpersonal relationships, such as anger, temper outbursts, and moods. Forces from without could be relied on for control in childhood, but the teenager makes efforts to free himself of such dominance and to establish his own internal authority on which he can rely, thus increasing his self-direction and asserting the individuality of his own personality. Growth in self-regulation is also needed for developing skills in interpersonal communication and for gains in popularity and leadership.

QUESTIONS FOR REVIEW

1 Explain the concept of puberty and relate it to the concept of adolescence.
2 Describe the role the thalamus-pituitary axis plays in precipitating accelerated growth during puberty.

3 Compare bone and muscle growth during puberty and discuss the contribution of each to physical fitness.
4 Describe sexual maturation at puberty and the role of the gonads in furthering it.
5 Give some reasons for early sexual maturation. Explain the social effects of early sexual maturation for boys and girls.
6 What are some typical new problems and conflicts arising and intensifying during pubertal changes?
7 What criteria are often used in estimating the level of sexual development during the pubertal years?
8 What emotional changes take place during puberty? How do emotional currents affect the attitudes of the individual?
9 Describe some health disturbances that frequently occur during the pubertal years. Give reasons for their occurrence.
10 List several developmental tasks of puberty and explain one of the major tasks.

REFERENCES

Selected readings

Cole, L., with Hall, I. N. *Psychology of adolescence* (7th ed.). New York: Holt, 1970. A general text on adolescent experience, motivation, and behavior, normal and deviant, including educational applications.

Conger, J. J. *Adolescence and youth: Psychological development in a changing world.* New York: Harper & Row, 1973.

Douvan, E., & Adelson, J. *The adolescent experience.* New York: Wiley, 1966. A national survey, based on interview data on about 3,000 adolescents, which discusses social attitudes, masculine and feminine crises, culture, values, and personality integration.

Jersild, A. T., & Alpern, G. D. *The psychology of adolescence* (3d ed.). New York: Macmillan, 1974. A general text on adolescence, emphasizing emotional development, social relationships, and the self.

McCandless, B. R. *Adolescents: Behavior and development.* Hinsdale, Ill.: Dryden, 1970. A popular text on various aspects of adolescent development and adjustment.

Specific References

Bayley, N. Tables for predicting adult height from skeletal age and present height. *J. Pediatrics,* 1946, **28,** 49–64.

DHEW, National Institutes of Health. *How children grow* (DHEW Pub. No. 73–166). Washington, D.C.: Government Printing Office, 1973.

Frisch, R. E. A method of prediction of age of menarche from height and weight at ages 9 through 13 years. *Pediatrics,* 1974, **53,** 384–390.

Frisk, M., Tenhunen, T., Widholm, O., & Hortling, H. Psychological problems in adolescents showing advanced or delayed physical maturation. *Adolescence,* 1966, **1,** 126–140.

Husni-Palacios, M., & Scheur, P. The high school student: A personality profile. *Proc. 80th Ann. Convention APA,* 1972, 565–566.

Kagan, J. Motives and development. *J. Person. soc. Psychol.,* 1972, **22,** 51–66.

Kiell, N. *The universal experience of adolescence.* New York: International Universities Press, 1964.

Konopka, G. Requirements for healthy development of adolescent youth. *Adolescence,* 1973, **8,** 291–316.

LaSalle, D., & Geer, G. *Health instruction for today's schools.* Englewood Cliffs, N.J.: Prentice-Hall, 1963.

Muuss, R. E. *Theories of adolescence* (2d ed.). New York: Random House, 1968.

Salzman, L. Adolescence: Epoch or disease? *Adolescence,* 1973, **8,** 247–256.

Selye, H. The stress of life. In M. L. Haimowitz & Natalie R. Haimowitz (Eds.), *Human development: Selected readings* (2d ed.), pp. 170–195. New York: Crowell, 1966.

Tanner, J. M. *Growth at adolescence.* Oxford: Blackwell, 1964.

Vermes, H. C. *The boy's book of physical fitness* (Rev. ed.). New York: Association Press, 1973.

Chapter 14

Dynamics of Adolescent Behavior

HIGHLIGHTS

Needs, desires, and activities change dramatically with sexual maturation and the intensification of heterosexual interests and concerns. Boys become fascinated by the sexual aspects of girls' appearance and behavior. Girls seek ways to be attractive to boys.

The need for physical, sexual, and social adequacy is a dynamic one, which presses the teenager to seek genuine acceptance by peers of both sexes and their groups. The girl strives to win and hold love; the boy seeks her company for companionship and sex as well as love.

Adolescent cognitive maturation presses the mind to deal with the ideal and creates strong desires to right the wrongs of the world. Many adolescents seek quasi-experimental proof of the validity of their assumptions.

Developmental tasks of adolescence include acceptance of the growing body as a symbol of the changing sense of self, a search for human models to identify with, the strengthening of self-control by assimilation of a mature moral system, and adult identity formation.

Self-projection beyond the confines of reality heightens the adolescent fantasies of uniqueness and heroism. Questions related to self-identity often elicit existential anxiety before acceptable solutions are found.

> *Personal and social concerns gain much in depth and persistence. The desire for self-regulation and self-improvement becomes a powerful motivation, as does the desire to help others who appear to need assistance. Narcissism and altruism alternate in dominance in a matter of days, if not hours.*

Adolescent motivation emerges from pubertal developments and experiences, which, in turn, are influenced by racial and socioeconomic background, resources, and childhood experiences. Physiological, sexual, emotional, cognitive, social, and value developments lead to changing behavioral tendencies. Environmental vicissitudes have a major modifying effect. Each adolescent's needs, desires, interests, attitudes, and problems spring from both inner dynamics and outer contingencies.

In the early part of adolescence, even a strong enthusiasm or a heightened concern for something often wanes quickly and is replaced by other interests, sometimes antagonistic, sometimes related, but always highly absorbing for some time. In the later phase of adolescence, some attractions, interests, and preferred activities become firmly established through practice and gain in depth and stability. These are usually closely related to personal endowments and gifts, status and environmental opportunities. Less oscillation and more consistency is shown as the years of adolescence pass.

ADOLESCENT NEEDS AND DESIRES

Needs, interests, and desires increase in complexity during adolescence. Gratification of somatogenic needs—e.g., oxygen, nutrients, and fluids—is necessary for the maintenance of organismic functioning. Besides these primary requisites for physical survival, there are locomotive and sensory drives that affect behavior and evoke new attitudes. Adolescents are eager to approach, explore, and learn about new objects and subjects. They experience stronger and more numerous aversions and dislikes than in childhood.

What many teenagers desire and appreciate most is the development of gender-appropriate behavior and sexual maturation. Sexual maturity implies the attainment of what Freud called "coition," which becomes the dominant sexual proclivity, usually remaining incomplete through adolescence. It is not to be confused with the earlier heterosexual interest or activity (Blos, 1973, p. 114). Knight (1969, p. 173) points out that our society cheats youth "by letting the sexual flower be picked when it is only a bud, thus never reaching the splendor of its full bloom." To most teenagers, sexuality is a major focus that captures their attention more and more often as adolescent years pass. At this age girls' legs are a source of fascination—not merely curiosity—for many boys. Most boys become very attentive to the sexual aspects of girls and women. Girls' sexual interests are more diffuse. Kuhlen and Houlihan's study (1965) suggests that the intensity of the heterosexual interests of puberty and adolescence is considerably greater for the present generation than for earlier ones.

The psychological dimension of human existence generates the need for affection, security, independence, and moral integrity. Sociogenic needs include group acceptance, identification, participation, and recognition. Cultural enrichment, intellectual understanding, and moral commitment also belong to the total structure of human needs. Evidence of a hierarchy of needs may be observed in the later part of adolescence, as a definite order of exigency begins to appear. Any unfilled need usually gains in motivating power; lack of love, for example, heightens the need for affection. When not successful in meeting needs, the adolescent compensates by overeating, hoarding, reveries, or other forms of problem behavior.

As the adolescent develops, his abilities and skills enable him to gratify fundamental and derived needs. The social structure and culture in which each adolescent finds himself or herself provide some means for satisying these needs; yet a poorly endowed or a handicapped adolescent at times finds it almost impossible, in spite of all efforts, to achieve gratification of needs or to develop successful strategies to compensate for lack of fulfillment.

Certain emotional and social needs affect the adolescent greatly and have far-reaching effects on behavior and personality. A brief analysis of some of these needs will shed light on adolescent dynamics and the accompanying drives, interests, and desires.

The need for *novel experiences* is a major force driving the adolescent toward activity and self-improvement. The adolescent is eager to join various groups, to plan and make trips; he or she is interested in adventures and new activities. Everyday experiences often appear monotonous, and the desire to escape into something sensational grows with the advancing adolescent years. Novel experiences add much to the adolescent's zest for living.

The need for *security* is another strong motivational force. During adolescence, security is greatly determined by the young person's estimate of himself—of his power and worth, his social status, and his moral integrity. Security depends on an attitude of self-confidence and self-control, which in turn result from the satisfaction of emotional and social needs. Because of earlier failures, many adolescents lack self-confidence. Consequently, their expectations are clouded with anticipation of danger and threat. Because of their ambivalence, it is often difficult for them to make decisions. They experience opposing desires and waver between independence and the need for support, selfishness and altruism, conformity and a desire for individuality. Such experiences create feelings of inadequacy and, in turn, of insecurity. Adolescent security is furthered by affectionate acceptance, concern, and tolerant consideration by parents as well as enthusiastic acceptance into peer groups. The experience of intimate friendship is also an important achievement in furthering security (Blatz, 1966, p. 42).

The need for *status* extends to family, peers, school, and community. The adolescent has a strong desire to be appreciated by parents as he is, to be accepted with his idiosyncracies, and to be dealt with on an equal and friendly basis. The very natural wish to share experiences with peers is fundamental, and the adolescent tries everything possible to be attractive to and on friendly terms with members of both sexes. Moreover, he wants adult rights and privileges. Desire for status influences him in school and other institutions. Membership insignia and clothing are used as external signs of his particular status.

The need for physical and social *adequacy* is a pressing one, and the young adolescent seeks to be fully accepted by his reference groups. If he is not, defense mechanisms become intensified and make practical adjustment difficult. With lack of social acceptance, tensions and conflicts are bound to arise; these, in turn, elicit strong anxiety and feelings of inferiority or loss of self-esteem, and sudden or gradual withdrawal follows. The loner broods over the aggressiveness and injustice of others. In this way, tendencies toward hostility and destruction begin, and violent assaults against others can result.

Related to the need for physical adequacy is the need for self-identity (Erikson, 1959). In the search for identity and standards, many adolescents show contempt for the values of their parents and adult society. Sons and daughters test parental values in many crude ways. They often count each parental cigarette and each drink or tranquilizer, hours "wasted" watching television and "white lies" told. If parents infringe on a law, their sons and daughters feel justified in going much further (Gaier, 1969). Sooner or later, teenagers engage in shoplifting, use of pep pills and other drugs, sexual experimentation, drinking, and truancy.

Although the teenage years are characterized mainly by transition and change, some basic attitude configurations apparently survive throughout childhood and adolescence. In her reexamination of the Berkeley Guidance Study data, Wanda C. Bronson (1966) postulates fairly salient and enduring *central orientations* (patterns of attitudes), originating from genetic material, physical-physiological events, and early interpersonal experiences. The central orientation determines the limits of the hierarchy of responses favored by the organism despite shifting developmental pressures. A core group of forty-five boys and forty girls representative of the Berkeley population from age five to age sixteen was repeatedly rated for a variety of behavior traits. The greatest degree of persistence and centrality was found along three dimensions: withdrawal-expressiveness, reactivity-placidity, and passivity-dominance. Reserved-withdrawn versus expressive-outgoing and reactive-explosive versus placid-controlled are styles of behavior related to the first and second dimensions.

There are distinct areas of concern for adolescent boys and girls, even when the concepts used do not differ. Douvan and Adelson (1966, pp. 343–350) feel that "the key terms in adolescent development for the boy in our culture are *erotic, autonomy* (assertiveness, independence, achievement), and *identity*. For the girl the comparable terms are the *erotic,* the *interpersonal,* and *identity*." The girl's concerns with identity and the erotic differ greatly from the same concerns of the boy. The development of interpersonal ties—the sensitivities and values of intimate associations with other persons—forms the core of the girl's identity, while the masculine identity "focuses about the capacity to handle and master nonsocial reality, to design and win for himself an independent area of work which fits one's individual talents and taste and permits achievement of at least some central personal goals." For the girl the erotic consists mainly of winning or maintaining love, but the boy is exposed to "turbulent instinctual struggles." Most adolescents, however, experience a strong desire for one another's company. Companionship often creates strong emotional currents and sexual stirrings leading to love and sexual intimacy. These topics are presented in the next chapter.

ADOLESCENT FANTASIES

During puberty the final phase of perceptive and imaginative development occurs. The adolescent enjoys many experiences vicariously—that is, he can go to various known and strange places, anticipate events, and engage in activities without being physically involved. When the adolescent suffers frustration or disappointment, he tends to compensate by recourse to fantasy, where he can cope more successfully with threatening situations and events. In daydreaming and reveries, he transcends the limits of his own powers and of time and space and enters upon experiences otherwise not attainable. Fantasy is often a means of escaping from mounting tension, anxiety, and frustration.

Persons of importance to the teenager form the "intrapsychic family" he thinks and daydreams about, the core of the adolescent's imaginary audience that observes him wherever he goes, so that his self-consciousness does not subside much even when he is alone. Moreover, his imagination is stirred by his personal fable—his belief in the uniqueness of his own experiences (Elkind, 1967). One major theme of pubertal reverie is the "suffering hero." Here the adolescent pictures himself as undergoing various trials and persecutions and eventually being vindicated. The "conquering hero" represents another frequent theme of imagination that the adolescent uses to compensate for his defects. Boys daydream about adult roles, travel, vocational success, possessions, friends, sexual encounters, homage and grandeur, and a variety of adventures. Girls daydream about their attractiveness, romances, singing, dancing, and attentions received from a handsome boy. They entertain fantasies about their friends, travel, and adult roles. Fantasy adds much to the development of "feminine intuition," as girls spend much time analyzing how they feel about others and how other persons feel in various situations and encounters (Lidz, 1968, p. 334).

An early empirical study by Symonds (1949, pp. 218–255), based on responses to pictures by forty subjects from twelve to eighteen years of age, reveals that themes of aggression (for example, violence and frenzied excitement) and themes of love (for example, dating episodes, driving together, and courting) are typical for adolescents. Other frequent themes of adolescent imagination include anxiety, guilt, depression, success, independence, happiness, conflict over good and evil, oedipal longings and conflicts, and dread of sickness and injury. At night, fantasies are manifested in dreaming. REM (rapid eye movement) cycles are closely associated with vivid dreaming; but when reports are obtained throughout the night, they indicate "considerable thought content occurring at all stages of sleep" (Singer, 1975).

The function of daydreaming is obviously to furnish an escape from the demands of reality, a retreat from self-restraint and criticism. This gives free rein to the operation of underlying needs, drives, attitudes, and emotions. The creation of "good times" and joyful events, though imaginary, serves as a useful outlet as well as an escape from the threatening confines of the present situation. Tasks and duties inherent in reality cannot be avoided by flights into the world of fantasy, of course, and as the years of adolescence progress, the vicarious experiences of daydreaming become more closely related to the adolescent's reality situation and fuse with his plans and aspirations. Vividly trying to imagine and anticipate coming events helps the young person to avoid a number of mistakes and errors. Excessive daydreaming, however, prevents him from engaging actively in the constructive learning activities necessary for the unfolding of abilities and skills.

COGNITIVE MATURATION

During puberty, in addition to perceptual and imaginative development, the adolescent attains his or her adult intellectual level. There is a great increase in ability to comprehend relationships, discern facts from assumptions, abstract what is essential, and use less concrete terms and symbols. The ability to learn—to absorb information and ideas—and to solve problems is near its highest point, as shown by the fact that adolescents attain almost the same scores on intelligence tests as adults.

The young adolescent begins to feel confident in his or her mental ability and enjoys intellectual activity. A vivid preoccupation with thinking, hypothesizing, and experimenting at thirteen or fourteen years of age leads to the acquisition of a theoretical and very critical attitude. Curiosity about existential and sexual problems is accompanied by a desire to seek out satisfactory solutions to problems through books, pamphlets, and encyclopedias. The teenager wants to formulate his or her own answers rather than rely on parents' or teachers' judgments as in the years of childhood.

The increased ability to abstract and to generalize is seen in the type of self-expression the teenager chooses. He now relies heavily on a conceptual rather than a concrete type of analysis. Less tangible relationships and roles are recognized. Some branches of science and philosophy begin to interest him. He can direct his attention to the scientific objectives of astronomy, cosmology, ethics, aesthetics, logic, and metaphysics.

At fourteen to fifteen years of age, many adolescents fully acquire *formal* thinking (i.e., they can deal with verbal propositions and can consider most of the possible combinations in each case) and propositional operations (i.e., they can combine various empirical associations on which multiplicative classes are based in many possible ways). Now they handle most of the formal operations, including disjunction, incompatibility, and various forms of implication and exclusion, successfully and are able to set up experimental conditions for verifying simple assumptions. Formal operations enable the teenager to construct a large number of possibilities in a system, including contrary-to-fact propositions. He can reason objectively about his thoughts. At this age the adolescent can successfully apply general principles to most specific situations (Elkind, 1967, 1974, pp. 99–104; Piaget, 1973, pp. 59–61).

DEVELOPMENTAL TASKS

The powerful desire to grow and to mature in order to secure acceptance into peer and adult society and culture is a mark of late adolescence. With time the adolescent acquires a clearer awareness of what is expected of him. Beginning with Erikson, many psychologists see identity formation as a principal developmental task of adolescence. Unless adolescents satisfactorily settle such questions as "Who am I?" and "Why am I?" they remain "at war" with themselves. If active questioning and turmoil about ideology and occupation continue, as well as uncertainty about their role in peer or adult society, they need a "moratorium"—additional years to find acceptable solutions—before they can embrace an adult style of life (McCandless & Evans, 1973, pp. 407–411). William Kay (1968, p. 251) speaks about the one supreme task of adolescence, the acquisition of a mature moral system to guide conduct. Adolescent maturation is incomplete unless one arrives at a universally applicable moral code. Robert J. Havighurst (1948/1952, pp. 33–71) was probably the first to enumerate distinct developmental tasks for adolescents. Achieving new and more mature relations with age-mates of both sexes, developing the intellectual skills and concepts necessary for civic competence, preparing for marriage and family life, and acquiring a set of values and an ethical system as a guide to behavior are some of the tasks of the adolescent.

Several developmental tasks pertain mainly to the middle and late phases of adolescence:

1 Accepting one's adult physique and its various qualities as something final and self-related.

2 Attaining emotional independence from parents and authority figures.

3 Developing skill in interpersonal communication and learning to get along with peers and other people individually and in groups.

4 Finding human models with whom to identify.

5 Accepting oneself and relying on one's own abilities and resources.

6 Strengthening self-control on the basis of a scale of values and principles, or *Weltanschauung.*

7 Outgrowing infantile, puerile, and pubescent modes of reaction and adjustment.

A person's ability to find sources and means of gratification for his or her needs and to master age-related tasks is the key sign of general adequacy. A great deal of exploration and learning are necessary to move ahead toward self-realization. The period of adolescence is a time of continual movement from the puerile level of functioning to that of adult maturity. Luella Cole (1964, pp. 4–9) has classified adolescent goals in nine categories: general emotional maturity, establishment of heterosexual interests, general social maturity, emancipation from home control, intellectual maturity, selection of an occupation, suitable use of leisure, a philosophy of life, and identification of self. Table 14-1 shows the fundamental goals of adolescent maturation.

LEADING INTERESTS AND CONCERNS

A major characteristic of most needs, abilities, and other driving tendencies is their activity-stimulating power, which evokes a variety of experiences for the adolescent. Only selected motivational tendencies and interests as they pertain to adolescence will be treated briefly here in order to understand their role in the life of the teenager.

Appearance

The adolescent's concern over physical appearance, including physical features such as face, chest, arms and hands, feet and legs, and bodily measurements, as well as voice and hair, clothes and ornaments, and use of cosmetics, stands out as a major dynamic factor. Experience teaches the adolescent that personal appearance plays a major role in social acceptability or lack of it, especially with members of the opposite sex. To be accepted or to gain popularity, the adolescent must make his appearance conform to the patterns and expectations of adolescent society, so his attention is often focused on himself. He examines critically his size and proportions, hairstyle, and other aspects of appearance in comparison with those of others of his age. Even a minute deviation becomes a matter of concern and worry. This concern is often so pronounced that teenagers, especially girls, are willing to undergo considerable inconvenience and discomfort in order to correct their defects. Some withdraw from group activities to avoid being exposed to unfavorable remarks and intensified feelings of inferiority or rejection. Since the postpubertal physical appearance is almost adult and resists change, the teenager has to learn to accept and adjust his feelings and attitudes toward it. Most aspects of physical appearance must be accepted as they are.

Changes of fashion and styles tend to produce additional worries as the adolescent learns about them from peers and from advertising, television, and magazines. To increase their charm is almost an obsession with many adolescent girls. Tallness, overweight, or a poorly developed figure are the more frequent feminine problems. Being shorter than average, too weak, or too heavy are characteristics feared by the male adolescent. Wanting to appear handsome is one of his preoccupations when masculine qualities begin to impress him.

Table 14-1 Developmental Goals of the Adolescent Period

From	Toward
Social and emotional maturation	
Intolerance and striving for superiority	Tolerance and feelings of adequacy
Social awkwardness	Social poise and gracefulness
Slavish imitation of peers	Interdependence and self-esteem
Parental control	Self-control
Feelings of uncertainty about oneself and others	Feelings of self-acceptance and sociability
Anger, temper tantrums, and hostility	Constructive and creative expressions of emotions; refinement of moods and sentiments
Growth in heterosexuality	
Acute awareness of sexual changes	Genuine acceptance of male or female sex identity
Identification with members of the same sex	Interest in and association with peers of opposite sex
Relationships with many possible mates	Selection of a mate for life
Cognitive maturation	
Desire for universal principles and final answers	Need for explanation of facts and theories
Acceptance of truth on the basis of authority	Demand for substantial evidence before acceptance
Many interests and concerns	Few, stable, and genuine concerns
Subjective interpretation of situations	Objective interpretation of situations and reality
***Weltanschauung*—philosophy of life**	
Behavior motivated by pleasure and the like	Behavior based upon realistic aspiration and conscience
Indifference toward ideologies and ethical principles	Interest and ego-involvement in humanistic ideologies and ethics
Behavior dependent on reinforcement	Behavior guided by moral responsibility and ideals

Source: Formulation draws extensively on the goals of the adolescent period presented by L. Cole with I. N. Hall in *Psychology of Adolescence* (6th ed.). New York: Holt, 1964, pp. 4–9.

Worries and problems about the complexion are frequent. Rapid growth, emotional turmoil, and conflicts may be accompanied by acne and other skin eruptions. Medication is often used to reduce skin disturbances. While it is difficult for a boy to cover up pigmentation and skin disturbances, girls can use creams, rouge, and powder to advantage. Adolescent girls spend a great deal of time before the mirror trying to achieve an attractive complexion. They apply mascara to make their eyes appear deeper and more colorful. Much interest is also paid to the general hygiene of the body, including dental care, control of perspiration, and manicure. "What makes a girl charming?" is a frequent question on the girl's mind.

The changes that occur at puberty affect the vocal quality of both the male and the female voice as the speech organs undergo final modification. When the voice pitch lowers rapidly, the voice cracks. Difficulties with voice are common among boys, whose voices drop nearly an octave. In girls the change is more moderate and is not likely to cause embarrassment. Realizing that attractiveness depends not only on appearance but also on the voice, girls in particular strive to acquire pleasant voices.

Apparel, jewelry, beads, and other ornaments are all important to appearance. The teenager has no difficulty in realizing the role clothes play among his peers and in society at large. In order to attract attention, some adolescents try to be very selective in their choice of clothes. As style-consciousness heightens, styles and fashions of the season are quickly adopted. A desire for novelty and surprise is an important factor in the adolescent's choice of clothing and ornaments. By the time he has taken on the characteristics of an adult and has discarded those of the teenager he has usually acquired skill and good taste in the selection of attractively styled clothing. By the end of adolescence most feel secure enough to abandon meticulous conformity in dress and hairstyle.

Self-regulation

Even children find great satisfaction in being competent and able to take care of themselves. The desire for self-direction usually becomes stronger as puberty nears completion and is one of the most outstanding characteristics of the adolescent. It manifests itself as a continually rising pressure to break away from existing family bonds and dependence on adults. The need to speak one's own mind and assert one's personality becomes strong after the completion of pubertal changes. Various forms of self-assertion, including aggressiveness in defense of one's own status, become prominent and frequently lead to friction between young adolescents and their parents. Differences of view in regard to selection of activities and companions, education, and vocation are frequent factors in parent-adolescent disputes. Many young adolescents begin to feel estranged from their parents for this reason. Feelings of being misunderstood and a desire to leave home arise if parents rely solely on their own experience or authority in maintaining restrictions and inflexible views. Many boys and girls find it difficult to achieve their vital right to self-regulation and feel pressed to rebel.

Parents and educators who tend to overemphasize their own advice, preferences, and controls have difficulty in realizing the importance of self-direction and self-reliance in the development of the teenager. The adolescent's desire for a spending allowance to use as he chooses and for privacy regarding telephone calls and mail, as well as his wish to have a room of his own, are disregarded by some parents. The adolescent feels hurt and distrusted when he is questioned about where he has been or what he has done. This overprotective practice of parents may be well-intended, but the adolescent must be allowed to make some errors as he strives to act on his own. Adolescence is the proper time to gain independence from parental influences and controls.

Vocation

Vocation is another major area of adolescent concern. In this phase of life, the adolescent understands the general need for a vocation. He is often aware of the vocation he would like to pursue, but the typical adolescent is not mature enough to make a serious choice. Douvan and Adelson's empirical findings (1966, p. 342) show that "for the most part the jobs boys choose represent modest advances over their father's position, and they are jobs with which the boys have had some personal contact." Over half the total sample of boys (N=1,045) named a professional or semiprofessional job they hoped to fill, while only one in fourteen thought his future was in semiskilled or unskilled work. A significant minority of the boys in this sample did not have any clear vocational plans.

Practically all studies of adolescent occupational preferences show a high percentage selecting professional vocations. This points to a lack of adequate self-appraisal, because many of those selecting professions will not qualify for them. Interest in glamorous and prestige occupations must be replaced by interest in practical occupations related to ability and economic resources. Achieving satisfactory occupational status contributes to economic security and personal independence. It is usually accompanied by improved adjustment, mature adaptation to adult expectations, and ability to marry and set up a home. Vocation and marriage are the two last milestones initiating adult life patterns. Marriage is no longer the primary vocation for a majority of girls. As a result, the career needs of girls decline only temporarily as marriage plans come into view.

Creative Self-expression

Communication with oneself, or self-reflection, precedes and often supplements advanced forms of communication with others. As self-centered baby talk precedes the acquisition of a common language, so imaginary and literary notions precede advanced forms of adolescent interpersonal communication. Literary and other forms of creative self-expression are fairly common among adolescents who have a strong desire to convey their feelings and ideas about personally significant events and to keep a record of them. Desire for self-expression in art is probably equally common, since many adolescents experience impulses toward pictorial or plastic creativity as well.

Letters, poetry, short novels, diaries, and autobiographical incidents in the form of short stories based on real life represent typical modes of expression at this time. The need to confide in an intimate friend is dynamic, but many teenagers do not as yet have such close friends or do not trust them enough to share all their personal hopes and concerns. Formulating their inner experiences and problems and putting them on paper can be a substitute for interpersonal communication. The diary often becomes the first silent confidant, to whom many secret desires, conflicts, problems, and ambitions can be told without reproach or embarrassment. It also serves as a repository for those experiences, feelings, and thoughts that adolescents consider important and want to treasure.

The writings of adolescents are primarily signs of emotional growth and cognitive maturation rather than indications of literary talent. From a psychological viewpoint, adolescent writings are equivalent to autistic forms of conversation. The tension of self-expressive drives finds relief from the feelings of isolation experienced in this phase of intensified psychosocial development. A higher degree of introversion during adolescence also seems to be indicative of vivid fantasy and creative productivity during the midphase of adolescence. Writing as a means of self-expression is used more often by girls than by boys. In addition, keeping diaries and similar forms of literary preoccupation are typically the activity of adolescents with superior intelligence. Keeping a diary occasionally becomes a habitual activity, and in some cases it is not abandoned with the attainment of personal maturity but continues into adulthood.

Watching movies and television plays of a dramatic or romantic nature and reading novels and stories from "real life" have similar ventilating effects on adolescent emotional tension and partially satisfy the need for novel experiences. Except for the creatively gifted, the need for creative and literary self-expression subsides during the early adult years.

Recreation

Play is the child's form of recreation and entertainment. During the course of pubertal development, child play is largely discarded as the need for teenage amusement and relaxation increases. Frequenting rock festivals, concerts, and motion pictures, watching television, reading selected magazines, parts of newspapers, and books, and listening to records are typical forms of adolescent recreation. Summer and winter sports, athletic and creative activities, trips to the beach and other youth gatherings, and dates and dancing parties are additional teenage diversions. Membership in various clubs not only satisfies the adolescent's social needs but provides recreational opportunities.

Interpersonal Communication

As the years of adolescence continue, interpersonal communication expands in varying degrees, depending on individual personality. Sharing personal experiences with peers is a fundamental requisite for adolescent adjustment. Many teenagers appear irritable and in low spirits when they are separated from their companions. When the adolescent is alone for a long while or far away from his close friends, a feeling of loneliness rises and conflicts and problems increase. Long telephone calls and letters usually satisfy the need for personal communication and are often adequate substitutes for face-to-face meetings.

Generally adolescents choose activities that offer opportunity for conversation. The desire to communicate personal views is often so strong in the adolescent that he is incapable of following any systematic procedure for talking in turn. Even in places where conversation is disturbing, such as movies or high school classes, whispering is frequent. Much leisure time is spent with friends lounging around corner drugstores—another outlet for conversational self-expression and social interaction. In late adolescence small-group conversations are usually quite free and frank.

Adolescent topics of conversation center on (1) boy-girl relationships, (2) travel and recreational activities, (3) athletic events and individual performances, (4) movie and television stars, (5) sex and morals, (6) parents and teachers and, (7) money, cars, clothes, fashions, and status symbols. Individual opinions generally receive a hearing by the group. Disagreements are infrequent and usually of little consequence.

Discussions with peers serve to improve the teenager's communication skills as well as to generate new interests and attitudes, broaden viewpoints, and amplify general knowledge, thus greatly enriching personality resources. Discussions prepare the ground for peer identification and emotional intimacy. Much of this communication promotes maturity and the development of social skills and graces.

Dating

Dating is the chief means for establishing close relationships with members of the opposite sex. Since urges toward heterosexual friendship exist, dating is used by the adolescent as a means of testing popularity with selected members of the other sex. Group dating usually precedes individualized dating. As soon as emotional acceptance by the partner occurs, dating tends to become steady. Going steady offers security, because the need for a companion is satisfied. Sharing of activities and experiences augments pleasure and fun. Table 14-2 enumerates the traits teenagers look for in a steady date.

Frequent dating leads to situations in which sexual exploration and intimacy are experimented with. Frequently the boy tries to get as much as possible and the girl yields as little as possible. The girl is expected to set the limit, and the boy is expected to conform. The adolescent girl often desires to strengthen her relationship to the boy, and she sometimes permits petting and intercourse in the hope of strengthening her ties to him. This can be a self-deceiving attitude, since she is often the object of sex exploitation. Restraint usually brings high dividends. Permissiveness may reinforce the boy's irresponsibility and "fun morality" and may degrade the girl in the eyes of the boy, who often turns elsewhere in his search for a more idealistic relationship. Most boys value girls who are adamant, although the ethics of young people in college seem to be undergoing many changes. Group outings

Table 14-2 Traits Desired in Companion and in Future Mate

Traits	Percentage	
	Boys	Girls
Companion		
Is dependable, can be trusted	92	96
Takes pride in personal appearance and manners	89	90
Is considerate of me and others	84	95
Shows affection	84	85
Acts his (her) age, is not childish	77	87
Has a pleasant disposition	77	84
Is clean in speech and action	72	82
Mixes well in social situations	51	59
Does not use liquor	49	49
Is popular with others	46	42
Future mate		
Desires normal family life with children	77	87
Knows how to budget and manage money	62	75
Is approved by my parents	56	74
Is independent of his or her parents	45	64
Has interests similar to mine	49	48
Has ideals similar to mine	46	52
Has a job	9	93
Knows how to cook and keep house	79	15
Is as intelligent as I am	27	42
Is started on a professional career	8	42

Source: Purdue Opinion Panel, Youth's attitudes toward courtship and marriage. *Report of Poll No. 62,* 1961. (Based on 2,000 high school boys and girls.)

and double-dating are often best for most of the adolescent years. Otherwise dating moves too fast toward steady dating, and the adolescent misses opportunities for varied experiences with a larger number of peers. Steady dating is difficult to stop, and its tendency to isolate the adolescent from other peers cannot be easily overlooked. Playing the field is an integral part of adolescent experience.

In steady dating the partners are exposed to many opportunities for early sexual intimacy. The girl should be aware of the sexual significance of bodily contact with a boy, especially in privacy. While a girl usually does not experience erotic sensations from sitting close to her date, placing arms around each other's waist, or many forms of necking, boys often become sexually aroused and interpret such activities as invitation to greater intimacy and further sexual advances. When the same person is dated frequently or exclusively, intimacies that would seem inappropriate on a first or second date appear quite natural, since they are the result of gradual advances over a period of time. Occasionally sexual aggressiveness becomes uncontrollable as the boy's impulsiveness is heightened. By permitting hugging and kissing, girls invite petting, and by permitting petting, they are drawn toward intercourse (Staton, 1963, pp. 325, 374). Chapter 15 discusses various aspects of adolescent sexuality.

Helping Others

The desire to help others intensifies throughout the adolescent years. It seems likely that the adolescent's sensitivity to the needs of others is related to his own problems and difficulties. His friends' problems are often treated as his own, but even a stranger in need or distress concerns the teenager. Most adolescents are capable of identifying themselves with a person or group in distress, becoming ego-involved and often throwing themselves enthusiastically into any action suggested. Altruism and charity permeate a major part of their activities and often outweigh the tendency to discriminate. Denial of fundamental human rights and oppression of ethnic or religious groups call forth an urge to assist the victims and a drive to reform the existing evils.

In applying their powers and resources to help others, adolescents are ready with advice and service and occasionally are prepared to sacrifice themselves. Many social services attract their attention. Since teenagers are quite naive in some respects, persuasive speakers can arouse their energies for radical causes as well. Without the total dedication of their youth, the success of Russian Communism, German Nazism, Italian Fascism, and the Chinese "cultural revolution" could not be explained. Youth fell prey to advanced propaganda techniques. Moreover, the extremist, with his ideology of upheaval and transformation, seems to have a specific appeal to adolescents. The desire for an ideal society can be appealed to and readily directed toward changing the environment: home and school, community and nation, and the world at large.

In studying adolescent motivation, one should bear in mind its dual source, which includes frequent and powerful promptings from within and the situational stimuli of the adolescent's social milieu, both of which mold and direct his or her energies into adequate or inadequate patterns of activity and experience. Conflicts and ambivalences are usually strong for the majority of teenagers. The following chapter on personality reorganization will elaborate on these concepts.

PREPARING FOR LIFE

Furnishing adequate educational facilities and providing occupational training opportunities for the nation's adolescent population is a challenging task for most communities—urban and suburban, large and small. The early 1960s saw a phenomenal rise in high school and college enrollments: 640,000 and 175,000 students each year, respectively. The late 1960s and early 1970s were marked by a leveling off of high school students, because the child population began to drop. Figures 14-1 and 14-2 and Table 14-3 show statistics for schools and related population statistics. The tendency for young people not only to complete high school but to continue their education in college augurs well for the future of science and technology, arts and the humanities.

Project TALENT (Flanagan et al., 1962; Flanagan & Jung, 1971;

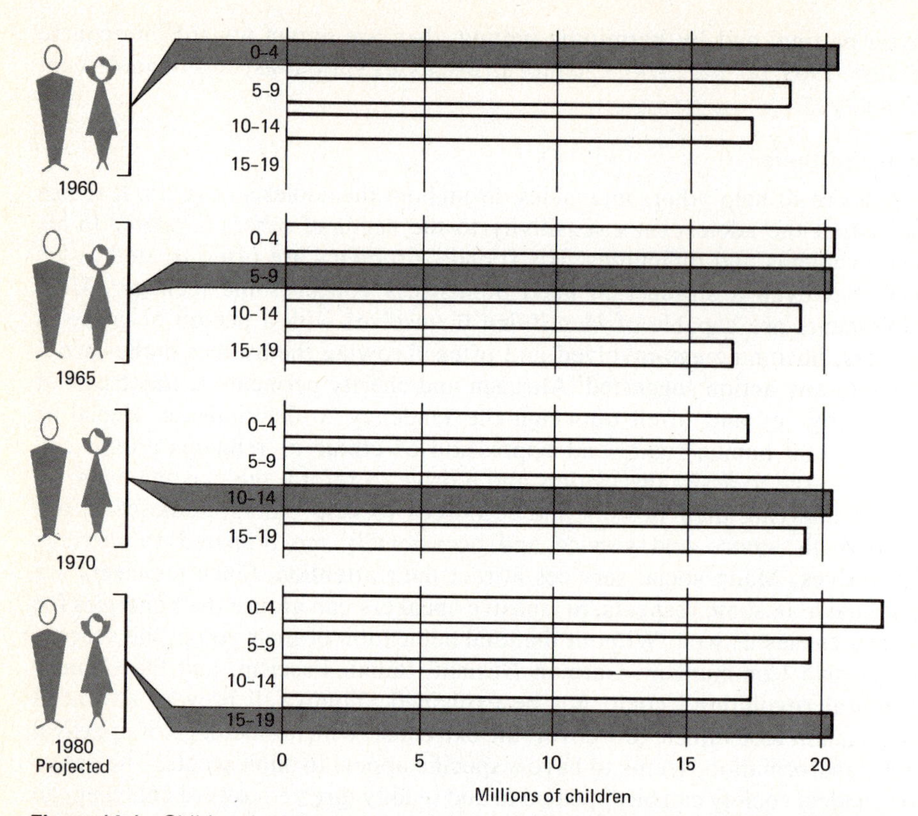

Figure 14-1 Child and adolescent population, by ages: United States, 1960–1980. [*Source: Children and youth: Their health and welfare* (Children's Bureau Publication No. 363), 1958; *Current population reports* (Series P-25, No. 321), 1966; U.S. Bureau of the Census, *Statistical abstract of the United States: 1973.*]

Table 14-3 Population of the United States by Age: 1960, 1970, and 1980

	1960		1970		1980†	
Age (years)	***N****	**Percent**	***N****	**Percent**	***N****	**Percent**
Total	180,684		205,395		227,510	
Under 1	4,112	2.3	3,595	1.8	4,241	1.9
1–5	20,162	11.2	18,013	8.8	19,881	8.7
6–13	28,773	16.0	33,353	16.2	28,614	12.6
14 and over	126,980	70.5	150,434	73.2	174,774	76.8

**N* in thousands.
†Series D projections, U.S. Department of Commerce, Bureau of the Census.
Source: Profiles of Children, 1970, p. 85.

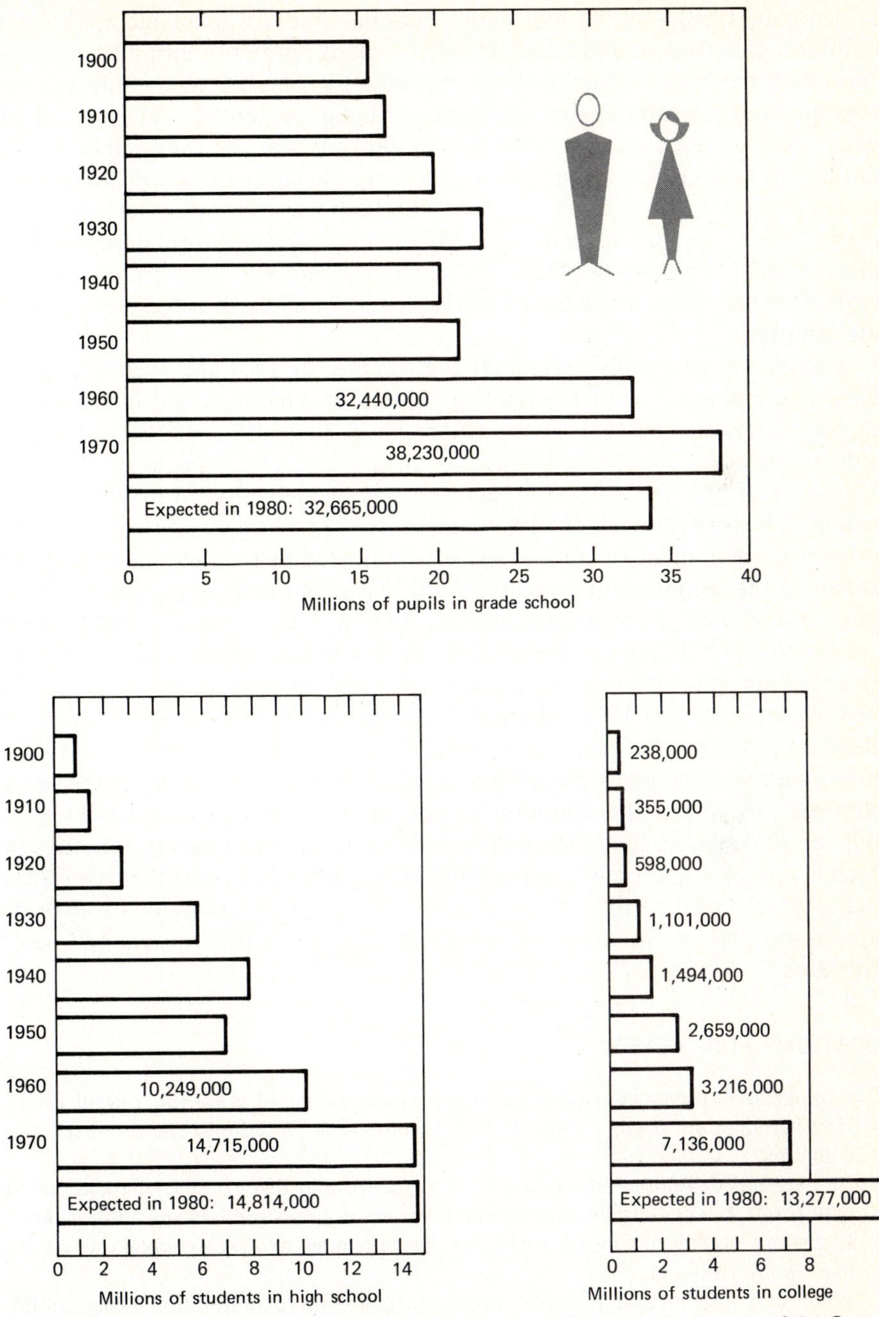

Figure 14-2 Schools in the United States are growing. [*Source:* U.S. Bureau of the Census, *Statistical abstract of the United States: 1973.*]

Schoenfeldt, 1968) was the first major national study of available ability and creativity. This study explored the relations among aptitudes, interests, motivations, and productivity, as well as the effectiveness of many educational programs and procedures for realizing individual potentials. The testing of nearly 500,000 high school students (5 percent of the total high school enrollment) was completed in 1960. Follow-up questionnaires will trace what has happened to the high school seniors of 1960—whether they have gone to college, taken jobs, or changed their earlier plans. Further information and data will continue to be collected from TALENT students one, five, ten, and twenty years after they have graduated from high school and will be compared with new samples.

Comparing nationally representative samples of 1960 and 1970, Flanagan (1973) claims that much of the reading content of what is taught in schools is grossly irrelevant. The 1970 survey (Flanagan & Jung, 1971) of eleventh-grade students showed a small improvement over the survey of ten years earlier, but 33 percent of the students still reported having difficulty with their school reading. (The percentage in the 1960 sample was 38.) The general inadequacy of the curriculum in meeting student needs is indicated by the fact that in 1960, 44 percent of the students surveyed reported that about half of the time or more often "I feel that I am taking courses that will not help me much in an occupation after I leave school." For 1970, the comparable figure was 45 percent. Flanagan (1973) feels that the principal defects of the educational programs of 1960 and 1970 alike can be traced to the traditional function of education as a means of training the elite. He speaks of being shocked "to find classes with homogeneous grouping in which all three levels are using the same textbook." He urges the adoption of educational goals to fit the individual needs of students, so that each will be able to apply his talents in activities in which his performance can be among the best. He suggests that the role of the teacher be changed to that of "an experienced guide, a continuous source of inspiration, and a valued companion in the student's search for self-realization."

QUESTIONS FOR REVIEW

1 Explain several psychogenic needs and indicate means of gratifying one of them.
2 How do human needs relate to the development of adolescent interests and concerns?
3 List several developmental tasks of adolescence and discuss one of them. What may result if necessary developmental tasks are not mastered during adolescence?
4 List some leading themes of adolescent daydreaming, and try to explain why such themes are frequent.
5 What new intellectual preoccupations does cognitive maturation bring to the adolescent?
6 What makes it difficult for the adolescent to attain self-regulation of behavior?
7 Discuss dating and the effects of early steady dating.
8 What are some of the frequent topics of adolescent conversation?

9 What role do interests and activities play in the adjustment of an adolescent?
10 Identify several forms of recreation and indicate the adolescent's interest in one or two of them.
11 Discuss the major national study entitled TALENT and indicate some of its findings.

REFERENCES

Selected Readings

Cole, L., with Hall, I. N. *Psychology of adolescence* (7th ed.). New York: Holt, 1970. A well-revised edition dealing with physical, intellectual, emotional, social, and moral developments during adolescence, including educational applications.

Hurlock, E. B. *Adolescent development* (4th ed.). New York: McGraw-Hill, 1973. A study of various aspects of growth and an analysis of factors in personality development. Includes chapters on socialization, social status, and teenage non-conformists.

Rosenberg, M. *Society and the adolescent self-image.* Princeton, N.J.: Princeton, 1965. Self-esteem, as an independent variable, is related to various neighborhoods. Its presence in the upper class, its lower level among minority groups, and the lesser degree of self-esteem among girls than among boys are discussed. Five thousand subjects from ten New York high schools were studied.

Specific References

Blatz, W. E. *Human security: Some reflections.* Toronto: University of Toronto Press, 1966.

Blos, P. The epigenesis of the adult neurosis. *Psychoanal. Study Child,* 1973, **27,** 106–135.

Bronson, W. C. Central orientations: A study of behavior organization from childhood to adolescence. *Child Developm.,* 1966, **37,** 125–155.

Cole L. *Psychology of Adolescence* (6th ed.). New York: Holt, 1964.

Douvan, E., & Adelson, J. *The adolescent experience.* New York: Wiley, 1966.

Elkind, D. Egocentrism in adolescence. *Child Developm.,* 1967, **38,** 1025–1034.

Elkind, D. *Children and adolescents: Interpretive essays on Jean Piaget.* New York: Oxford University Press, 1974.

Erikson. E. H. Identity and the life cycle. *Psychol. Issues,* Monogr. No. 1, 1959.

Flanagan, J. C., et al. *The talents of American youth. Design for a study of American youth.* Boston: Houghton Mifflin, 1962.

Flanagan, J. C. Education: How and for what. *Amer. Psychologist,* 1973, **28,** 551-556.

Flanagan, J. C., & Jung, S. M *Progress in education: A sample survey (1960–1970).* Palo Alto: American Institutes for Research, 1971.

Gaier, E. L. Adolescence: The current imbroglio. *Adolescence,* 1969, **4,** 89–109.

Havighurst, R. J. *Developmental tasks and education.* New York: Longmans, 1952. (Originally published, 1948.)

Kay, W. *Moral development.* London: Allen & Unwin, 1968.

Knight, J. E. *Conscience and guilt.* New York: Appleton-Century-Crofts, 1969.

Kuhlen, R. G., & Houlihan, N. B. Adolescent heterosexual interests in 1942 and 1963. *Child Developm.,* 1965, **36,** 1049–1052.

Lidz, T. *The person: His development throughout the life cycle.* New York: Basic Books, 1968.

McCandless, B. R., & Evans, E. D. *Children and youth: Psychosocial development.* Hinsdale, Ill.: Dryden, 1973.

Piaget, J. *The child and reality: Problems of genetic psychology* (Arnold Rosin, Trans.). New York: Grossman, 1973.

Schoenfeldt, L. F. *A national data research for behavioral, social, and educational research.* Palo Alto: American Institutes of Research, 1968.

Singer, J. L. Navigating the stream of consciousness: Research in daydreaming and related inner experiences. *Amer. Psychologist,* 1975, **30,** 727–738.

Staton, T. F. *Dynamics of adolescent adjustment.* New York: Macmillan, 1963.

Symonds, P. M. *Adolescent fantasy.* New York: Columbia, 1949.

Chapter 15

Personality Reorganization and Patterns of Adjustment

HIGHLIGHTS

Numerous developments occurring during puberty further major modifications of various personality components during the adolescent years—indeed, until an adult style of life takes hold.

The magnifying of awareness elicits critical self-evaluation and rejection of puerile behavior and "kid stuff." Unless the young person is supported by rationally advanced information, he or she may reject childhood beliefs and may, with peer support, adopt an extremistic ideology.

Poor organization of defenses, conflicts with parents and peers, sex and aggression, —all induce strain, guilt, and anxiety in various degrees for most adolescents. In more severe cases neurotic symptoms or misdemeanors appear.

Even before the advent of adolescence, many young people explore alcohol, tobacco, and stimulating drugs. Emotional turmoil and peer pressures impel many teenagers to sex, alcohol, nomadism, and dangerous drugs. Adjustment problems are compounded with each wayward step.

Personality traits mirror the physiological, sexual, emotional, social. cognitive, and value developments of the individual. By *personality* is meant a dynamic system of traits, attitudes, and habits producing a varying degree of consistency in the total response repertoire of an individual. Adolescence is a highly significant phase of personality development and integration. A fully mature personality is possible only when all the major areas of growth have had an opportunity to develop toward their near-maximal level. It is of paramount importance that a person complete each successive level of development in order for a sound, integrated personality to be attained. The personality patterns established during childhood are modified greatly during adolescence, when new developments and experiences complicate the adolescent's life.

Briefly, some of the new factors and experiences that appear in adolescence and alter the personality are (1) acquisition of adult physique, (2) sexual maturation, accompanied by new drives and emotions, (3) greater self-awareness, resulting in a heightened desire for self-direction and a reevaluation of standards, goals, and ideals, (4) the need for companionship, with prime emphasis on heterosexual friendships, (5) treatment by parents and peers, and (6) conflicts arising from the adolescent's transition from childhood to adulthood.

No two adolescents have equal endowments or potentialities, either physical or cognitive, nor are acquired abilities and skills the same for all teenagers. Aspirations formulated during the early phase of adolescence are usually high. Many young persons strive toward goals and achievement levels that are unrealistic in view of their actual endowments and acquired abilities. The goals and ambitions of adolescents are comparable to neurotic and prepsychotic strivings. If their aspirations are not lowered to realistic levels, they feel that they are failing themselves and experience deep disappointment and even despair, often accompanied by feelings of inadequacy tinged with depression or defeatism (Konopka, 1973a; Turiel, 1974). Lack of adequate behavioral differentiation can be the result of having a low biological and cognitive endowment or remaining too long in any of the childhood or adolescent stages of development because of frustration or merely insufficient stimulation for growth and adjustment. Any regression or fixation restrains flexibility and learning efficiency, thus lowering the accumulation of personality resources. Some minor difficulties in adjustment during childhood, when not remedied, produce more serious problems in adolescence. Early traumatic events, privations, frustrations, and severe conflicts make lasting imprints, which later magnify anxiety and lead to withdrawal or hostility and destructive tendencies.

OUTGROWING CHILDISH MOTIVATION

During the pubertal period the young person removes himself or herself psychologically from family and neighborhood. The adolescent begins to value his or her personal aspirations more fully and is eager to realize some of them. The young person is challenged to strive toward responsible behavior, unlike a child, who runs away from many duties. Pleasure often conflicts with the notion of duty or obligation.

The increasing desire to show off skill or prowess leads the adolescent to participate in individual and group contests. Competitive sports, tasks requiring physical strength, and striving for popularity or for scholastic or other honors are some of the more frequently sought fields of adolescent self-assertion. Any type of achievement, if acknowledged by others, increases one's feelings of self-importance and adequacy. Any failure is deeply discouraging and motivates withdrawal from the activity. Occasionally the resulting conflict leads to depression and self-blame or to anger toward and aggression against others associated with failure. The opposite also occurs; the young person's level of aspiration rises toward unrealistic goals as he or she tries to compensate for earlier failures. Perfectionist strivings are not infrequent during puberty and adolescence. These are often marked by repeated attempts at self-improvement or attainment of high athletic or scholastic standing and an increase in popularity within peer groups of the community and school.

Emancipation from home ties consists mainly in gaining emotional and social autonomy from parents. The earlier feelings of tenderness and affection toward parents during childhood are now readily directed toward individuals of one's own age. The young adolescent becomes emotionally and socially distant from his parents, making many attempts to gain and hold the affection, confidence, and esteem of his age group and to seek security in peer identification. The deep-seated adolescent need for exchange of personal experiences, thoughts, and desires is best satisfied by his peers in both dyadic and group situations. Parental claims to intimacy and dependence meet with resistance and resentment, even open defiance, from the adolescent. Parental controls and restrictions are often seen as barriers to outside associations and group activities. Unlike a child, the adolescent will not readily compromise. By testing the parents' limits, many teenagers move beyond their own. When their parents rescind controls prematurely, "the adolescent is left without the support and protection he needs—albeit sometimes from his own desires and impulses" (Lidz, 1968, p. 330).

IMPRINT OF VALUES

Pubertal gains in emotional and cognitive sensitivity press the adolescent to question his or her early beliefs and practices. A time comes when many childhood concepts, attitudes, and practices of a moral and religious nature are minutely examined in the light of abstract and propositional hypothesizing. In most cases, doubts about some moral or religious tenets and practices arise. The critical attitude forces rejection of what has been feared and respected before. This is especially true for those whose previous religious or moral instruction lacked developmental planning. Since most young persons are not exposed to any comprehensive moral or religious guidance commensurate with their rising level of feeling and understanding, crises occur and often lead to denial of faith. For the rational justification of belief and emotional acceptance of a religious practice, a higher level of religious instruction is needed during early adolescent years. Acceptance of any particular faith rather than rejection of it aids the adolescent in the process of humanization during this critical phase of reorganization of values for life.

Many adolescents are perplexed and ambivalent when asked about their belief in God or an afterlife. If they answer "yes," they are probably repressing a "no"—and vice versa. Their answers change from day to day, if not from hour to hour, until a life philosophy for early adulthood is achieved. Modern religious counselors can often help settle the doubts the young person has about relations to other people, to church, and to God. This personal communication helps to reduce the existential despair that often emerges before the adolescent years are over.

Parents have a part to play in value formation during this period. By explaining the beliefs they hold as adults, they help the young adolescent to realize the role religion plays in adult life, thus giving his earlier beliefs new meaning. Church and school also ought to contribute to the moral and spiritual development of young people. Religious books and magazines may be helpful in gaining a deeper understanding of adult aspects of religious faith and morality. Without some progress in moral-religious self-evaluation, the young adolescent will fail to define himself and to formulate a philosophy of life with ultimate meaning (Rode & Smith, 1968). Konopka (1973a) formulates a developmental theory of values in which the adolescent period is seen as the most significant phase of value formation. Value formation is an emotional and intellectual process highly influenced by human interaction. It is recognized that in modern society every person has to deal with conflicting value systems. Highly influenced by Kant's ethic stressing "the moral law within you," Konopka accepts two absolute values: (1) the importance of the dignity of each person and (2) the responsibility of human beings for one another. Konopka feels that young people in the United States value highly (1) participation in decision making, (2) honesty—admission of one's own feelings to oneself as well as to others, (3) acceptance of a variety of values, and (4) spiritual values marked by rejection of worldly goods and competition. High value is placed on feelings and concerns over living up to one's value system.

As has been pointed out, a primary source of information concerning meaning and values is the social environment of the adolescent. Values and meanings are taken in from significant persons such as parents, teachers, and peer-group leaders. Peers begin to rank high as an adolescent strives to free himself from excessive parental and adult influences. Occasionally, life-determining decisions result from intimate talks with close friends. In the later years of adolescence, societal and cultural norms and expectations gain substantially in their conditioning power. Peers and other social agents praise and reward certain forms of behavior and censure and punish others. In line with the former, the adolescent tends to develop positive approach values, and with the latter, negative avoidance values (Seidman, 1963).

Gordon W. Allport (1961) maintains that our national values, some of the finest mankind has ever formulated, are deteriorating badly. Unless Judeo-Christian ethics are revitalized, "our youth may not have the personal fortitude and moral implements that the future will require." Many adolescents' problems can be traced to the fact that our society has no set hierarchy of values, and the young are often confused about the importance of moral and cultural values in their lives.

Moral awakening in adolescence is often accompanied by the rise of sensitivity to moral issues of poverty, injustice, discrimination, and other perverted practices of society. A degree of scrupulosity is experienced by many teenagers, and impulses of asceticism are not foreign to them. Social service becomes impelling, though often for only a short time. In many cases there is little encouragement and even less reinforcement for engaging in tasks of this kind. Most young persons grope their own way as a concern for meaningfulness rises. Despite their efforts, many fail to reach humanistic goals. Deficiency in the growth of higher values is a prevalent phenomenon. Without the values that outstanding educators and philosophers have formulated, genuine progress in humanization is crippled (Konopka, 1973a).

Kohlberg and Kramer (1969) have described five stages in the development of morality. During stage one, the child differentiates right actions from wrong actions by their consequences. An act is understood to be wrong if punished and right or acceptable if praised; deference to authority is emphasized. During stage two, the child sees an act as right if it satisfies a need for him. Those persons in his life who give him most are seen as "good." During stage three, behavior which pleases others becomes most important. Other people come to be valued for their good intentions and their kindness. Stage four is characterized by strict adherence to rules. Correct behavior is equated with maintenance of the social order and respect for authority. Acts which do not meet these criteria are seen as wrong, regardless of their other characteristics. For example, stealing would be seen as wrong irrespective of circumstances (e.g., extreme poverty or starvation). During stage five, right and wrong actions are defined in terms of their effect on the rights of the individual. Rational consideration is given to the social good, but it remains somewhat selfish in that it is not universalized (pp. 100–102).

For adolescents, moral disequilibrium frequently produces conflict and inconsistencies or even sweeping denials of morality. Many college students reject Kohlberg's fourth stage of morality but cannot move on completely to the fifth. And they are still unable to reach the highest level of moral functioning—an orientation to universal ethical principles in which human rights are reciprocal and equal, and right actions are defined as those taken in accordance with consistent, universally applicable principles freely chosen by the individual.

During the 1960s especially, many young people transcended the materialistic aspects of life and pursued altruistic goals. For lack of practicality their efforts were ineffective in changing the adult society, but their ideals brightened future prospects. A different hierarchy of values was envisioned by young people. This may be partly responsible for the rebellion that adolescents found themselves experiencing in the late 1960s and early 1970s. The principle of equal rights sometimes conflicts with the quest for personal distinction and hedonistic self-gratification. The development of a full human nature implies a scheme of philosophy oriented to the higher values (Konopka, 1973a; Morgan, 1972).

Alienation is another recent problem among college students on large campuses. A sense of personal identity and moral values is lost in the mass of students and the dearth of personal interaction with influential adults—parents and professors alike. Many adolescents become defiant when they feel they are reduced to a set of symbols. They cannot become mere labels or IBM scores. They resist when pressed into particular programs or plans (Morgan, 1972). Frequently deterioration of beliefs and morals is accompanied by depersonalization and loss of standards. The pendulum may swing toward immorality before it swings back to the opposite side. Uncurbed hedonism, the experience of rootlessness, and loneliness—all lead to mass alienation. "The alienated person is not only out of touch with other persons but also out of touch with himself" (Gersten, 1965). There is a general consensus that excessive cultural homogeneity and uniformity prepare the soil for developmental estrangement as one becomes less and less able to apply oneself creatively and feels pressed toward criticism and repudiation of society and culture. Alienation is now chosen by youth in large numbers as its basic stance toward society (Keniston, 1965, p. 3). Apparently there is a deep estrangement from both the *ordo socialis* and the *ordo divinis*. For the majority, this estrangement is temporary, lasting about three to five years; for others it continues throughout early adulthood.

Although the teaching of humanistic values ideally should permeate the entire educational process and the school's resources should be amply used to teach moral values and altruistic orientations, efforts in this direction would probably meet with serious obstacles. Despite the rise of the ecumenical spirit, the major American faiths probably differ too much to agree on a common course for teaching moral and religious living; a basic course suitable to all faiths would be much too vague to be of any significance to the individual student. But to omit from the classroom all reference to religion and the institutions of religion is to neglect an important segment of American life and culture. Knowledge of religion is essential for a full understanding of our literature, art, law, and history, as well as many current issues.

The adolescent increasingly expects and demands the privileges of an adult. Important among these are the use of the family car until he purchases his own car, the selection of his own clothes, resistance to any early curfew, and permission for extended visits of his friends. Although it is true that the adolescent's demands may not be in keeping with the level of maturity or responsibility attained, it is also true that some parents try to maintain their authority and exercise extensive control over their children's lives beyond the proper age. The adolescent's rights and privileges must go well beyond those of a child (Conger, 1971; Montgomery, 1973).

HETEROSEXUAL RELATIONSHIPS

A marked heterosexual interest appears at the age of thirteen or fourteen in girls and at about fifteen or sixteen in boys. This interest and curiosity is closely related to sexual maturation, which occurs at the peak of changes. Girls wish to attract boys' attention, and vice versa. A desire to make social contacts and find companionship also appears at this age and continues to increase as the years of adolescence pass. From the age of fifteen or sixteen, many girls want to be romantically desirable. Boys are eager to deserve admiration and entertainment from girls. After the major pubertal changes are completed, adolescent pleasure motivation centers heavily on members of the opposite sex. Since most schools are coeducational, young people make a variety of contacts with members of the opposite sex during their school years and learn much about the characteristics and interests of the opposite sex.

For Erikson (1963, p. 262), adolescent love is basically "an attempt to arrive at a definition of one's identity by projecting one's diffused image on another and by seeing it thus reflected and gradually clarified. This is why so much of young love is conversation."

When one or two romances fail, the girl tends to become cautious and more selective before "falling in love." In late adolescence, the heterosexual interest usually centers on one particular person. The oncoming adult sex drive and erotic sentiment fuse into a single and powerful force of heterosexual motivation, leading toward an increased desire for associating fully and intimately with a chosen one, and the desire for sexual consummation and marriage intensifies.

During the middle stages of group-conscious adolescence, dating is frequent, leading to observation and a deeper insight into the overt personality traits of the opposite sex. If the impressions are positive, they stimulate affection and love, referred to as infatuation, or "puppy love." Teenagers in this period are not as yet emotionally and socially mature enough for a long-term intimate association. Through their expectation of finding an ideal companion, they are readily "hurt." Jealousy and envy lead to arguments and arouse bewilderment and other antagonistic emotions. Stress ensues. The breaking up of each intimate cross-sexual association causes feelings of inadequacy or depression and sometimes a total reappraisal of the self-concept. Many adolescents experience five or six romantic attachments before they marry. Teenage marriages resulting from infatuation show the highest divorce rates, because most adolescents are not yet mature enough to choose properly and to assume full marriage responsibilities. Insufficient financial resources compound the problems of young couples, since most parents are against early marriages and reduce their emotional and financial support.

Psychologically, the sex hormones are nonspecific and the sex drive is undifferentiated, at least before sexual habits are formed. When heterosexual stirrings begin, the primary urge is for togetherness and a pleasant sensation of bodily contact. Social and emotional warmth is desired. Most postpubertal girls desire affiliation, affection, and love—not sex. Purely sexual urges appear earlier in boys than in girls. At the lowest socioeconomic level, however, some girls are seduced by sexually sophisticated men and permit coitus at eleven or twelve, have first babies at fifteen or sixteen, and may become afflicted with venereal disease. Morally and emotionally (a complete repression of both moral and emotional factors is nearly impossible), premarital sex is disastrous for many girls, since what is natural for girls is a desire to be romantically desirable. The recent tendency to have sexual experience earlier than in the past carries a risk of acquiring an unhealthy sexual life (Godenne, 1974).

After fifteen or sixteen years of age, many girls, once sexually aroused, are practically as urgently driven to sex activity as boys. After a high degree of sexual excitement, the urge to consummation rises rapidly and climax becomes inevitable for most boys. Androgen is probably the erotic arousal factor in both boys and girls. This hormone, then, helps to explain the higher sexual excitability of boys as compared with girls.

Today sexual information of all kinds is more readily available than ever before. Ranging from "school sex-education programs to hardcore pornography, the erotic atmosphere is a given condition of existence for contemporary adolescents growing into adulthood at a time when Western man is fighting to purge himself of his long-held guilt over his own sexuality" (Maddock, 1973). Live sex acts—peep shows and X-rated movies—are exhibited in many places. Youth is threatened by practically unbridled obscenity. Many bookstores and drugstores provide pictorial illustrations of sexual activities, including many perversions. Debasing as it is, many older children and young adolescents are driven to such places by sheer curiosity. By disregarding their better selves and reasonable moral standards, they venturesomely get into some aspects of the prurient exploitation of sex. Is it any wonder that sex-related problems and perversions are more ubiquitious than ever before? Apparently, young people need some safeguards against excessive sexual stimulation.

At sixteen or seventeen many boys are eager to test their sexual potency. They do not require any affection toward a girl to make sexual relations with her attractive, especially after R- or X-rated movies or a couple of beers. "Sexual expression without some form of commitment is both exploitative and meaningless" (Maddock, 1973).

Sex activity has a powerful self-reinforcing factor that enhances sex in all its variety of forms. A major national survey of thirteen- to nineteen-year-olds (Sorensen, 1973, pp. 58–60) indicates that sexuality is a mystery to adolescents. This is probably why 50 percent of the boys and 39 percent of the girls agree, "There isn't anything in sex that I wouldn't want to try, at least once." A sexual encounter with another person is never dull to the teenager, since it "offers infinite means by which the adolescent can explore himself; sexual activities permit two people to share that experience of exploration and self-definition." About 85 percent of this sample of teenagers (*N*=411) believe that "in general . . . my sex life is pretty normal for a person of my age."

It has been noted that marijuana users tend to fantasize and enjoy masturbation in "substantially greater proportions than nonusers"; in varying degrees, guilt, concern, or anxiety over masturbation is experienced by 81 percent of all teenagers (Sorensen, 1973, pp. 138–140). Adolescents frequently apply the situational ethic to sexual activity: "they will do what they believe is best to meet the requirements of a given situation." Sorenson (1973, pp. 363–364) recognizes that adolescents need personal values to apply to sexually exciting situations in order to reduce any possible damage springing from otherwise indiscriminate engagement in sex.

Despite a decline of sex morality, the majority of girls reach the age of twenty as virgins (Luckey & Nass, 1969). Justification of premarital sex by a strong emotional bond or love is a frequent rationalization of sexual excesses among teenagers today. Sexuality is intimately related to personal identity. To break the bonds of dependence, the teenager has to master his own sexual impulses to a high degree; he must become able to refuse sex as well as to accept it under his own standards. Sex is a nonhomeostatic drive. There is no regular decline with its satisfaction. Rather, the opposite is true: the more one engages in sex, the more one desires it. This is especially true of adolescent sexuality. The existence of "the pill" and other contraceptive means makes pregnancy avoidable, but early loss of virginity elicits feelings of major loss in many girls.

Many teenagers read about and see colorful illustrations of the techniques of sex and seduction. This, however, does not teach them how to love or why personal commitment is necessary for long-term love. Excessive sexual awareness produces more harm than good for anyone, just as excessive awareness of one's heartbeat often disturbs the natural rhythm of the heart. Hypochondriasis focused on sexual organs is just as undesirable as a morbid concern with the heart, the brain, or any other organ or organ system.

One hesitates to speak of a sexual revolution among young people today, since comparisions with previous generations are difficult; yet the rising incidence of venereal disease, which in the early 1970s had reached epidemic proportions in many metropolitan areas, including inner-city high schools, and the spiraling rate of pregnancies among unmarried girls are signs of increased sexual activity by teenagers (Godenne, 1974).

In many colleges "girls who want to stay virgins find it hard to hold the line. Many boys refuse to date them and some girls treat them as squares. . . . Some girls yield, not out of inner need but out of pressure." For many college boys, sexual conquests are a symbol of maturity and masculinity. Often boys play with love as a means of getting sex. For many girls sex is a safeguard against loneliness (Ginott, 1969, pp. 159–166).

With Erikson's theory of identity formation in mind, it may be reasoned that since most teenagers have a long way to go in establishing a firm sense of identity, sex relations are rarely intimate relations, although they may be accompanied by affection. Since most boys experience intense sex tensions, coitus without intimacy is highly acceptable to many of them, while "petting with affection" is acceptable to many girls (Smart & Smart, 1973, pp. 143–144). The most usual code for girls is "petting with affection"; for boys it is the double standard (p. 151). "The custom of petting-with-affection combines the sex-affection ideal with the standard of abstinence, since the couple stops short of full sex relations" (p. 143).

To avoid feelings of guilt, the illusion of love is created by many girls. Many young people rush too soon into heterosexual activity and pay for it by becoming emotionally shallow and not establishing their own sex identity (Blos, 1962; Godenne, 1974). After early seduction, many girls remain puppets manipulated and exploited by boys unless the girls develop proper defenses against sexual abuse. Excessive knowledge furthers heightened self-consciousness about even minute aspects of sex—a poor gift for most teenage boys, who readily worry about the size, shape, and other aspects of their penises. They know more than physicians knew a generation ago but often deny or repress much sexual knowledge and some aspects of sexuality itself. Sexuality is a dynamic power that does not readily yield to easy handling. Sorensen's survey (1973, pp. 373–376) has shown "an unfortunate tendency toward premature intercourse among adolescents" seeking "genital satisfaction rather than love." "Does he or she exist only to gratify my urge?" If the answer is "yes," a person is oblivious of the humanistic and loving aspects of sex at the age of adult sexual readiness.

Erikson (1970) feels that contemporary teenagers need a psychological moratorium for dramatizing and experimenting with infantile, pubertal, and adult patterns of behavior. Going one step further, it may be added that the sexual *polymorphism* of puberty often evokes a wide range of behavioral exploration possibilities, from the purely narcissistic to the clearly heterosexual. A reenactment of the urges of a premoral id is arousing the worst in young people—a blatant regression in "blowing off accumulated steam" and "sowing wild oats."

The appearance of all forms of sex and violence on the screen and of many forms in real life raises a new issue: the legitimacy of obscenity. Regressive trends appear to be gaining the upper hand here too. Pornographic pictures and X-rated movies are widespread. Most boys of fifteen know where to get hard-core pornography. Many clinical reports indicate that obscenity and pornography debase adolescent sexuality and provoke various forms of sexual regression. "Streaking" is just one of those regressive self-expressions of pubertal minds when body and sex awareness rise rather sharply. Denial of taboos and recurrence of engagement in oedipal relations can be predicted as the final and probably fatal step a young mind may move into when internal controls are not strengthened at this age, because there are so few encouragements and reinforcements of moral and idealistic stirrings and impulses (Maddock, 1973).

ADOLESCENT CONFLICTS AND PROBLEMS

Adolescent behavior often has a surface appearance of gay and carefree activity marked by rollicking antics and enthusiasm for living. Beneath the shiny veneer of adolescent self-expression, however, the trained observer detects marks of the many anxious thoughts and the uncertainties young persons are undergoing in this decision-making and problem-solving period of development. In learning to adjust to his own changing body and motivation, assailed by insatiable and perverse drives and desires, the adolescent is often struggling within himself. Life is offering new goals and views, and he is becoming increasingly aware of new relationships with parents and peers. Problems in adjusting spring from many sources, including new abilities and sexual urges, intense feelings of love and hate, adult needs and childhood limitations. Table 15-1 shows fifteen leading adolescent problems and interests and their changes from 1935 to 1957 and 1967. Money, mental hygiene, and sex adjustment ranked highest as personal problems in 1967 for adolescents in the Detroit area, and sex adjustment, money, and personal attractiveness were the problems of foremost interest for the sample.

Ambivalence springs from the presence of contradictory feelings toward the same person, object, or situation. The young adolescent is often torn between immaturity and a facade of maturity, attraction and repulsion, frenzied activity and idleness. His bipolarity of emotion and thinking points to a lack of harmonious fusion among his various psychobiological drives. Ambivalence is especially frequent when the sexual drive becomes involved before sexual and emotional maturity is attained. Lack of perspective and moderation seem to reinforce the states of doubt and conflict. The faith that "it won't happen to me" plays hide-and-seek with the fear that it will. The feeling of omnipotence tangles with the feeling of helplessness (Konopka, 1973b). Frequently adolescents (and many adults) lack the ability to make important decisions by themselves. Often their closest friends make decisions for them. In late adolescence, with normal gains in ego strength, there is usually a decline in the intensity of ambivalent tendencies.

Ego Strength and Self-Defenses

The advanced level of adolescent development allows for the acquiring of numerous defensive responses. By means of these defenses the person attempts to reestablish a temporary balance between internal superego regulations and external stimulants and pressures. They are compromises between internal needs and emotions and the demands of reality. A weak ego needs defenses to protect it against additional damage; plain denials and spontaneous

Table 15-1 Rankings of Adolescent Life Problems and Interests: 1935, 1957, and 1967

	As personal problems						As problems of interest					
	1935		1957		1967		1935		1957		1967	
Issues	Mean rank	Rank order	Mean rank	Rank order	Mean rank	Rank order	Mean rank	Rank order	Mean rank	Rank order	Mean rank	Rank order
Health	6.61	2	8.9	12.5	7.79	11	6.1	2	6.7	1	6.09	4
Sex adjustments	10.0	15	8.9	12.5	5.09	3	9.3	12	7.1	7	3.84	1
Safety	8.6	12	9.6	14	8.30	15	8.5	10	9.5	12	8.18	14
Money	6.5	1	6.4	2	4.39	1	7.6	7	7.4	9	4.63	2
Mental hygiene	8.5	11	7.6	5.5	5.06	2	9.2	12	8.8	11	6.89	7
Study habits	7.1	4	5.7	1	5.31	4	9.0	11	9.6	13	7.81	12
Recreation	8.3	10	10.1	15	8.08	14	5.2	1	6.8	3	7.04	9
Personal and moral qualities	7.2	5	6.9	3	5.84	5	7.6	7	7.2	8	6.25	6
Home and family relationships	8.2	8.5	8.0	7	6.39	6	8.4	9	6.8	3	7.36	11
Manners and courtesy	7.9	7	8.1	8	7.98	12	6.9	4	8.6	10	7.21	10
Personal attractiveness	7.0	3	7.3	4	7.99	13	6.8	3	7.0	6	6.01	3
Daily schedule	9.2	14	8.5	11	7.63	10	10.4	15	11.2	15	9.25	15
Civic interests, etc.	8.7	13	8.2	9	7.05	8	9.4	14	9.8	14	8.03	13
Getting along with others	8.2	8.5	8.3	10	7.42	9	7.6	7	6.8	3	6.19	5
Philosophy of life	7.5	6	7.6	5.5	6.47	7	7.5	5	6.9	5	6.95	8

Sources: Percival M. Symonds, Life problems and interests of adolescents, *School Rev.*, 1936, **44,** 506–518; Dale B. Harris, Life problems and interests of adolescents in 1935 and 1957, *School Rev.*, 1959, **67,** 335–343; Sharon Kroha and Justin Pikunas, Adolescent life problems and interests in 1967 (unpublished study; $N = 424$).

repressions are not enough at this level of life. Some of the more frequently used dynamisms are compensation and substitution, rationalization, displacement, introjection and projection, fixation, regression, repression, and reaction formation. Nietzsche and Sigmund and Anna Freud are the original formulators of many of these defenses. Those commonly employed by adolescents are briefly discussed here.

Rationalization is a dynamism that appears in the early years of childhood and often continues throughout life. It is an attempt at self-justification by finding reasons to excuse oneself from disapproval, criticism, or punishment. The actual facts in a situation are misinterpreted by the person in order to protect his or her own self-concept from the adverse opinion of others. Rationalization is commonly known as a defense to save face.

Projection refers to a dynamism by which personal weaknesses and undesirable qualities and traits are attributed to other persons and external sources. Undesirable elements of the self are unconsciously treated as though they existed in another person and not in oneself. For example, dishonesty and hostility are often projected; when this occurs, lying and aggressiveness are seen by the individual as characteristics of others, against which he must protect himself.

Any evidence of return to an earlier, less mature level of functioning is referred to as *regression* rather than as a situational response. When an individual is exposed to a frustrating or anxiety-provoking experience, his mature manner of adjustment may fail and avoidance or some more primitive reaction is used to protect the self. If the source of frustration or stress is not removed or resolved, this response pattern, properly called regression, becomes habitual. The level of regression is judged by the level at which one functions or in terms of the number of years a person's behavior regresses. When regression results in the use of infantile ways of handling needs and problems, this level of self-defense is referred to as *infantilism.* When the adolescent's behavior can be characterized by a frequent need of assistance, extreme dependence, and inability to delay the gratification of his needs or desires, he is then displaying segments of infantile behavior along with segments of more mature functioning.

Neurotic Tendencies

Conflicts with parents and peers, sexual and moral stresses, change in residence or school, revision of standards and values—all strain the adolescent for a time. Televised sex and violence narcotize him. His appearance and health concern him. His past fears and concerns return often to bother him. Fierce and prolonged stress situations cause many adolescents, especially those who have experienced childhood trauma, to suffer emotional disturbances and an inability to integrate the personality into a sound functional system. As a result, anxiety intensifies and forms a basis for a neurotic pattern of behavior marked by chronic symptoms. Of the various forms of neurosis, the adolescent appears to be most susceptible to conversion reactions, anxiety attacks, and obsessive-compulsive reactions. More extreme cases of maladjustment may result in psychotic reactions such as hebephrenic or catatonic schizophrenia.

Persistent irrational ideas or actions which the subject often recognizes as illogical but which he nevertheless attends to because they reduce anxiety or tension, constitute obsessions. As obsessions intensify, they are partially released in various forms of compulsive behavior. *Obsessive-compulsive* reactions afflict a significant minority of adolescents. The tension brought about as a result of trying to ignore or suppress these urges is almost unbearable to the young person. The underlying cause of this behavior is usually a deep sense of inferiority or guilt that creates tension and expresses itself symbolically in the persistence of the same ideas and behavior patterns.

Adolescents often experience guilt over the processes and urges that arise as a result of pubertal changes. Sexual arousal and autoerotic expressions, as well as menstruation in girls and nocturnal emissions in boys, sometimes produce guilt that in some cases leads to obsessive-compulsive responses. This is especially true when knowledge of these phenomena is incomplete or lacking or when negative attitudes toward matters pertaining to sex have been deeply instilled.

Frequent and unnecessary hand washing is a compulsion usually resulting from a deep sense of guilt, often attributed to sexual fantasies and activities. This guilt is often referred to the future—that is, the hand washing or any other compulsion may be either an immediate defense or an attempt to prevent actions that the adolescent fears he or she will commit.

Obsessions are irremovable ideas that often plague a person's consciousness and seriously disturb his efficiency and adjustment. The content of these ideas may be almost anything: fear of mental breakdown, sexual fantasies, hatred for parents or members of the opposite sex. These or similar ideas easily arise in adolescence, in part because of a lack of success in heterosexual attraction, the need for independence from parents, and an abundance of sexual stimuli. Lack of complete sexual information or the concomitant moral instruction necessary for satisfactory sexual adjustment, overprotective or dominating parents, lack of self-knowledge—all contribute to obsessions and consequent maladaptive behavior in adolescence. If discovered early enough, obsessions or compulsions can be successfully treated by reeducative psychotherapy. But often they are not made known in time and persist throughout adolescence and into adulthood. In these cases long-term psychotherapeutic treatment is necessary, and its success is less certain.

Anxiety is an unrealistic and unpleasant emotional state in which threats and dangers to the life of the person are vividly anticipated. Pathological anxiety is out of proportion to any danger stimulus and is often undifferentiated and diffuse. Adolescents, because of the multitude of adjustments and decisions they must make and because of their immaturity and lack of experience, are an easy prey to this type of emotional disturbance. The tensions resulting from physiological changes, lack of confidence in themselves, and indecision and ambivalence in many areas of their lives readily lead to the experience of intense anxiety. When these tensions and strains accumulate, anxiety attacks may be the result. A person experiencing intense anxiety becomes terrified and sweats profusely, and his heart palpitates. This state of near-panic usually subsides in two or three hours. Freud views anxiety as the core of all psychopathology.

Calm and deliberate reassurance is often of great help in allaying anxiety. If the adolescent can be convinced that his fears are shared by others and that others are surmounting these fears and adjusting satisfactorily to new situations in spite of them, he may be helped. If the anxiety is deeply rooted in the personality structure, counseling and reeducative psychotherapy are needed to help the adolescent rid himself of this affliction. A person showing anxiety symptoms may be considered neurotic if anxiety attacks or other striking symptoms occur periodically, but not if a single anxiety attack occurs, for example, a final examination or a similar situation of intense strain.

Conversion reactions are essentially symptomatic externalizations of inner conflict and anxiety that are usually not recognized by the subject. The conflict is usually of such a nature that it is unacceptable to the conscious mind and its tension is expressed in physiological symptoms. These symptoms are of many types, including anesthesia, writer's cramp, hysterical paralysis, and neuromuscular convulsions. There is no organic basis for conversion symptoms; they are not physically or neurologically caused. Rather, they are purposeful, unconsciously adopted defenses for resolving conflict and reducing anxiety. Cardiac disturbances, severe pain, and nausea are unconsciously utilized by some adolescents as fair and honorable methods of escaping stressful situations.

The prevention of conversion reactions begins in childhood. The child must be taught to face reality and deal with problems as they arise. Once a pattern of evasion and escape is built up, the problems and conflicts of adolescence prove too much for an already precarious mode of adjustment. Dealing with these problems after such a pattern has taken hold is difficult, because the subject is often unwilling or unable to bring the conflict to the surface, where it might be understood and dealt with effectively.

Delinquent Trends

Delinquent trends are often the reactions to continual stimulus privation, need frustration, and prolonged lack of success. "Battered" children usually become aggressive teenagers. Many adolescents tend to respond to situational deprivation by verbal or physical aggression. Aggressive behavior involves some form of attack, such as using abusive language or provoking or striking another person. In some instances the immediate response to frustration is withdrawal and toleration of the situation, yet the suppressed aggressive tendencies pile up and are likely to show themselves later. Thus aggressive reactions are only temporarily delayed, repressed, disguised, displaced, or otherwise deflected from the original source. It is probable that some aggressive energies can be redirected or compensated for during this period of increased conflict and frustration, but some adolescents will turn to delinquent activities as a means of discharging aggression.

Two of the most common delinquent activities are truancy and stealing. Truancy is a form of withdrawal from the reality of school in order to avoid classroom tasks that are perceived as unpleasant. It often precedes other typical forms of delinquency such as uncontrollable misbehavior at home and at school and offenses against others and against society. In the school situation, refusing to submit to the regulations of the teacher causes added difficulty. For some adolescents, however, probation and the resulting frustrations can serve as valuable assets for building ego strength and discovering sound solutions, as well as for the development of more mature personal judgment. This frequently happens when the adolescent has the advantage of a good home.

Cheating and stealing at home and at school, destruction of property, shoplifting, association with "rough gangs," and a tendency to get involved in fights are typical forms of delinquent behavior resulting from frustration and tension. Shoplifting became something of a fad in recent decades. When control of aggressive impulses is inadequate, cruelty and sex offenses also occur.

Neurotic and delinquent tendencies often mix with abuse of dangerous drugs. How far it is safe for the teenager to go in exploring various drugs is a question that is difficult to answer. Much depends on the conflicts, tensions, anxiety, and difficulties he faces at home and in school. The other side of the story is related to the peer pressure he will be exposed to before and after dances, concerts, and movies until he returns to the relative safety of his home. Association with other drug users is the only stable variable found in the majority of teenage addicts. This points to a strong psychosocial basis for taking drugs (Lipsitt & Lelos, 1972). Other important factors in the making of an addict are use of drugs by parents and a weak or absent father—conditions that reduce the inclination to face reality (Lieberman, 1974; Smart & Fejer, 1972).

For a long time adult American society has been chemophilic—overmedicated and pill-oriented. This orientation is increasingly affecting the younger age groups. Since most adults use alcohol and tobacco, most teenagers explore them early in adolescence if not before. Since many adults use tranquilizers and sleeping or diet pills, they should not be surprised that their sons and daughters explore drugs. As a result, many teenagers are attracted to certain amphetamines or barbiturates. Moreover, "turning on" is often used as a defense against fierce aggressive and sexual impulses. Many surveys report that parents who regularly use alcohol, tobacco. and other mood-changing drugs unintentionally favor a tendency in their children to experiment with drugs; apparently, many teenagers model their drug use after their parents' (Smart & Fejer, 1972). Intensive dependence on drugs is often strengthened by medical practitioners who prescribe drugs to cope with emotional problems, thereby avoiding psychosocial approaches (e.g., Daytop Village, Inc., or

Phoenix House in New York) that might be better for young patients (Lennard et al., 1971).

A spectacular surge of marijuana smoking (*Cannabis sativa* with tetrahydrocannabinol) among American teenagers occurred in the 1960s. Pleasure motivation is obviously a major factor, since a euphoric high is a prominent experience of "pot" smokers. Symbolic status aspects also play a major role; it is the "in thing" to do to escape tension and up-tightness. Marijuana intake is accompanied by initial excitement or anxiety, then by inattention and flight of ideas, blurring of vision and floating sensations, heart palpitation, weakness, and finally a tendency to sleep (Hollister, 1971; Nahas, 1972).

Marijuana became widely available first on campuses and in inner cities, then in high schools, and finally in grade schools and most neighborhoods. Its use spread by leaps and bounds. In the early 1970s, most boys of ten or eleven knew how to get "grass," usually in a matter of minutes. By 1975, nearly nine million Americans were reported to be smoking marijuana and many more millions were exploring it. It is one of the "soft" but illegal and hazardous drugs. There is a growing body of evidence indicating that the effects of marijuana are dose-related and cumulative. Its hazards include probability of chromosome damage, disruption of cellular metabolism, irreversible brain damage, debilitation of lungs and bronchial tract, sexual impotence and growth of female-like breasts in men, and personality reverses, including sharp deterioration of mental health. Some of these reverses increase susceptibility to viral diseases (Maugh, 1974). Chronic use of marijuana does not lead to physical dependency, nor cessation to any serious withdrawal symptoms, but psychological dependency often gains in strength over a period of time. Use of larger amounts or of potent varieties produces acute panic and clearly psychotic reactions, often with brain-wave changes and disorganization of cognitive and social judgment. The Canadian Commission (1972, p. 274) stated that "the long-term heavy use of cannabis may result in a significant amount of mental deterioration and disorder." Additionally, the Commission observed that marijuana plays a role in spreading use of many drugs by "stimulating a desire for drug experiences and lowering inhibitions about drug experimentation."

Chronic use of "pot," or "hash," as mentioned above, leads to a moderate degree of psychological dependence on it—another crutch many young persons begin to need in social situations and later in private situations as well. Instead of solving internally arising conflicts, many adolescents "smoke them out"—create a moratorium in limbo. For a minority of adolescents there is no stop: they move from one drug to another. Chemical options do not solve their conflicts or problems; their "lifts" are only apparent, not real. Experimentation with narcotics, if prolonged, "hooks" many individuals on dangerous drugs like methaqualone and morphine or its derivative, heroin.

The danger of extreme dependence on narcotics is often overlooked by young persons who seek release from stress or tension or who find it difficult to secure thrills or fun by other means. Educational efforts ought to aim at showing the teenager all the available facts honestly. At the same time, student leaders should be enlisted to help to discourage unlawful use of drugs. Rehabilitative programs in which methadone is substituted for heroin lead only to new blind alleys, since the diverse human conditions creating chemical addiction remain untreated. Large-scale introduction of any new potent psychoactive drugs creates new dangers for many young persons who need psychotherapeutic strategies for the solution of their deep intrapsychic problems.

More active and aggressive teenagers, when confronted with difficulties at home, run away from their parents. Nomadism was never as popular in the United States as it is today. Most teenagers express a desire to travel, and many are permitted to do so by their benevolent parents, who sooner or later realize that refusal spells other troubles. The youth exodus begins late in May when the school year ends. Faraway "glamour" cities continue to attract large numbers of those who run away. Disenchanted offspring of the middle class form the bulk of the flow. Runaway havens, public and private, have been set up in many cities to accommodate the homeless young. The average age of runaways continues to drop from the late to the early teens. In 1970, about 600,000 minors ran away from home, most of them aged thirteen to seventeen and nearly as many girls as boys. Many teenagers willingly go home once they realize that their parents care enough to look for them or to send money for the trip home. Frequently their romantic *naïveté* about life in communes and other runaway shelters, and about their inhabitants, changes rapidly into astonishment, fear, and even shock. This makes them ready to turn back before they burn all bridges. Each September the majority of the runaways are back in school, often readier to adjust to its limitations and demands. Yet the strains of the nuclear family often rise again, and the problem of running away will remain for many years to come. Regression to nomadic adventure is a frequent adolescent escape from the demands of reality.

ADOLESCENT IDENTITY AND THE PATTERN OF LIFE

Even for experts in human development, it is difficult to make conclusive statements about adolescent identity. The formulation of the future adult identity is foreshadowed in childhood by the parents and others on whom the adolescent is dependent, and he has little choice but to build his life pattern on this base. The adolescent task of forming his identity within the family framework, yet separate from it, implies disagreements and conflicts with most parents, especially the critical and the overprotective ones. A warm parental relationship, if not genuine understanding, helps the adolescent to attain a stable and realistic identity (Ginott, 1969).

The adolescent finds that "within himself he is fractionated, divided into pieces. He senses, perhaps vaguely, a lack of wholeness and inner harmony of life as it is lived in social context." He needs "responsible relatedness to the past as he lives in his present life space yet is motivated by future goals" (Montgomery, 1973). The teenager needs opportunities to look outward and test himself or herself in a great variety of settings. Experimentation with ideas, roles, and people is necessary before meaningful commitments are possible (Konopka, 1973a). Television and motion-picture stars are often imitated by adolescents, and professional athletes are frequently idolized by pubertal boys. The Beatles were popular in the early sixties, the Monkees in the middle sixties, and the Rolling Stones in the late sixties; The Who and Elton John climbed to the top in the early seventies. Who can predict the characteristics that will appeal to the next generation?

A teenage culture has a retarding effect on maturation insofar as it furthers resistance to wholesome inner growth and self-improvement. The hazards of late maturation have been well recognized—alienation from established adult society and alienation from self are the Scylla and Charybdis of the passage to adulthood. Turmoil continues as the young "muddle through," and crises of development occur until they "settle down" in marriage or work.

The family framework is too limited for most adolescents to try out new images and roles. Peers and peer groups serve the purpose of providing feedback as the adolescent acts out a role and sees how it affects others and what feelings arise within himself. In the heterosexual role the teenager gets involved in bisexual groups and forms dyads within the larger peer structure. There is a need to view himself through the eyes of a selected member of the opposite sex. Adolescent identity is thus formed through the resolution of conflicts regarding appearance and sexuality, vocation, and ideological choice. When tested on family members and peers, the identity is modified and made compatible with peer groups and often with the family as well. Once the adolescent gets a firm grasp on the major aspects of his identity, he is ready to function in peer and adult society without suffering neurotic anxiety or magnified frustration at every obstacle.

Late adolescence is a major phase of identity development, largely influenced by a person's experiential background and the direction taken by his dispositions, endowments, abilities, and other resources. Self-identity has its roots in early childhood experiences in which the parents' directing influence is very strong. Childhood experiences that affect the concept of self and identity are conflicts encountered by the child, such as repeated failures in school, which lead many children to truancy, dishonesty, rationalization, and contempt for authority. Pleasurable and satisfying events also affect self-development and usually promote desirable traits if the pleasure is lawful and not excessive. The neighborhood and its mores and standards also make an impression on the child or adolescent, as he tends to assimilate what he is exposed to. Responsiveness to what is good must be reinforced throughout childhood in order to foster an appreciation of ethical principles and higher values during adolescence.

In late adolescence, self-direction becomes a major force in the preservation of identity. Strivings for independence, combined with the anticipation of approaching adulthood, make the adolescent weigh carefully his behavior, abilities, and weaknesses with the future in mind. He searches for long-term meanings and goals and decides what kind of person he wants to be. Of crucial concern are his relations with his peers and the adult world. Much depends on how he sees others in the fabric of society. He considers the ethics of his community and the larger society, the wishes of his parents, and the attitudes of his friends. He attempts to use self-control in cultivating the traits he desires. While self-control necessitates personal effort, self-appraisal depends largely on the estimation of others. Occasionally the casual opinions of peers, parents, and others greatly disturb a young person. After considering the views of others, he consciously or unconsciously undertakes alteration of himself. The search for self is often completed with the design of a moral code to live by, and still more often by a choice of vocation and marriage, which represent standards attributed to the adult pattern of life.

Feminine identity is highly reflected by a girl's attractiveness, interpersonal competence, and success in the search for men (or the man) by whom she wishes to be sought. The feminine identity is thus uncompromisingly linked to the male identity and leaves little for itself. As a result, the adolescent girl encounters frequent frustration when she decides to "be someone," yet her society provides no clearly charted guidelines for her autonomy (Erikson, 1968). The woman's liberation movement is evidence that social attitudes are changing in favor of autonomy and equal career opportunity for girls. "There are already many women in the world who perform valuable services outside the family, yet retain a feminine way of feeling and relating" (Smart & Smart, 1973, p. 147). Marriage will no longer be as crucial a factor in ending and solving the woman's identity crisis as in the past.

During adolescence the teenager begins to develop a philosophy of life. The majority of adolescents construct their *Weltanschauung* on the basis of religion. Most religions offer a comprehensive authoritative perspective for a philosophy of life. Large numbers of persons adopt an ideology in which the nation or the economy takes the place of religion, while others attempt a plan of life based on other philosophies, such as hedonism (seeking pleasure as an end in itself). Because of the confusion of our present culture, the process of *Weltanschauung* formation usually extends well into the years of adulthood. The adolescent's progress toward an ideology indicates his or her approach to an adult level of maturity. The next chapter will attempt to define signs indicating personal maturity at the postadolescent level of life.

QUESTIONS FOR REVIEW

1 What are some major causes of magnified self-awareness in adolescence?
2 Why do many adolescents set unrealistic goals for themselves?
3 Discuss the concepts of personality and the self, including major relationships between them.
4 Give some reasons why a child's motivational structure has to be revised during adolescence. Why are slight modifications unsatisfactory?
5 Why are close peer associations desirable at this age?

6 What does an adolescent need to further his or her value development? What factors further deeper moral and religious concerns in adolescence?
7 Explain how changes in motivation and ability affect the self-concept and personal adaptation to reality.
8 In what ways do heterosexual relationships change during the period of adolescence?
9 What are the key adolescent problems? Relate sexual maturation to the rising problems of this period.
10 List and discuss major findings of Sorensen's survey of adolescent sexuality.
11 Define anxiety and its effects on adolescent adjustment.
12 Give some major reasons for drug abuse by teenagers, and analyze one of them.
13 What social conditions contribute to misdemeanors and delinquency among teenagers?
14 What factors contribute greatly to the adolescent's identity definition? Discuss two or three of them.

REFERENCES

Selected Reading

Gersh, M. J., & Litt, I. F. *The handbook of adolescence: A medical guide for parents and teenagers.* New York: Stein and Day, 1971. A compact information source on sex and venereal disease, drugs, "nervous breakdowns," acne, weight, and birth control.

Konopka, G. Requirements for healthy development of adolescent youth. *Adolescence,* 1973, **8,** 291–316. A very comprehensive statement about requirements for sound development. Includes obstacles hindering normal development and suggests foundation of a federal office for youth affairs.

Nesselroade, J. R., & Baltes, P. B. Adolescent personality development and historical change: 1970–1972. *Monogr. Soc. Res. Child Developm.,* 1974, **39** (II, Ser. No. 154), 1–80. The authors believe that cultural moments, rather than sequential unfolding of universal behavior patterns, are major influences.

Sorensen, R. C. *Adolescent sexuality in contemporary America.* New York: World Press, 1973. A national survey of attitudes and activities related to sex among adolescents aged thirteen to nineteen.

Specific References

Allport, G. W. Values and our youth. *Teachers Coll. Rec.,* 1961, **63,** 211–219.

Blos, P. *On adolescence: A psychoanalytic interpretation.* New York: Free Press, 1962.

Commission of Inquiry into the Non-medical Use of Drugs. *Cannabis.* Ottawa: Information Canada, 1972.

Conger, J. J. A world they never knew: The family and social change. *Daedalus,* Fall 1971, 1105–1138.

Douvan, E., & Adelson, J. *The adolescent experience.* New York: Wiley, 1966.

Erikson, E. *Youth: Change and challenge.* New York: Basic Books, 1963.

Erikson, E *Identity, youth and crisis.* New York: Norton, 1968.

Erikson, E. Reflections on the dissent of contemporary youth. *Internat. J. Psychoan.,* 1970, **51,** 11–22.

Gersten, W. M. Alienation in mass society: Some causes and responses. *Sociol. soc. Res.,* 1965, **49,** 143–152.

Ginott, H. G. *Between parent and teenager.* New York: Macmillan, 1969.

Godenne, G. D. Sex and today's youth. *Adolescence,* 1974, **9** (33), 67–72.

Hollister, L. H. Marihuana in man: Three years later. *Science,* 1971, **172,** 21–28.

Keniston, K. *The uncommitted: Alienated youth in American society.* New York: Harcourt, Brace & World, 1965.

Kohlberg, L., & Kramer, R. Continuities and discontinuities in childhood and adult moral development. *Human Developm.,* 1969, **12,** 93–120.

Konopka, G. Formation of values in the developing person. *Amer. J. Orthopsychiat.,* 1973, **43,** 86–96 (a).

Konopka, G. Requirements for healthy development of adolescent youth. *Adolescence,* 1973, **8,** 291–316 (b).

Lennard, H. L., Epstein, L. J., Bernstein, A., & Ransom, D. C. *Mystification and drug abuse.* San Francisco: Jossey-Bass, 1971.

Lidz, T. *The person: His development throughout the life cycle.* New York: Basic Books, 1968.

Lieberman, J. J. The drug addict and the "cop out" father. *Adolescence,* 1974, **9** (33), 7–14.

Lipsitt, P. D., & Lelos, D. Relationship of sex-role identity to level of ego development in habitual drug users. *Proceedings 80th Ann. Convention APA,* 1972, **7,** 255–256.

Luckey, E. B., & Nass, G. D. A comparison of sexual attitudes and behavior in an international sample. *J. Marr. Fam.,* 1969, **31,** 364–379.

Maddock, J. W. Sex in adolescence: Its meaning and its future. *Adolescence,* 1973, **8,** 325–342.

Maugh, T. H., II. Marihuana: The grass may no longer be greener. *Science,* 1974, **185,** 683–685.

Montgomery, L. J. From personal identity to social integrity—The educator's role. *Adolescence,* 1973, **8,** 179–188.

Morgan, C. A. Assignment infinity: Exploring some aspects of human potential. In E. H. Williams & D. R. Schrader (Eds.), *Tenth Annual Distinguished Lectures Series,* pp. 20–39. Los Angeles: University of Southern California Press, 1972.

Nahas, G. G. *Marihuana: The deceptive weed.* New York: Raven Books, 1972.

Rode, A., & Smith, S. S. The God-seekers: A group portrait. *Adolescence,* 1968–69, **3,** 441–446.

Seidman, J. M. Psychological roots of moral development in adolescence. *Cath. psychol. Rec.,* 1963, **1,** 19–27.

Smart, R. G., & Fejer, D. Drug use among adolescents and their parents: Closing the generation gap in mood modification. *J. abn. Psychol.,* 1972, **79,** 153–160.

Smart, M. S., & Smart, R. C. *Adolescents: Development and relationships.* New York: Macmillan, 1973.

Sorensen, R. C. *Adolescent sexuality in contemporary America.* New York: World Press, 1973.

Turiel, E. Conflict and transition in adolescent moral development. *Child Developm.,* 1974, **45,** 14–29.

Part Seven

Adulthood

Achieving adult status is the main task of the third decade of life. Each prior level of development made significant contributions toward the attainment of this goal. Each aspect of growth provided part of the foundation for the qualities, traits, habits, superiorities, and shortcomings that will make up the adult person. Although childhood and adolescence will be outgrown, adulthood remains a person's status for the remainder of life. Becoming adult is a process that involves choosing a vocation, selecting a spouse, and integrating into the personality the sociocultural structures and dynamics of the society in which the person is to function. Full maturity and adult status come with the consolidation of the personality structure, the development of one's identity, and self-actualization, especially as these are related to the vocational and marital roles that a person will assume. For the early and middle period of adulthood, a *stability* model seems plausible for most variables of interest to the psychologist, including cognitive abilities and many personality variables. For later years of adulthood, a *decrement-with-compensation* model which permits both ontogenetic decrement and compensation by external support systems seems to be appropriate (Schaie and Gribbin, 1975). The four chapters in this section present many of the psychosocial and cultural aspects of maturing and living in the early, middle, and later years of adulthood.

Chapter 16

Achieving Adult Status

HIGHLIGHTS

Most persons face many difficulties in their attempts to abandon behaviors that are puerile or pubescent. The legal age of majority serves as a reminder of what adolescents ought to become—adults adapted to the modern demands of life.

The criteria of adult maturity include differential responsiveness, participative activity, sensitivity to the needs of others, a unifying philosophy of life, and willingness to assume adult responsibilities. A mature person is one who has acquired a personal identity and has made progress in integrating his or her total personality into a smoothly functioning system.

Preparation for vocational activity and marital life is a part of the maturing process. A school dropout or a hippie has usually failed to assume an adult pattern of life. Making progress in marital adjustment and in vocational advancement is a task demanding personal resourcefulness and persistent effort.

Although recent Supreme Court decisions and state laws have changed the legal age of majority, or adulthood, to eighteen years, there is obviously much more to achieving adulthood than attaining a certain age. While American society lacks specific rites of passage to mark the transition to adulthood, it has a number of milestones along the road of maturation. Being permitted to drive a car at sixteen, to vote and marry at eighteen, to drink liquor in public at eighteen or twenty-one—all are indications of adult status, rights, and privileges. Our society expects a high level of cognitive functioning and efficient learning, vocational selection and preparation, increasing autonomy, and strong attachment to the peer group as signs of adulthood (Coleman et al., 1974, pp. 98–110). When the developmental tasks of adolescence have been mastered and the acquired abilities and skills have been integrated into a smoothly functioning system, a person becomes an adult capable of mature behavior. There is still need, however, for constant adjustment during the early years of adulthood. Childhood and adolescent experiences provide the developmental matrix within which the forces that will determine the adult pattern of behavior interact. The thrust toward the future becomes more powerful as a young person approaches adulthood. Values, skills, habits, and coping mechanisms learned and practiced at earlier levels may facilitate or hinder the person as an adult.

Without a continual willingness to learn and to act and without a satisfactory store of knowledge and skills, a sound pattern of adult living will not be acquired. Without sufficient interests and incentives, learning loses its efficiency. For the teenager, accomplishments without peer approval or praise tend to lose meaning. A strong sense of personal adequacy and self-reliance must be established before various areas of human endeavor can be effectively explored. A maximal use of abilities and skills is possible only when the person is not disturbed by anxiety or fears and is capable of being realistic about his or her assets and liabilities. Any undue emphasis on liabilities and deficiencies acts as an impediment to adult adjustment. The ability and the willingness to assume adult activities and responsibilities must be implanted within the adolescent as he or she approaches the twenties. Balancing daring and aspiration with endowments, abilities, and assets is also important in promoting integrity and adjustment (Coleman et al., 1974, pp. 108–111).

OVERCOMING IMMATURITY

The ability and the motivation to respond in mature ways under varying circumstances is an important adult developmental task. A mature response implies the overcoming of puerile and pubescent tendencies and vicissitudes. Frequently seeking help is an indication of helplessness. Frequent irritability, moodiness, and emotional outbursts as responses to everyday disappointment or frustration are adolescent reactions. Constant search for excitement, fun, the "high," or the sensational are also signs of adolescent motivation. If an adolescent level of motivation becomes deeply entrenched within the personality structure of a person, he or she will be handicapped in assuming adult responsibilities later.

As in early childhood, frustration may be one of the major factors in producing positive development. When the larger society frustrates the adolescent's immature wishes, he or she is often driven to explore more adult behavior. Conversely, if the adolescent behavior proves gratifying and the important persons in the young person's life reinforce immature behavior, there is little motivation to change to adult ways of living.

During the postadolescent years, many learn to exhibit an external "facade of maturity" but internally are still frequently moved by anxieties and by ambivalent feelings. Time and again they avoid facing their problems and use fantasy, rationalization, or illness as means of escape from unpleasant situations and challenging events. Moodiness and emotional oscillations are also signs of a preadult level of living. In the early twenties many persons find it difficult to exercise internal controls, yet loss of control may result in tragic developments later. The lack of readiness to assume one's sexual role or an inability to foster deep and permanent human relationships at the start of the adult years are signs of immaturity.

Experimenting with various roles and relationships is necessary for the realization of one's limits and hidden strengths. Most adolescents enthusiastically enter into new relationships and assume the roles offered to them. They are eager for intellectual, emotional, and social enterprises that help them to experience various forms of life as well as suitable job opportunities. Certain tendencies toward particular roles and interpersonal relationships seem to be determined by attitudes acquired about earlier roles during the years of childhood. As discussed in an earlier chapter and as shown in Bossard and Boll's classic study (1955), the younger children of a family find some roles already taken by siblings and must seek unoccupied roles. In early adulthood, then, certain programs, roles, and positions, though highly desirable to a person, may be unavailable; he or she must then search for suitable substitutes. A premedical student who fails to be accepted in medical school must reexamine his or her goals and aspirations and must look for alternative training programs to find a niche in professional life.

Satisfactory progress toward an adult level of functioning is often disrupted by anxiety, ambivalence, conflicting drives, and resulting excessive use of defensive methods of coping. Too frequent use of defense mechanisms under normal conditions is an index of immaturity and a barrier to exploring newer methods and means of adjustment. The growing person is destined to become, but not to remain, an adolescent. As the period of early adulthood progresses, the normal doubts about identity and self-worth subside. This occurs because the person receives from other adults confirmation of his or her acceptance as a peer. Feelings of alienation and fear of "going crazy," if they were present in the adolescent years, diminish.

For many persons there is too much continuity of early childhood traits and attitudes and too little revision. A major longitudinal study at the Fels Institute (Kagan & Moss, 1962, p. 266), based on eighty-nine subjects, summarized its most consistent finding as follows: "Many of the behaviors exhibited by the child during the period three to six were moderately good predictors of theoretically related behaviors during early adulthood." Apart from revisions and modifications of behavior, the threads of continuity show themselves throughout adulthood. If development turns into deviation, however, parents, educators, and society are often responsible for it. "It is not evil babies who grow up into evil human beings, but an evil society which turns good babies into disordered adults, and it does so on a regimen of frustration." (Montagu, 1955).

To achieve an adult pattern of behavior, the young person must surmount the obstacles to further growth present in some features of adolescent personality functioning. In a thoughtful article, James A. Knight (1967) identified some of the typical features of the adolescent, such as egocentricity and the reactivation of narcissism; a high degree of ambivalence involving many people and issues; marked indecisiveness on many points; rebellious drive for independence and escape from adult standards and authority; great need for the protective coloration of the peer group; various degrees of anxiety, guilt feelings, and disorganized behavior; the need for freedom and the need for restraints and controls; and lack of self-confidence and commitment. Knight emphasizes that it is most difficult for the adolescent to accept his finiteness, for whatever he does, "he is determined to succeed. Thus, he overdoes most everything, and thereby often appears clumsy and awkward." There is much to outgrow and to revise before the person becomes a full-fledged adult.

MATURITY: CONCEPTS AND CRITERIA

The previous chapters have discussed the many aspects and factors of the uneven process of development. The first two decades of life are a preparation for maturity. In growing up, the organism and personality reach and begin to operate on progressively advanced levels of maturity. One should bear in mind that an adult acts at the highest level of his abilities only when the situation calls for it; for example, in a political debate among civic leaders. When the same adult finds himself among children, he may speak simply and act in a somewhat juvenile manner. A teacher often has to use a lower level of self-expression to make himself understood or to initiate activities based on the interests of children. Responding to a situational demand, he merely faces reality and acts appropriately; this naturally does not mean that he regresses into immature forms of behavior. Ability to act at a child's level is very helpful in showing full understanding of children.

Children and adolescents often use direct imitation of adult models in their striving for mature behavior. In smaller children, this modeling behavior is often amusing, as when a girl of five dresses up in lipstick, nail polish, and her mother's shoes or when a boy uses shaving lotion and flounders around in his father's overcoat. The powerful effect of identification and nonverbal learning is also illustrated in the child's adoption of negative parental characteristics such as dishonesty and bigotry and the adolescent's imitation of his parents' drinking or reckless driving.

The definition of maturity poses many problems. It must take into account individual differences in abilities and self-expressiveness, social expectations, ethnic, cultural and subcultural variations, the unique circumstances of each individual life, and a host of other factors. Table 16-1 presents concepts of the

Table 16-1 Models of Adult Maturity

Model/Theory	Concept of man or behavior	Contending forces	Standard for maturity	Leading exponents
Psychoanalytic	(a) Determined early in life; psychosexual; instinctive	Unconscious vs. conscious motivation; id and superego vs. ego	Genital sexuality; sublimated expression of most sexual and aggressive urges	Freud
	(b) Psychosocial	Polarized conflicts of social origin	Strong ego and capacity for intimacy	Erikson
Learning theory/ S-R theory	Determined by external contingencies	Internal drives vs. external reinforcements	Anxiety-free hierarchies of response	Hull, Skinner
Self theory	Self-actualizing; basically good	Organismic impulses and drives vs. self-concept	Self-acceptance; internal locus of evaluation	Rogers
Personalistic	Individualistic and unique	Propriate striving vs. environmental limitations	Extension of the sense of self; unifying philosophy of life	Allport
Existential	A stranger in the world	*Eigenwelt* vs. *Dasein* (anxiety, despair, death)	Authentic existence	Heidegger, Binswanger
Humanistic	Indeterminate; gifted but need-bounded	Lower needs vs. higher values and universal ideals of authentic humanness	Self-fulfillment; life with commitment to highest values	Bühler, Maslow

human being and models of maturity advanced by major personality theorists. The diversity of standards illustrates some of the basic problems reaching a consensus on the definition of adult maturity.

Criteria of Maturity

It is somewhat easier to set and discuss the criteria of maturity than to apply them in individual cases. Maslow (1971, pp. 182–186) defines maturity in terms of "the ascendency of the humanistic orientation" where society is a potential hindrance to humanistic development of the individual. The lower needs for physiological survival, safety, belonging, and esteem and the higher needs for self-actualization and cognitive understanding interact, but only the latter represent genuine growth motivation toward maturity. It is the interaction between need-determined behavior and the contents and constraints of the person's environment that determine his or her style of life (Maddi & Costa, 1972, pp. 162–163).

The self-actualizing person transcends the lower needs and engages in self-actualization and cognitive development; he shows self-respect, relatedness to others, and willingness to continue to grow as a human person. He gains in autonomy from others close to him but also from his total environment and culture. In Rogerian terms, he is becoming a fully functioning person who trusts himself and accepts his own experiences. He finds what he needs to adapt himself to all the aspects of reality he faces (Maslow, 1967, 1971, pp. 113–116, 302–303).

Other methods of appraising the level of maturity are less bound to one theory of personality. These criteria are the attainment of a particular level of functioning in different sectors of personality. Many of them are classified as ego functions in dynamic theories of personality. Following is a discussion of several global criteria of maturity.

Differential Responsiveness Intellectual development and, in particular, various avenues of learning enable the child and the adolescent to expand and improve his or her understanding of the many realities of life, their dimensions, and their relationships. The child's early forms of exploration and subsequent modes of questioning and reading are important means of acquiring knowledge. The variety of experiences to which the growing person is exposed contributes substantially to the extension of familiarity with the many details of his environmental and cultural matrix. A child may discover that certain types of antisocial behavior, such as unwillingness to share or attempts to dominate, lead to unpopularity with his playmates. As a result, he learns to control such aspects of behavior for the sake of preserving friendships. Some adolescents deliberately introduce a single behavior variable such as flirting or sulking to test the reaction of others, thereby employing a form of experimental procedure and reasoning.

Lack of differential responsiveness on the part of youths and adults is frequently indicated by certain popular misconceptions, such as the inability to interpret "doctor" as anything but "physician," or "music" as anything but "melody" or "song." Progressive improvement in sensitivity and refinement of conceptual interpretation is a sine qua non in such a process. Accumulation of a variety of experience and knowledge represents a capital gain for feelings of adequacy and self-reliance. Maturity of response in various situations depends on previous experience and the range of one's information pertaining to each situation. Summarizing the research on intellectual development, Nancy Bayley (1970, p. 1201) wrote:

> In following the processes of mental growth from birth to some time near 30 years, when growth in most mental test scores finally appears to level off, it becomes clear that mental abilities are complex both in their nature and in their causes. With growth, abilities change in nature as they become increasingly complex and as they move from concrete to abstract processes.

Interdependence Growth in autonomy and independence from significant persons in one's life is a kind of psychosocial weaning. Many steps are involved in this process.

Late infancy and preadolescence are characteristic in this respect. At about the two-year level a new attitude becomes dominant, marked by excessive resistance to parental control and suggestion, stubbornness, and attempts at contrary behavior. The infant begins to feel that he has a mind and will of his own and starts to exercise them. On the other hand, the pubertal adolescent transfers emotional affiliation from his parents to peers. This gain in independence is a major sign of maturation, yet the young person may remain fully dependent on a clique or a "best friend" and feel lost when separated. In approaching maturity, however, the adolescent must break away from dependence on the peer group in particular and "peer culture" in general in order to integrate himself into adult society and culture as a self-reliant individual.

In young adulthood, a satisfactory level of autonomy will not be achieved if a mere transfer of dependence occurs. For example, if a husband or a wife relies too much on the partner for emotional support, neither can continue to grow emotionally. Conversely, total independence is not a mature goal; it can be attained only as a form of severe withdrawal, as in certain schizophrenic reactions

Participative Activity Without active and personal engagement, little can be learned or accomplished. Passivity, spectatorship, and an attitude of "let George do it" act as restrictions on the constructive use of energy, and unused energy can lead to physiological tension. The mature adult is able to act on his or her initiative, set goals, and involve the whole person in activities. Individuals, on the whole, respond better to challenge and stimulation than they do to routine. Self-knowledge, active engagement of abilities, and experimentation with one's potential can help the person to mold himself or herself to an appropriate model.

Application of Knowledge and Experience In the process of formal and personal education, constant self-examination is necessary to improve discrimination in terms of what is worth knowing and how to apply knowledge. In evaluating the possible implications and consequences of a decision, a person must study alternatives and strive for a wider perspective. Previous experience and knowledge can be valuable assets, if they are sources of learning rather than determinants of current behavior. The mature person establishes the locus of evaluation within himself. Following the phenomenology of Edmund Husserl and Marx P. Scheler, Rogers (1964) emphasized "a letting oneself down into the immediacy of what one is experiencing, endeavoring to sense and to clarify all its complex meanings." Because the mature person can pause before important decisions to evaluate the present in terms of prior experience, he is better able to perceive and react to danger signals. This in turn, facilitates better choices based on long-range goals rather than short-term expediency or impulse.

Communication of Experience Personal adequacy and adjustment are enhanced by the ability to relate experiences satisfactorily, especially emotionally significant experiences. It is a frequent observation of psychotherapists and marriage counselors that many adults continue to have difficulty developing this skill. A limited ability or willingness to communicate with significant others can severely limit one's relationships and consequently the satisfaction that could be derived from these relationships.

Sensitivity to the Needs of Others Infants and children are sensitive to only their own needs and interests. Although daily disregarding the feelings of others, children react with great volume and depth of feeling to any frustration of their wishes. Their world revolves about them, and their concern for others extends only as far as the other is important to them. A missing toy or an absent mother may produce virtually the same reaction.

Sensitivity to the needs of others gradually develops during childhood but frequently does not reach any depth before adolescence. The adolescent preoccupation with self is usually transformed into an examination and consideration of others. This observation of others often leads to deep insights relative to the needs of others. Yet an adolescent's activities for the gratification of others' needs are frequently interrupted by the emergence of his own acute desires. His own needs usually take priority. A young adult may, however, attain a level of control that permits constant service to the needs of others.

Sensitivity to the needs of others tends to decrease with advancing age. In old age, self-concern deepens and usually constricts the direction of personal interest in others. Then the needs of others are often disregarded and forgotten in favor of preoccupation with self and gratification of the person's own drives and inclinations.

Dealing Constructively with Frustration One of the major signs of maturity is the increasing ability to delay the gratification of psychological needs and to control or tolerate considerable amounts of disappointment, deprivation, anxiety, and frustration in general. In recalling and examining past frustrations, one should draw positive lessons for future activities and investigate the possibility of prevention where possible.

The adolescents' standards and ideals on the one hand and emotional and sexual needs on the other often do not readily fuse into an integrated and consistent behavior pattern; therefore disillusionment and conflicts occur, which, if intense, are likely to disturb the young person in a variety of ways and situations. As the adolescent becomes better able to cope with and solve these conflicts, he advances toward the attainment of maturity. He learns ways to express his drives and emotions without hurting others or himself.

Willingness to Assume Adult Responsibilities A young adult has to develop his abilities and advance his readiness to assume personal responsibilities pertaining to his status, duties, and obligations. There should be progress in anticipating and in setting long-term objectives during the years of adolescence and early adulthood. Frequently willingness to assume responsibilities involves sacrifice and courage on the part of the young person. He must learn to overcome fear of failure, disregard moods and feelings of disgust or apathy, and ignore cutting comments from his peers when his responsibilities must be met and solved in a rational manner, before he can develop a personality that is reasonably reliable in fulfilling duties.

Moral Character During the years of childhood, adolescence, and adulthood, a person is exposed to a variety of societal, cultural, and moral factors and has a long-term task in integrating his behavior with them. It may be recalled that at any moment in time one is "linked to all that has gone before and to what will follow" in his own life experience (Cox, 1970, p. 14). Moral maturation commences when a child begins to show some obedience to the dictates of his conscience. It is further advanced when an individual becomes his own judge in daily activities and in the organization of a responsible pattern of living. When a number of ethical and moral principles are assimilated and start acting as effective behavior organizers, people begin to show character, which is one of the ultimate indicators of advanced maturation and of adult personality. Moral character adds much to the humanization of the adult and to the control of his own destiny—he comes to the realization that "the decisions I make must be valid for the rest of my life" (Allport, 1964).

Lawrence Kohlberg (1971, p. 71), reviewing the relationship between the development of character and methods of facilitating this development, states:

> The goal of moral education is the stimulation of the "natural" development of the

individual child's own moral judgment and capacities, thus allowing him to use his own moral judgment to control his behavior. The attractiveness of defining the goal of moral education as the stimulation of development rather than as the teaching of fixed rules stems from the fact that it involves aiding the child to take the next step in a direction toward which he is already tending, rather than imposing an alien pattern upon him.

Progress in moral maturation is marked by an increase of freedom to choose what the individual recognizes as good. Habitual choice of what is good points to moral maturity, and it is an ideal for adolescent and adult alike. Hence the morally mature person guides himself in terms of assimilated moral standards and goals. For control of his instinctual urges, impulses, and undesirable emotions, he utilizes selected signals to himself to initiate substitution and sublimation. For instance, a person uses self-control by substitution, replacing an unacceptable thought or desire with a more acceptable one, or by suppression, forcing a thought out of consciousness by distraction. He may use neuromuscular control by refusing to give vent to an impulse such as striking someone when he is angry. In any such use of control, self-imposed standards and values take precedence over his own convenience and gratifications in satisfying biological or emotional drives.

Weltanschauung "Unifying philosophy of life," Gordon W. Allport's term (1961), or "rational design for living," the phrase employed by Magda B. Arnold and J. A. Gasson (1954), both refer to an important criterion of the mature person. A rational design for living helps to establish a coherence between the natural and the supernatural, between the person's self-imposed ethical principles and his religious experience. Allport (1961, p. 304) comes to the conclusion that "an integrrated sense of moral obligation provides a unifying philosophy of life whether or not it is tied to an equally developed religious sentiment."

DEFINING THE MATURE PERSON

A mature person structures the environment and is able to perceive himself or herself and others objectively. He or she has acquired a personal identity and integration of the total personality. In the process of living the mature person carries out the developmental task for his or her level of life and develops an ever-increasing number of abilities and skills for coping with the present and the future.

In terms of the previously presented criteria and their basic implications, a mature individual is a person of chronologically adequate physiological, sexual, cognitive, and ego development who:

1 Has the ability to respond differentially in terms of his needs and the outside factors operating in his situation

2 Channels his tensions, impulses, and emotions into constructive behavior and directs his behavior toward achievement of positive long-term goals, yet retains the basic sensitivity, openness, emotional driving strength, and high degree of satisfaction and enjoyment of late adolescence

3 In reference to his parents and peers, establishes interdependent patterns of relationships and is able to impress and influence them and to maintain his role and respond with flexibility

4 Is satisfied by and derives enjoyment from his status and occupation; continues to develop and expand a reservoir of abilities, skills, and viewpoints; learns to recognize his own assets and limitations; and seeks compromise and creative solutions

5 Is at home with reality in most of its aspects; has learned to relate effectively with persons of all ages and to see himself and others with respect, patience, and humor

6 Values and considers the alternatives and consequences of his actions; finds ways of contributing to his community, his nation, and humanity

7 Feels whole and satisfied with himself and his own design for living and is committed to his internalized values and ideals

In terms of later adulthood, the mature person is one who has incorporated into his self-structure the Eriksonian attributes of trust, autonomy, initiative, industry, identity, intimacy, generativity (creative productivity), and integrity. His dynamic yet integrative pattern of adjustment makes him capable of working effectively and loving wisely.

Major Implications

It is well to notice that the criteria of maturity presented above are related and overlap to an appreciable extent and that therefore, for practical estimation, one needs to consider only three or four of them to come to accurate conclusions about a particular person. The criteria need to be considerably modified for a child or an adolescent. Moreover, any strict application readily leads to a perfectionist interpretation. Striving for a higher level of integration, consistency, adequacy, and progress in maturity during adulthood does not make one perfect; mediocrity usually prevails. There is always much to be desired in terms of human potential for creative and integrative activities when one closely observes the spectrum of interpersonal, intergroup, and international behavior. Gardner Murphy's *Human Potentialities* (1975, pp. 129–137, 243–252) points to many possible extensions of creative activity into new dimensions and fields by developing and applying powers inherent in the "three human natures," namely, biological, social-cultural, and creative. Maslow (1971, pp. 41–50) adds "peak experience"—which is transcendent and illuminates values—as the climax of self-actualization toward "fully human existence."

Maturity and Adjustment These two attributes of life appear to be inseparable. Without satisfactory progress in maturation, adequate adjustment to oneself, to others, and to one's culture is impossible. Age-related maturity is a determining disposition toward proper responses and relationships, even when situations involve frustration, deprivation, or other adversity. Mature persons, as a rule, recover more rapidly than immature persons from conflict, strain, or frustration.

It is conventional to indicate that maturity promotes mental health and general welfare since it safeguards a person from prolonged conflicts and unrealistic fears or anxieties. Maturity provides stimuli for familial and communal improvements, because a mature individual acts as a corrective influence in his broad environment, in some ways at least. Adult life-styles marked by genuine maturity are what every society ought to aim to develop in its education of the child and the adolescent.

ACQUIRING A VOCATION

One of the important steps in the adolescent's striving toward adulthood is in the area of vocational training. Today, because of automation, the need for unskilled labor continues to decline. Moreover, most vocational training takes longer than it did in the past, and changes in occupational choices cannot be made without encountering some difficulty or financial loss. Vocational choice involves a vital decision since, especially for boys, it will affect the whole life. The adolescent realizes that a vocation is needed for the economic independence for which he is striving. Ability and vocational preference must be reconciled at the outset of vocational training.

Many factors must be considered in the selection of a training program, and several decisions have to be made. The question of whether or not to continue in high school is one of the early decisions. Despite the fact that a great majority of parents—more than four-fifths—urge their adolescent boys and girls to complete their high school education, over 30 percent of the adolescents in the United States were high school dropouts in 1970. The process of choosing a career is the first significant confrontation with the sense of maturing, if not aging. The individual senses that such a decision is fateful because it determines how the rest of his or her life will be "filled in." The fact is that the person must make a choice; yet he may wonder if society will grant him his first or even second choice, for more and more candidates are being rejected by professional and graduate schools (Sarason, Sarason, and Cowden, 1975).

There are indications that the United States is approaching realization of the ideal of universal education. Despite increased employment, larger numbers of youths are going to school, not only because the population is growing but also because a larger proportion of school-age children continue their education into the high school and college levels.

Over the last several decades, the proportion of young people at work has declined sharply. Labor statistics indicate that in 1970 to 1972 only less than half of the sixteen- to nineteen-year-olds were employed. The increase of certification requirements in the last two decades has effectively barred most young people from professional and semiprofessional occupations. Moreover, trade unions fear any large-scale incursions by young people. The resultant deferment of higher responsibilities and of productive opportunities reduces sharply the options young people have in seeking employment. Schools add their share by segregating young people from adult workplaces (Coleman et al., 1974, pp. 128–133). The minimum wage, now set at a relatively high level, discourages the employment of the young, whose productivity is not yet sufficiently high to merit it (p. 168).

Table 16-2 shows employment and school status by age, sex, and race from sixteen to twenty-four years of age. During the 1960s, both the employment rate and the school attendance rate continued to rise, with a new high for both. In October 1970, students made up about 31 percent of the sixteen- to twenty-four-year-old labor force, up nearly ten percentage points from 1960. This increase of students in the labor force reflected both a rise in the proportion of young people in school and the increasing tendency of students to work. Rates of participation in the labor force were generally higher for whites than for blacks and other races. As seen in Figure 16-1, the number of college students sixteen to twenty-four years old attending college full time has more than doubled during the decade, rising to 5.2 million in 1970 from 2.5 million in 1960. Recent statistics from the Office of Education show that in the fall of 1973, total school enrollments reached 59.4 million, with 59 percent in elementary, 26 percent in high school, and 15 percent in college.

Child labor laws, including the Fair Labor Standards Act (1938) and the Child Protection Act (1966), protect children and young adolescents from work likely to hinder well-rounded development; thus they increase the adolescents' chances of remaining in school longer. The Fair Labor Standards Act sets sixteen years as the minimum age for work during school hours and for occupations in manufacturing at any time. At present, most states permit children and teenagers to work after school hours.

Many young people who withdraw from school feel pressed to take full-time jobs. Part-time employment satisfies students who are partly supported by their parents. Withdrawal from school at times entails hesitation on the part of parents to supply an older adolescent with funds. Availability of jobs is of crucial importance in this connection and will largely determine whether the person can find work suited to him and make a satisfactory adjustment to his occupation. The accessibility of trade schools is also an important factor in the acquisition of the technical skills sought by industry.

Bridging the gap between school and work is a significant challenge for youth today. First of all, young people must be adaptable if they are to fit into the rapidly changing world of employment. They must be well educated and emotionally prepared for jobs consistent with their education and talents. National Science Foundation studies indicate that, nationally, less than half of those in the top third of their high school graduating classes go on to graduate from college. Those who do not graduate take jobs below their potential capacity and deprive other young people with less educational ability of the jobs they could have handled adequately. From 1960 to 1970, 26 million young

Table 16-2 Employment and School Status (United States): Population Sixteen to Twenty-four Years Old, October 1970* (Numbers in Thousands; Percent Not Shown Where Base Is Less Than 75,000)

	Labor force, enrolled in school				Labor force, not enrolled in school			
Age, sex, and race	Population	Number	Percent of population	Unemployment rate	Population	Number	Percent of population	Unemployment rate
White, male:								
16–24	6,612	2,945	44.5	12.7	5,790	5,397	93.2	10.1
16–19	4,711	1,955	41.5	14.6	1,540	1,345	87.3	14.9
20–21	1,030	458	44.5	9.4	1,263	1,184	93.7	12.3
22–24	871	532	61.1	8.5	2,987	2,868	96.0	6.9
White, female:								
16–24	5,375	2,148	40.0	11.3	8,463	5,104	60.3	9.2
16–19	4,223	1,550	36.7	13.7	2,169	1,104	61.4	15.1
20–21	719	337	46.9	5.9	2,272	1,332	63.6	7.8
22–24	433	261	60.3	3.4	4,002	2,328	57.9	6.7
Negro and other, male:								
16–24	808	236	29.2	26.7	1,050	891	84.9	17.5
16–19	648	170	26.2	32.9	325	235	72.3	25.1
20–21	100	31	31.0		258	225	87.2	21.3
22–24	60	35			467	431	92.3	11.4
Negro and other, female:								
16–24	811	205	25.3	24.9	1,340	776	57.9	20.2
16–19	667	149	22.3	30.9	373	189	50.7	37.6
20–21	98	29	29.6		378	220	58.2	15.9
22–24	46	27			589	367	62.3	13.9

*Members of Armed Forces and inmates of various institutions excluded.
Source: Monthly Labor Review, August 1971, **94** (8), p. 15, table 2.

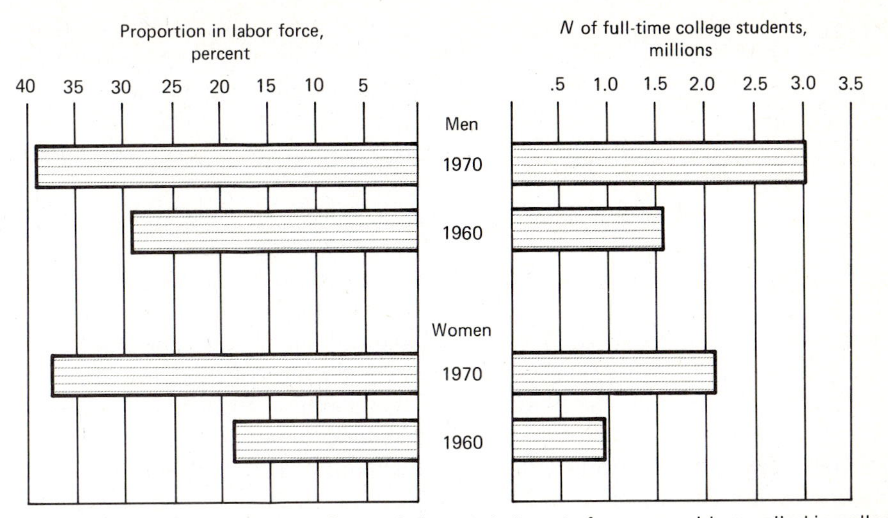

Figure 16-1 Status in the labor force of sixteen- to twenty-four-year-olds enrolled in college. [*Source:* U.S. Bureau of the Census, 1971.]

people entered the labor force. The poorly educated segment of this population will compete with machines for employment on somewhat unequal terms, since the machines being produced today have skills in many ways equivalent to a high school diploma.

> In the developing cybernated system, potentially unlimited output can be achieved by systems of machines which will require little cooperation from human beings. As machines take over production from men, they absorb an increasing proportion of resources while the men who are displaced become dependent on minimal and unrelated government measures—unemployment insurance, social security, welfare payments. (Ferry et al., 1968, p. 311)

Despite steadily increasing educational requirements for employment, nearly 30 percent of all young people drop out of school before finishing high school. Moreover, about 40 percent of all students entering college withdraw before completing a four-year program. Desirable employment is highly probable only for those who complete college. As shown in Table 16-2, the fairly high unemployment rate of 1970 for people with limited education is likely to rise as many occupations for the unskilled decline slowly in the years ahead. It is also noteworthy, that, according to census figures, the high school dropout earns only 12 percent more than the eighth-grade graduate, while the high school graduate earns 15 percent more than the high school dropout. The college graduate earns 42 percent more than the high school graduate and 83 percent more than the eighth-grade graduate. Since the American picture of success is related to both education and earnings, school and college are the major stepping-stones for advancement. Investments in education will go on paying dividends. Apparently the teenagers of today are growing up in an era where only lifelong education will prepare them to profit from the opportunities and advantages of rapidly changing industry and business. Figure 16-2 indicates that the average educational level of men and women, white and black, is rising and is currently well into high school for blacks and slightly beyond it for whites.

The existence of over 100,000 distinct occupations in the United States makes vocational choice complex. There is apt to be some trial and error unless a person has formed definite vocational goals based on a realistic appraisal of self and of the requirements of a given occupation. Limited knowledge of requirements adds stress to the vocational aspirations of many adolescents.

As the percentage of high schools having vocational guidance services steadily increases, it becomes more likely that young Americans will choose occupations commensurate with their abilities and the opportunities at hand. An adolescent is likely to put considerable value on results of aptitude tests, since his vocational interests are often confused. E. K. Strong, Jr. (1943) found that only after the age of twenty-five are vocational interests well crystallized. It is unfortunate that a majority of people, including those who attend college, have to make a vocational choice before this age. During early adulthood most persons, frequently after two or three trials and changes, settle down and maintain the same occupation.

Employment counseling programs were established on a nationwide basis in 1945. Since the National Defense Education Act of 1960, there has been rapid growth in the guidance services offered by secondary schools. In 1960–1970, there were 46,189 counselors and 5,622 psychologists in the public sec-

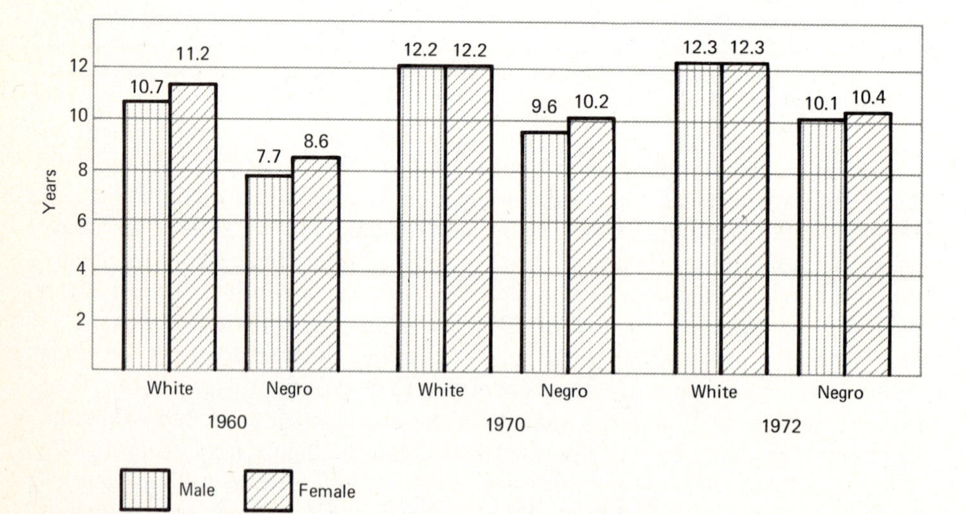

Figure 16-2 Median school years completed by persons twenty-five years old and over: United States, 1960, 1970, and 1972. [*Source:* U.S. Bureau of the Census, 1970 and 1973.]

ondary schools. Most school dropouts and at least 40 percent of high school graduates need both aptitude testing and employment counseling (DHEW, 1972). David A. Goslin (1963, pp. 53–54) estimates that each year between 200 million and 250 million standardized ability and aptitude tests are administered in the United States. He emphasizes (p. 189) the need for assessing ability through aptitude testing.

SELECTING A MATE

At the same time that vocational opportunities are attracting the young adult, so are the selection of a spouse and the establishment of a home. Despite the fact that these accomplishments involve difficulties and risk, they contribute substantially to the achievement of adult status and adjustment. The average age for marriage has become lower during the course of this century, but it seems to have remained relatively constant since 1950, as indicated by Figure 16-3.

In late adolescence or early adulthood, a strong and lasting attraction toward a peer member of the opposite sex occurs. Such attachments are frequently mutual and offer deep ego gratification as well as promoting feelings of adequacy and security. The percentage of companionships and romances that lead to mutual love and marriage rises steadily and reaches its peak in early adulthood. The universal desire to love and be loved and feel needed finds its complete satisfaction in marriage. While many persons in the early twenties are sufficiently mature to assume marital and parental responsibilities, a significant number are not yet ready. Quite a few, some of whom have been damaged in their earlier development by the attitudes of their parents, later find the partners they select incompatible. The desire for personal security, as well as

Figure 16-3 Ages of brides and bridegrooms: United States, 1910–1972. [*Source:* U.S. Bureau of the Census, *Statistical abstract of the United States: 1973* (94th ed.); *Current population reports* (Series P-20), 1973.]

parental pressure, influences some girls to contract marriage, particularly during the later part of young adulthood. Pregnancy is another fairly frequent reason for early marriage. Thus the voluntary acceptance of this new role can be questioned in many marriages—early marriages in particular.

The time needed to complete adolescent cognitive, emotional, and social maturation has increased with the complexity of our technology and culture and the rising standard of living. In American society, completion of pubertal growth does not imply adult competence as it does in most primitive societies. During the twentieth century, the duration of adolescence has increased with each generation. Since success in marriage depends on a certain level of maturity, each increase in the period of adolescence implies a longer period of delay before marriage. Since the early 1950s, each generation has married at an early age: Apparently many married before they were psychologically ready to do so successfully. Teenage marriages are frequently contracted because of sexual involvement and pregnancy. Marriages are often preceded by too short a period of courtship, thus increasing the chances of incompatibility between the individuals involved. These inadequate preparations for marriage are contributing causes of discord, separation, and early divorce. Statistics, though incomplete, show that during the first half of the century the rate of divorce was actually increasing from generation to generation. Many teenage marriages are dissolved by divorce. Emotionally immature teenagers find it difficult to postpone marriage. In most states teenagers cannot marry without parental consent, and the laws determine at what age such consent is unnecessary. However, elopements are still frequent. Table 16-3 shows the increase in the rate of divorce from 1940 to 1972. The total rate has grown from 25 to 52 persons per 1,000 married. Related statistical findings indicate that divorces are more frequent in states with lenient divorce laws; among city families,

Table 16-3 Divorce Rates: United States, 1940–1972

	1940	1950	1955	1960	1965	1970	1971	1972
Total number (thousands)	264	385	377	393	479	715*	768*	839*
Rate per 1,000 population	2.0	2.6	2.3	2.2	2.5	3.5*	3.7*	4.0*
Rate per 1,000 married women 15 years old and over	8.8	10.3	9.3	9.2	10.6	15.0†	15.9†	(NA)
Percent divorced, 18 years old and over:‡								
Male	1.4	1.8	1.9	2.0	2.5	2.5	2.9	2.8
White	(NA)	(NA)	(NA)	2.0	2.4	2.4	2.8	2.7
Negro and other	(NA)	(NA)	(NA)	2.2	3.4	3.4	3.4	3.2

NA—Not available.
*Preliminary.
†Preliminary rate for women fourteen years and over.
Sources: U.S. Bureau of the Census, *Current Population Reports,* Series P-20, 1973; U.S. National Center for Health Statistics, *Vital Statistics of the United States,* annual. (Beginning 1950, median age at first marriage based on sample.)

especially those of lesser education; in the laboring class; and in teenage marriages and mixed-faith marriages. Divorces are most frequent within the first few years of marriage.

The expectation of marital happiness is often frustrated on psychological grounds, one being the dominance relationship. If each partner has a strong tendency to dominate the other, congruity is difficult to establish. An example of incongruity may also be seen in intense striving for independence on the part of both partners. Lack of similarity in interests and activities presents a major obstacle to the mutual sharing of marital life. Neurotic behavior and doubts about the spouse's faithfulness often evoke anger or jealousy and jeopardize marital stability. Finally, ineffectiveness of communication makes other problems insoluble.

Prerequisites for a Successful Marriage

According to Simmel's early theoretical assumptions (1950) and subsequent research by sociologists and psychologists, race and religion have usually been the most decisive factors in designating males and females as acceptable or unacceptable for courtship and marriage. Experience gathered from thousands of families indicates a difficulty in sharing married life without sharing a basic identity in appearance or in faith. Difference in socioeconomic status also adds many difficulties to some marriages.

Similarities in socioeconomic background, age, intelligence, talents, and acquired abilities contribute to the permanence of a marital relationship. These factors are helpful in promoting mutual understanding and enjoyment of the same activities, especially if love is to be more than infatuation or mere companionship. Table 14-2 (page 264) shows traits indicated as desirable in a companion and in a future mate by 2,000 high school boys and girls. The second group of characteristics suggests that students realize that physical attraction and affection alone are not enough for marital success. The first group identifies traits desired in a companion whether marriage is being considered or not.

Mutual gratification of needs at the deepest psychological level is the essence of marital stability (Yorburg, 1973, pp. 125–127). A dissimilarity between the personality needs of one and the corresponding drives of the other will thus assist in deepening a relationship beyond the level of mere companionship. For example, persons with high assertiveness tend to be attracted to persons having receptive traits as dominant characteristics. Persons with strong strivings toward independence fit the needs of those preferring dependence and submission better than persons with similar strivings. Robert F. Winch (1958, pp. 96–98, 101–103) originated and presented supporting data for the hypothesis that complementary rather than similar need patterns promote marital adjustment. Reciprocity in need of gratification furthers love and makes possible the fullest possible personality development within the family structure. Nevertheless, in terms of basic interests, abilities, and values, it appears that similarity of partners is a major force contributing to marital adjustment.

Moreover, persons who are competent in interpersonal relationships, mature, and well adjusted are capable of contributing to marital success.

Happiness in marriage also depends on the degree of similarity of the partner and one's concept of the ideal mate. The image of the cross-parent plays a role in building such a concept. The positive qualities of the mother and other women in his life deeply impress a boy and subconsciously influence his image of the ideal mate. As the "dream model" crystallizes during adolescence, it begins to act as a definite influence on the boy-girl relationships that follow. The girl's image of the father's desirable qualities and traits causes her to look for these characteristics in male companions.

The selection of a mate is greatly influenced by popularity among members of the opposite sex and by attitudes concerning the appearance and personality of a life partner. It has been found that men rank appearance, contiguity of interests, and cheerfulness much higher than women do; to women, intellectual abilities, educational status, and social ease are of prime importance. Of course, the intensity of love tends to outrank these and many other considerations, especially among young persons contemplating marriage. When in love, marital partners are usually successful in complementing one another. When love declines, other factors gain in significance.

Recent studies of factors in mate selection and marital happiness reflect the changing nature of people's attitudes and expectations of marriage. In *The Intimate Experience,* Arlene Skolnick (1973) aptly summarizes the prevailing negative expectations for continuation of marriage and the family. Standard texts on marriage and family adjustment, however, continue to emphasize the possibilities for happiness if both spouses have similar interests and backgrounds, a happy family life during childhood, compatible expectations of each other, of the self, and of parenthood, and significant investment in each other's happiness (Kelly, 1969). When all data are considered, a successful marriage is most likely achieved when the woman is twenty-five, the man is twenty-eight, and children come two to three years after marriage. The popular age for marriage is twenty-three for men and twenty for women, a fact which may contribute to the high divorce rate (Lasswell, 1974).

In most young families strain is inevitable, even if the husband and wife are close in background, training, personality, and values. Some grinding of the gears still takes place as they try to mesh their life-styles together into a family style and as the family style changes when children come. Some disruptive effects come from in-laws, occasionally from peers or neighbors, more often from mass media. Under the best of circumstances individual members, possessing conflicting needs and desires, standards and ambitions, are bound to grate against one another from time to time. Family friction often intensifies when wives seek fulfillment through careers. Sexual frustration of husbands and wives is often a symptom of unresolved conflict in other areas of family life. Willingness to compromise is necessary to avoid unrelenting family friction and disaster (Wernick, 1974, pp. 120–123).

Need for Marital Counseling

Marriage represents a major transition that challenges personal maturity and adequacy. Personal problems, if present, readily produce family problems. Growing obligations, an incongruity of leading traits, lack of preparation, and sexual incompatibility are other obstacles to marital adjustment. Friction between the parents during childhood, heightened self-defensiveness, and financial difficulties add to the discord and disappointment of many couples. Marriage counseling is needed for most such cases to avoid deterioration of the marital relationship. A marriage counselor usually realizes that marital conflicts spring either from severe incompatibilities or from personal problems of one or both parties. Lack of maturity is a frequent source of difficulty. A husband, for example, may continue to be attracted by other women. Jealousy results and may remain a permanent obstacle to mutual understanding. Perfectionist strivings on the part of the husband may be expressed in remarks indicative of dissatisfaction with the wife's actions or criticism of her methods of doing routine tasks. Frequently these lead to more disruptive arguments and general discord within the budding family structure. Conflict is sometimes related to previous vocational or other aspirations of the wife, who regrets substituting homemaking for them. The wife's financial demands and pressure on the husband to achieve a better-paying position are other sources of early discord. Generally several undesirable factors or traits work together to disrupt equilibrium or make it impossible to establish. Professional counseling aids familial integration by discovering the underlying factors of discord and establishing new lines of communication.

Higher levels of education and socioeconomic background or achievement lend themselves to extensive social participation and adaptability to others, in marriage relationships as elsewhere. Although sexual and economic factors are often a source of family friction, they are primarily symptomatic of deeper causes of marital discord. General lack of preparation and undesirable personality traits, such as impulsiveness, a critical attitude, domineering behavior, and neurotic tendencies often cause family dissension. The belief in romantic love for the "one and only" tends to produce disappointment and frustration as well. Sexual incompatibility is usually recognized as symptomatic of unresolved marital discord, but in some cases it is a major cause of marital disintegration. (Wernick, 1974, p. 123).

INFLUENCES ON PSYCHOLOGICAL STATUS

Many factors affect in countless ways the status of the adolescent and young adult among his peers and others. The pattern of adult society may be unattractive to a person in the teens or even twenties. Some have a strong desire to continue the life-style set by adolescent peers despite the fact that chronologically they have reached the age of majority. A large number of young adults remain engrossed in typically adolescent activities and in other ways oppose what is conventional in adult society. They show contempt for authority through criticism of those directing social and cultural affairs.

The majority of young adults, however, appear to have little difficulty in accepting societal and cultural norms and acting in accordance with expectations based on these guides. The need for self-reliance, however, is still great, and many postadolescents experience anxiety in acting on their own judgment. The need for support from others is not easily discarded, and it takes time to become comfortable amidst the complexities of adult society and culture. Appearance and level of maturity are some of the factors that rank high in estimating developmental status. Physique, as it compares with that of peers, is one of the major influences on status. Adolescents and adults differ more in height than children of the same age; the maturing and the physically mature differ greatly in height, weight, and many other quantitative features. To begin with, as discussed earlier, some girls as well as boys begin pubertal transformation sooner than their age-mates. In a typical eighth-grade class many girls are adolescents, while others lag behind in development and are only entering the pubertal phase of accelerated growth. There may be several postpubertal boys in the class, along with others who are still children and conspicuously small by comparison. Many have blemishes and skin eruptions, in addition to voice changes. Embarrassment and tensions mount when attempts at self-assertion fail more frequently than before. Some are upset by unforeseen sexual phenomena. Doubts about personal adequacy emerge more frequently as social acceptance decreases. The situation may become grave when these adolescents enter high school and find unexpected difficulties in making new friends. Physical appearance is important in attracting friends throughout life.

Intrafamily frictions often develop as the adolescent girl tries to become womanly and charming by the use of cosmetics and women's dress and the adolescent boy tries to show his manliness by attempting to smoke and to drink. Selection of friends, late-evening social life, money, and the use of the family car are frequent points of conflict with parents. The atmosphere of the home, its educational and socioeconomic level, and its morale all have a major bearing on the total family relationship.

Adolescents need many opportunities before they can make adult commitments. Some developmental psychologists go beyond the idea of a mere "moratorium" and suggest a postadolescent stage of life in which the adolescent is basically mature but not committed to the traditional roles of either marriage and homemaking or occupation. Such a postadolescent stage, up to the age of about twenty-seven or thirty, remains uncharted; ties to society may be unsettled and crises and emotional turmoil may run unchecked until eventually the person settles down in marriage or in an occupation. At any age, but most often in twenties, there are many "half-developed" persons who do not fill any adult role or position adequately.

In the modern nexus of culture and scientific advance, excessive reliance on science is common. Science is often seen as the road to all knowledge. There is danger of a partial or complete exclusion of other key ways of knowing, such as religion, philosophy, and the arts. Lack of perspective results, which in turn promotes compartmentalization and denial of important dimensions of reality. As a consequence, personality development may remain incomplete in its structural aspects, with a lack of maturity evident in many responses to the fundamental questions of life.

Personality growth usually continues throughout early adulthood, and in some respects it progresses until the late years of senescence. If ordinary developments have taken place in childhood and pubertal conflicts have been resolved, self-direction and objectivity increase rather sharply at the entrance into adulthood. The search for identity often continues until well into adulthood, when one fully answers the twin questions of "Who am I?" and "What am I?" One must know where he stands and what he believes. It is difficult to take social roles without establishing an identity; it is barely possible to make crucial adult decisions without a clear idea of who one is (Lantz & Snyder, 1969, p. 314).

It is advantageous for the maturing person to capitalize on the preceding phases of development. The genuine smile of infancy; the curiosity and experimentation of the preschool years; the affiliative trends and vivid emotions of childhood; and the zest for adventure, courage, and idealism of adolescence—all may be incorporated into an adult personality even though now expressed in new ways. During the years of adulthood, much should be modified and added. Personal initiative for achievement, continuity of effort, sensitivity to the needs of others, and a vision of alternatives in planning are all needed for an adult style of life. Self-reliance comes not only from identifying with mature models but also from knowledge of religion, philosophy, social sciences, and the arts. It will be supplemented by information and counsel gained from other adults of professional competence.

GUIDING THE MATURING PERSON

Few would argue against a need for continued guidance of adolescents and young adults, yet not many would agree on an exact approach. Since the adolescent is no longer a child, parental and educational demands should be different from those placed on a child. Baumrind (1968) correctly observes that if the imposition of authority by use of power is legitimate in childhood, it is not applicable in adolescence, because "the level of cognitive and moral development of the adolescent is such as to require that he be bound by social contract and moral principles rather than by power." Several ideas should be kept in mind in attempting to clarify the young person's need for guidance. First, he or she is in a state of rapid development. This development is pervasive and encompasses practically all dimensions of the personality. Whenever change or reshaping is in progress, any detrimental influence can deform it more readily than after the young person has acquired an advanced level of personality organization. Certain undesirable influences are therefore incongruous with what has been developed up to this age. The maturing person's past experiences and internal controls must also be respected. He or she usually needs many good influences to support strivings for a higher level of humanization of emotions and drives, values and aspirations. Lawrence K. Frank (1951) points out also that "the best preparation for tomorrow is to live adequately today, to deal with today's requirements so as to be able to go forward without too much 'unfinished business.'"

Modified informed permissiveness seems to be the best approach in helping adolescents and young adults alike, especially in the complex situations of modern life. Since the young person's desire to learn often runs high, his attention is usually satisfactory or good and he will benefit from appropriate information about alternatives in a permissive atmosphere. A positive attitude of respect for his freedom of expression and choice usually appeals to him and helps him to reciprocate. Permissiveness is modified by his needs and internal control, about which the guidance worker, parent, teacher, counselor, or camp leader need competent information from reliable sources. Without proper guidance, growing into adulthood is not a process of enlightenment or certainty. Percival M. Symonds's intensive study (1961, p. 206) of a small sample of young adults (N=28) shows that "their maturing was on the whole a blind, trial-and-error process."

It is important that proper guidance be available to all growing persons who need information and advice. The maturing person should be encouraged to come out of himself and to accept greater responsibility in making decisions for his own advantage and that of his family. Finally, in guiding maturing persons, Goethe's maxim must be taken seriously: "If we take people as they are, we make them worse. If we treat them as if they were what they ought to be, we help them become what they are capable of becoming."

QUESTIONS FOR REVIEW

1. What adolescent achievements indicate a readiness to enter adulthood?
2. What are the typical difficulties the adolescent faces before attaining an adult pattern of life?
3. Why do ambivalence and defense mechanisms often indicate a lack of maturity?
4. Define adult maturity and present a theoretical model for it.
5. Identify several criteria for testing maturity and evaluate one of them.
6. In what ways does a philosophy of life promote maturity?
7. What are the outstanding characteristics of a mature person? How will such a person cope with conflicts and adverse situations in his or her life?
8. Indicate some relationships between (a) maturity and adjustment, (b) maturity and mental health.
9. List and explain some major factors operating in the vocational choice of a maturing person. Consider educational achievement.
10. Enumerate several major factors contributing to marital success. Analyze one of them.
11. List factors contributing to marital friction and explain two or three of them.
12. What kinds of parental guidance and professional counseling does a maturing person often need?

REFERENCES

Selected Reading

Allport, G. W. *Pattern and growth in personality.* New York: Holt, 1961. Chap. 15. Extension of the self through security, self-objectification, and a unifying philosophy of life is discussed.

Cox, R. D. *Youth into maturity: A study of men and women in their first ten years after college.* New York: Mental Health Materials Center, 1970. Presents the patterns of maturing after college, stressing the normal and mentally healthy ones. Includes experiences in work and marriage.

Heath, D. H. *Explorations of maturity: Studies of mature and immature college men.* New York: Appleton-Century-Crofts, 1965. A systematic research monograph exploring the meaning of maturity and reporting results of testing a small sample of college students for maturity.

Lantz, H. R., & Snyder, E. C. *Marriage: An Examination of the man-woman relationship* (2d ed.). New York: Wiley, 1969. Presents marriage and the family in historical and cross-cultural context, emphasizing premarital relationships and the changing role of women.

Specific References

Allport, G. W. *Pattern and growth in personality.* New York: Holt, 1961.

Allport, G. W. Crises in normal personality development. *Teachers College Record,* 1964, **66,** 235–241.

Arnold, M. B., & Gasson, J. A. *The human person.* New York: Ronald, 1954.

Baumrind, D. Authoritarian vs. authoritative parental control. *Adolescence*, 1968, **3,** 255–272.

Bayley, N. Development of mental abilities. In P. H. Mussen (Ed.), *Carmichael's Manual of child psychology* (3d ed.). New York: Wiley, 1970.

Bossard, J. H. S., & Boll, E. S. Personality roles in the large family. *Child Developm.*, 1955, **26,** 78–81.

Coleman, J. C., et al. *Youth, transition to adulthood.* Report of the Panel on Youth of the President's Science Advisory Committee. Chicago: University of Chicago Press, 1974.

Cox, R. D. *Youth into maturity: A study of men and women in their first ten years after college.* New York: Mental Health Materials Center, 1970.

DHEW, Office of Education. *Statistics of state school systems, 1969–70.* Washington, D.C.: Government Printing Office, 1972.

Ferry, W. H., et al. The triple revolution. In A. E. Winder & D. L. Angus (Eds.), *Adolescence: Contemporary studies,* pp. 308–321. New York: American Book, 1968.

Frank, L. K. Personality development in adolescent girls. *Monogr. Soc. Res. Child Developm.,* 1951, **16,** No. 53.

Goslin, D. A. *The search for ability: Standardized testing in social perspective.* New York: Russell Sage, 1963.

Kagan, J., & Moss, H. A. *Birth to maturity.* New York: Wiley, 1962.

Knight, J. A. The profile of the normal adolescent. *Ann. Allergy,* 1967, **25,** 129–136.

Kelly, R. K. *Courtship, marriage, and the family.* New York, Harcourt, Brace, 1969.

Kohlberg, L. Stages of moral development as a basis for moral education. In C. M. Beck, B. S. Crittenden, & E. V. Sullivan (Eds.), *Moral education: Interdisciplinary approaches.* New York: Newman, 1971, pp. 23–92.

Lantz, H. R., & Snyder, E. C. *Marriage: An examination of the man-woman relationship* (2d ed.). New York: Wiley, 1969.

Lasswell, M. E. Is there a best age to marry? An interpretation. *Fam. Coordinator,* 1974, **23,** 237–242.

Maddi, S. R., & Costa, P. T. *Humanism in personology: Allport, Maslow, and Murray.* Chicago: Aldine, Atherton, 1972.

Maslow, A. H. *Self-actualization and beyond.* In J. F. T. Bugental (Ed.), *Challenge to humanistic psychology.* New York: McGraw-Hill, 1967.

Maslow, A. H. *The further reaches of human nature.* New York: Viking, 1971.

Montagu, A. *The direction of development.* New York: Harper & Row, 1955.

Murphy, G. *Human potentialities.* New York: Viking Press, 1975.

Rogers, C. R. Toward a modern approach to values: The valuing process in the mature person. *J. abnorm. soc. Psychol.,* 1964, **68,** 160–167.

Sarason, B., Sarason, E. K., & Cowden, P. Aging and the nature of work. *Amer. Psychologist,* 1975, **30,** 584–592.

Schaie, K. W., & Gribbin, K. Adult development and aging. *Ann. Rev. Psychol.,* 1975, **26,** 65–96.

Simmel, G. *The Sociology of Georg Simmel* (K. H. Wolf, Trans.). Glencoe, Ill.: Free Press, 1950.

Skolnick, A. *The intimate environment.* Boston: Little, Brown, 1973.

Strong, E. K., Jr. *Vocational interests of men and women.* Stanford, Calif.: Stanford, 1943.

Symonds, P. M., with Jensen, A. R. *From adolescent to adult.* New York: Columbia, 1961.

Wernick, R. *The family.* New York: Time-Life Books, 1974.

Winch, R. F. *Mate selection: A study of complementary needs.* New York: Harper & Row, 1958.

Yorburg, B. *The changing family.* New York: Columbia University Press, 1973.

Chapter 17

Developmental Tasks of Early Adulthood

HIGHLIGHTS

Early adulthood is the period of maximum consolidation of developmental progress. It ushers in the adult's life work, marriage, and child rearing.

Specific tasks of early adulthood include achieving emotional, social, and economic interdependence, along with marriage, parenthood, and establishment of the home.

A person at this time of life is establishing the pattern of his or her adult personality, enhancing self-actualization, and continuing the process of development and adjustment.

Following the conflict-ridden stage of adolescence, the maturing person in our society is faced with yet another series of tasks and problems, those of integration into adult society and culture. The challenges and responsibilities that must be met and accepted are numerous, and the potential hindrances to satisfactory development and adjustment are varied. If a person is to participate fully as a peer in the adult culture, he or she must vote intelligently, keep informed on international, national, and local issues, and be willing to assume responsibilities in civic, religious, and educational organizations. This chapter attempts to analyze the major developmental tasks and problem areas in the early adult years.

ACHIEVING INTERDEPENDENCE AND RESPONSIBILITY

Among the criteria of maturity mentioned in the previous chapter are the achievement of interdependence and willingness to assume adult responsibilities. These two features are crucial attributes in distinguishing an adult from an adolescent. The process of becoming interdependent and responsible is particularly important in the emotional, social, and economic areas of life. Each of these areas will be examined separately. As used here, *interdependence* refers to a balance between dependence and independence and between giving and taking. Dependence is a mark of infancy and childhood; independence is often a characteristic of an eccentric person or an aloof schizophrenic rather than of a mature adult. It may be added that as regards the family or peer group, young adults show both dependence and sharing as well as autonomy and support. They both give to and receive from others.

Emotional

Emotional interdependence is best understood as the progression from dependence first on parents and then on peers to a level of relative autonomy. Although maintaining close emotional ties with others, a person becomes less susceptible to anger or despair when others disagree or are displeased with him than he was as an adolescent. Emotional autonomy is the most important and most difficult area of interdependence to achieve. The young adult must attain a level of affective development in which emotional needs are best satisfied by peers rather than by parents. Too strong an attachment to one or both parents will create severe problems, especially in marital adjustment. If a person derives his or her greatest satisfaction from pleasing parents or from being with them rather than with peers, particularly of the opposite sex, that person's emotional development is retarded or distorted.

In most cases, the process of freeing oneself from emotional dependence on parents begins during preadolescence. The increased social life of adolescence and new friendships with members of both sexes aid in the transference of emotional ties. During late adolescence and early adulthood, feelings toward parents should become more adult in tone. That is, the affection should be coupled with mutual respect and less like the dependent attachment of childhood. Evaluating decisions, actions, or persons in terms of parental opinions or acting only with parental approval are characteristics of an earlier level of social-emotional development.

Attaining emotional independence from parents does not imply complete self-sufficiency. Rather, one always remains interdependent in relation to others. Affection, security, status, and related needs are now satisfied primarily by the marriage partner, peer group, and occupational associations. The person, in turn, contributes to the gratification of the emotional needs of others in the family and reference groups.

Emotional interdependence is not achieved by a mere transference of emotional dependency from parents to peers. The young adult who is as dependent on his peers as he once was on his parents is still far from being emotionally mature. Freedom from parental control, in terms of emotional attachment, is usually recognized by both parents and young adults as a natural step in the growing-up process. However, emotional maturity means a certain degree of freedom from group domination as well. A young adult who feels lost when he is not with his particular peer group, or one whose decisions are dictated by his peers, is still emotionally dependent. Similarly, the husband who merely shifts his emotional dependence from mother to wife has made little progress toward emotional maturity. Establishing proper emotional ties with others is a comparatively slow process, and the adolescent and early adult years serve this purpose well.

The capacity to love someone other than oneself is another integral factor in adult emotional interdependence. Excessive self-love and the inability to give of self are signs not only of emotional immaturity but also of a personality disorder. A successful marriage demands a giving of self, and one who is incapable of this emotional giving will have grave difficulties in adjusting to marriage. Marriage demands a deep and permanent emotional cathexis and self-involvement. It flourishes on genuine partnership and mutual support from sharing responsibilities (Bee, 1959).

Social

Social interdependence refers primarily to the acceptance of a person in adult society. He must demonstrate decidedly adult traits and qualities or, despite his age, he will not be treated as an adult. The adoption of mature social characteristics is generally not too difficult, and this criterion of social maturity is met by many who do not strive for leadership.

Some young adults, however, are not satisfied with mere acceptance in an adult group. Having been leaders in their peer groups, they strive for positions of power and prestige in groups or organizations composed of older people. Often their attempts meet with rebuff or a wait-and-see attitude; in a sense they are on probation, and adults will not, as a rule, grant them positions of leadership until they have proved that they are also good followers and are able to contribute something worthwhile to the group.

A young adult should try to contribute to the group and in that way benefit both the group and himself. Volunteering for some of the necessary routine tasks and performing them well will gain him the respect of the adult group along with a heightened feeling of achievement and confidence in his dealing with his seniors.

A decidedly more difficult task, and one that is allied to emotional maturity, is the achievement of self-direction rather than group domination. The socially mature adult is in large measure inner-directed rather than group-controlled. His decisions and behavior flow from personal conviction based on his own values and standards. The group is not a constrictive force binding him with ties of dependency. Decisions of the group that run contrary to his convictions are not readily accepted by an adult who has achieved social

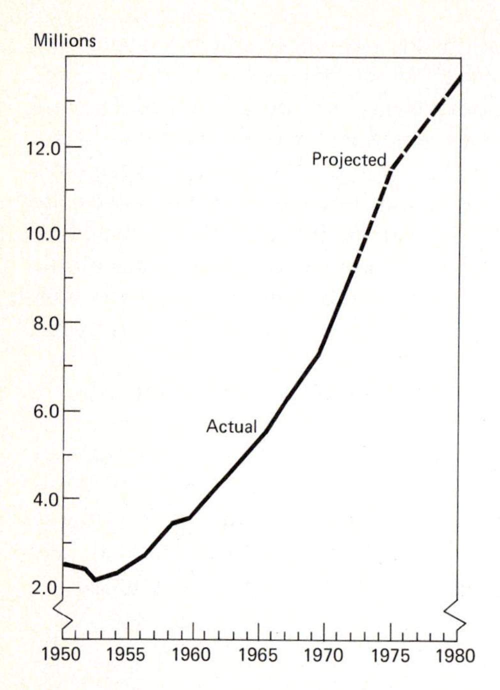

Figure 17-1 College enrollment: United States, 1950–1980. [*Source:* U.S. Bureau of the Census, *Current population reports* (Series P-20, No. 149, and Series P-25, No. 338), 1966; U.S. Office of Education, *Projections of educational statistics to 1980–1981,* 1974.]

maturity. He has reached the point at which his feelings of personal adequacy and security are such that they do not need constant reinforcement, which implies insecurity or dependence. The adult accepts a democratic consensus as binding, while also recognizing that communication of one's beliefs is part of group interaction.

Social interdependence carries with it social responsibilities. To partake of the privileges of adult life, a person must assume the obligations of an adult. Occupational, educational, civic, and religious areas are but a few of those in which the young adult must assume responsibilities and be willing to do his part.

That many young adults are preparing for their future responsibilities by continuing their education through college training is evident in Figure 17-1. From 1960 to 1970, college and professional-school enrollment has more than doubled; that is, it rose from about 3,570,000 to 5,675,000 students, excluding those over thirty-four (U.S. Bureau of the Census, 1972).

Economic

Integration into adult society usually presupposes economic autonomy, that is, being able to support oneself and one's family. This, as a rule, precludes living with a parent, whether or not the young adult is economically self-supporting. American culture, with its emphasis on additional higher education, makes the attainment of economic independence difficult. Without special training the number of occupational opportunities is severely curtailed; but in order to secure the necessary education for a specific occupation, young people are, in most cases, forced to prolong their financial dependence on parents. Parents usually encourage further training and the concomitant extension of financial dependence, and their partial authority over their children is therefore often extended into the mid-twenties. The rising number of federal, state, and other scholarships, grants, and loans for undergraduate and graduate studies facilitate financial autonomy for students.

The young adult is faced with yet another facet of this problem. To continue training is, in many cases, to postpone marriage. Is the period of marriage postponement "beyond the achievement of physical maturity . . . one of stress and strain, emotional incompleteness, frustration, physical restraint, and intellectual confusion for myriad young people, among whom the college student is one of the most common and most poignant sufferers," as Henry A. Bowman (1975), pp. 309–310) pictures it? Probably not, since many young persons are emotionally not ready for an exclusive long-term companionship, still less for parenthood and its responsibilities. Marriage is not a solution for adolescents who think themselves in love but are not ready to assume the contingent responsibilities.

Many young people, finding postponement of marriage and family life too great a sacrifice, try to combine marriage with further education. The result is, in most cases, a severe strain on the marriage relationship or an even more prolonged program of training or education, because the student is often forced to take a job and relegate education to evening classes. The birth of children complicates an already strained situation. This is not to imply that some young married couples cannot successfully combine marriage and extended training. The fact remains, however, that the first years of marriage demand from both partners a great deal of cooperation and concentration. Attempts to combine education and marriage put a strain on both endeavors. However, the great number of stipends and the large amount of housing for married students on many campuses encourage such prolonging of education. The percentage of the young married population completing college education continues rising at a fairly sharp rate.

Sidney L. Pressey (1965) questions the need for education so prolonged that half a lifetime may go by before one can begin his or her life work. Since the body of knowledge increases so rapidly, it might be better to start professional work earlier and then enter refresher courses and adult programs later to keep up with expanded knowledge and technology. A lifelong education, as suggested in Chapter 3, is a practical alternative.

Economic independence demands, along with freedom from financial dependence, an acceptance of financial responsibility. Young adults, especially in establishing a home, usually must go into debt. The prompt payment of bills and debts demands a degree of perspective and maturity. Usually payment of large bills demands some self-denial in the form of putting off other purchases and curtailing entertainment. However, chronic indebtedness in large amounts is a sign of financial mismanagement and a symptom of personal dependence. In recent years apartment living has become popular. It reduces greatly the efforts needed in regular homemaking but it also requires a steady income. During the 1970s, job stability declined greatly as compared to the 1960s.

Establishing a Home

Perhaps the largest problem confronting young couples is the establishment of their homes. The economic, interpersonal, and social problems are many in such an undertaking. The interpersonal adjustments that must be made will be discussed in the next section. Other problems are primarily economic. However, this may be considered as merely symptomatic of a deeper problem, namely, that many young couples desire at the beginning of their marriage the same standard of living that their parents have attained after twenty or thirty years of work, or a higher one.

MARITAL ADJUSTMENT

For most people, marriage is their deepest personal and social commitment. The constructive channeling of forces that did not participate in bringing the two persons together is a task of great magnitude. These forces often start disturbing the marital life before the end of the first year of marriage, unless one has high reserves of self-control and flexibility for readjustment. The marital role per se involves the acceptance of another role, that of parenthood. A desire for children by at least one partner, social expectations that married persons will have children, and accidental pregnancies are some of the factors operating in this regard. Temporary use of selected contraceptives helps in planning a family when there is a desire for offspring. The probability of conflict and divorce is doubled when a couple has children during the first year of marriage (Wernick, 1974, p. 131).

A young married woman is likely to entertain fantasies of holding a baby in her arms or having someone little to love her. Unless she works she may also want someone with whom to spend the hours her husband is at work. The less pleasant aspects of pregnancy, childbirth, and the many sleepless nights when the baby cries or is sick are less likely to come to mind or to impress her deeply at this early stage of married life. The passing years of married life usually intensify her desire to have children.

Family life involves fundamental learning in every area of living, both before and after children arrive. The demands of sharing in daily chores, conversation, and other activities and of thoughtfulness for others and respect for their claims and peculiarities become more complex with the addition of each new member to the family. It is impossible to discuss all the possible forms of relationship and interaction that occur within the framework of the family. Excluding extremes, each family represents life in most of its aspects. Family life is a stage of further growth for each member. Friction and rivalry within the total constitution of the family, even between parents, often serve as a necessary "training ground" for learning new and more mature responses and better ways of settling interpersonal differences. The need and opportunity to utilize additional potentialities is often present. As a rule, family living maximizes the opportunity for personal self-realization, although this is less true for the wife and mother who, busy with her home and children, has too little time left for her own education and cultural concerns.

Since the parents come from different familial constellations and often have quite different backgrounds, it is normal that they should have some differences of opinion about the rearing of children and that their attitudes toward children should be different. This constitutes a source of friction in their efforts to train and educate their children. A significant difference of opinion often appears concerning methods of discipline. Leniency by one parent is counteracted by strict adherence to certain modes of punishment by the other parent, and the sympathy expressed by the first parent may make a child antagonistic to the other. The tendency toward egalitarianism in the American family makes the role differentiation between father and mother difficult. As a result, many young persons feel somewhat confused as to how to organize their families and how to share in the care and management of their children (Winch, 1971, pp. 428–429).

It is apparent from the statistics (Table 17-1) that many marriages are not succeeding. The reasons are varied and the causes difficult to ascertain, but the fact is that an increasingly large percentage of marriages are failing. Table 17-2 shows the distribution of single, married, widowed, and divorced men and women. In 1972 the percentage of men who were divorced was 2.8, while for women the figure was 4.3. When divorced, men apparently remarry at a faster rate than women.

ADJUSTMENTS TO PARENTHOOD

The decision to have children is often made in some ignorance of the realities of parenthood. Unforeseeable contingencies may later distort the desire for parenthood even more and arouse feelings of rejection and hostility toward the offspring, especially if the child belongs to the sex other than that desired or appears to be retarded or handicapped. Since there is also likely to be tender sentiment and love, a mental conflict of considerable intensity develops with some mothers, and difficulties in adjusting to motherhood are frequent. Most young mothers, however, have sufficient control and stamina to overcome initial burdens and begin to gain satisfaction as the child develops and exhibits his or her personality. Their desire to be a good mother usually gains in strength, especially if the husband-wife relationship is satisfactory.

Considering overpopulation and economic realities, the modern family needs some form of birth control, but the method to be used is often at issue. After giving birth to two or three children in four or five years, most mothers need a longer rest period between pregnancies to prevent biological exhaustion and to safeguard their family's welfare as well as their own.

With many sources as a frame of reference, including fragmentary recollections of their own childhood, young parents begin to associate their role with maintenance of the physical and emotional welfare of their child. Care for the child's physical needs involves providing not merely age-related food and

Table 17-1 Percent Distribution of Single and Ever-married Men and Women by Marital History, Year of Birth, and Race: United States, 1971

(As of June; for Persons Born between 1900 and 1954)

Race and year of birth	Men			Women						
		Ever married in 1971			Ever married in 1971					
	Single in 1971	Once	Twice or more	Single in 1971	Once	Twice or more	Never divorced or widowed	Divorced	Divorced after first marriage only	Widowed
Total	23.2	66.5	10.3	17.5	85.5	14.5	73.9	14.9	13.7	12.3
white	22.4	67.4	10.1	16.6	85.9	14.1	74.8	14.3	12.9	11.8
Negro	29.2	58.3	12.5	24.5	80.8	19.2	64.4	21.1	18.6	16.9
other	30.9	61.6	7.2	22.4	91.7	8.1	81.2	9.4	8.8	9.5
Year of birth:										
1945–1954	67.2	31.7	1.1	50.9	94.3	5.6	90.2	9.2	8.8	0.7
1935–1944	12.2	79.2	8.6	7.1	87.7	12.3	82.0	16.2	15.3	2.0
1925–1934	6.4	80.7	12.9	4.5	84.6	15.4	78.4	17.2	15.6	5.3
1915–1924	5.4	79.2	15.3	4.3	82.2	17.8	70.7	17.0	15.4	13.7
1900–1914	5.5	76.4	18.1	6.2	80.8	19.2	53.9	13.9	12.5	36.1

Source: U.S. Bureau of Census, *Current population reports,* Series P-20, No. 239.

Table 17-2 Marital Status of the Population, by Sex and Age: United States, 1972
(In Thousands of Persons 18 Years Old and Over as of March, except Where Indicated as Percent)

						Percent distribution				
Sex and age	Total	Single	Married	Widowed	Divorced	Total	Single	Married	Widowed	Divorced
Male	64,228	12,558	48,054	1,834	1,781	100.0	19.6	74.8	2.9	2.8
18–19	3,629	3,336	290	—	3	100.0	91.9	8.0	—	0.1
20–24	8,247	4,690	3,455	1	101	100.0	56.9	41.9	(z)	1.2
25–29	7,117	1,375	5,539	4	199	100.0	19.3	77.8	0.1	2.8
30–34	5,913	732	4,996	12	173	100.0	12.4	84.5	0.2	2.9
35–44	10,988	904	9,578	55	451	100.0	8.2	87.2	0.5	4.1
45–54	11,212	613	10.057	154	390	100.0	5.5	89.7	1.4	3.5
55–64	8,848	452	7,769	390	321	100.0	5.1	87.8	3.5	3.6
65–74	5,410	314	4,391	593	112	100.0	5.8	81.2	11.0	2.1
75 and over	2,862	141	1,980	708	33	100.0	4.9	69.2	24.7	1.1
Female	71 557	9,856	49,047	9,601	3,052	100.0	13.8	68.5	13.4	4.3
18–19	3,783	2,912	846	6	20	100.0	77.0	22.4	0.2	0.5
20–24	8,992	3,275	5,464	21	231	100.0	36.4	60.8	0.2	2.6
25–29	7,361	900	6,059	35	357	100.0	12.4	82.3	0.5	4.9
30–34	6,126	412	5,347	58	309	100.0	6.7	87.3	1.0	5.0
35–44	11,614	522	10,069	335	689	100.0	4.5	86.7	2.9	6.0
45–54	12,142	514	9,926	991	711	100.0	4.2	81.7	8.2	5.9
55–64	9,984	550	6,935	2,056	443	100.0	5.5	69.5	20.6	4.4
65–74	7,046	470	3,395	2,962	219	100.0	6.7	48.2	42.0	3.1
75 and over	4,509	292	1,007	3,138	73	100.0	6.5	22.3	69.6	1.6

Dash represents zero; (z) indicates less than 0.05 percent.
Source: U.S. Bureau of the Census, *Current population reports,* Series P-20, No. 239.

clothing but also many comforts, age-mates, and toys for play activity, and an adequate education. The emotional welfare of the child is built on many expressions of affection and empathy, of "I know how you feel." Continuing love of the child implies frequent help and sympathy in time of trouble, tender handling, and patience with desires and whims. The parents' activity centers on their children for long periods of time. In such instances, parents serve as role models for their children, teaching them—primarily nonverbally—the kinds of personality characteristics, attitudes, and values with which the growing child will identify.

Parental attitudes and approaches in child rearing tend to differ in some ways with each child. The firstborn child often undergoes "experiments" in caring for a baby resulting from lack of information and experience in child management. The advice of other people is sought and followed with relatively little skill. Occasionally the firstborn baby is a victim of parental overprotectiveness—all his needs and whims are responded to. Experience with the firstborn proves invaluable when the second child arrives. The mother often shows more confidence in herself and less concern about the baby. The second child may be less often stimulated or even neglected and many of his whims and cries disregarded. Difficulties with the firstborn are likely to arise if the second-born is given too much attention.

The third child is no novelty unless the first two were of the same sex and different in sex from the third. Care is more casual and tension-free. The parents have learned to delimit gratification of children's demands. The third and following children are likely to suffer from this from the very beginning. With the exception of the last, they will be less dependent on their parents and more affected by their older siblings.

The children in the middle are likely to show less maladjustment in their childhood and adolescent years than the firstborn or the last. Good adjustment on the part of parents and their adequacy in dealing with the specific needs of each child are key factors in childhood adjustment to the demands of reality.

In adulthood, being a member of a family typically involves sacrifices in personal freedom. Parents of young children are restricted in their private and social life. While most kinds of occupational activities permit the father to be away for a considerable time, the mother is obliged to spend most of her time caring for, stimulating, and entertaining her children. Since the extended family has moved definitely into the past, high-quality substitution for the mother is difficult. Baby-sitting by someone else tends to produce much concern on the part of parents until they develop a more relaxed attitude toward the realistically heavy responsibilities of child rearing. The recent growth in the number of day-care centers has proved their acceptability to working mothers.

REMAINING SINGLE

A significant minority of young adults in our culture remain single. Recently, the percentage of the adult population remaining single has increased slightly. The reasons for remaining single vary. A daughter may feel responsibility for the care of parents, especially if they are aged or infirm, and consequently deny herself chances of marriage. For other persons the opportunity for mature love and an acceptable marriage may never present itself. Disappointment in love may result in bitterness and militate against accepting new intimacies. Table 17-2 shows a higher percentage of single males than of single females at every age up to fifty-five.

Whatever their reasons, those who remain single face special problems in adjustment. Their personal adjustment is often difficult, owing to a feeling of aloneness, especially if they are not living with close relatives. They must perform the "total role" of breadwinner and homemaker. The problems of companionship, of receiving love, and of maintaing emotional balance are, for a time at least, acutely felt. Socially they are somewhat out of place in adult gatherings of married couples. Additionally, they are often urged by their married friends to consider new cross-sex associations and are often the object of matchmaking attempts by well-meaning friends, most of whom fail to realize that many unmarried persons *choose* to remain single.

The opportunities for additional education and for community, religious, or fraternal service as well as for personal advancement are, in many cases, greater for single persons than for those with the responsibilities of a family; whether they use their free time for their personal growth depends on their values and aspirations.

ENHANCING SELF-REALIZATION

During the years of early adulthood, people generally come to a realization of their total nature. The unrealistic ambitions of adolescence have yielded to more practical goals. Self-knowledge in many areas is deepened by the improved reality testing of this period. Assets and liabilities have been more clearly delineated by occupational experience, and personal qualities have been brought to the fore by the adjustments necessary in marriage. Planning ability has been tested by the reality of family finance, and the ability to provide has by now been adequately demonstrated. Interests have crystallized somewhat, and the self-concept is more or less stabilized at least in terms of self-acceptance and self-esteem. Reliance on the habitual begins to gain in motivational power.

It is at this stage that a person is presented with enviable opportunities for self-realization. Unless he faces too many stressful situations, consolidation of the gains made up to this point and a valid assessment of self lead to real progress in applying himself to the attainment of goals. After a long period of preparation for adult life, he is now more or less on his own to move ahead and to use his endowments completely. The returns of such diligence and ego involvement differ greatly from person to person.

Setting the Pattern of Life

Toward the end of early adulthood, a person is in a position to predict his or her future rather accurately and to adjust the pattern of life accordingly. Plans are made for the attainment of some long-range goals, such as educating the children or ensuring the financial welfare of the family. The trial-and-error experiences of raising children usually solidify into a definite pattern for their guidance through childhood and adolescence. The person's philosophy of life is often altered in the light of experience and in anticipation of future goals and responsibilities. Most people settle down during this period of life, and adolescent intensity subsides.

QUESTIONS FOR REVIEW

1. Discuss *interdependence* as it is used in this chapter with reference to dependence and maturity.
2. Explain how emotional interdependence is attained.
3. Discuss the problems of attaining economic independence.
4. Present some basic considerations in planning a family.
5. Explain the relationship between emotional maturity and marital adjustment.
6. Discuss some basic factors leading to marital happiness.
7. Characterize some general attitudes of parents toward their children.
8. Identify some desirable qualities and traits of mothers and fathers.
9. Discuss some major problems of the single person.
10. What developments in early adulthood help to set the adult pattern of life?

REFERENCES

Selected Reading

Bee, L. S. *Marriage and family relations.* New York: Harper & Row, 1959. An analysis of the major factors influencing success in marriage, with emphasis on personality patterns and interrelationships.

Bowman, H. A. *Marriage for moderns* (7th ed.). New York: McGraw-Hill, 1974. Discusses dating, partner selection, readiness for marriage, adjustments to marriage and family living, divorce, and other aspects of marriage.

Clemens, A. H. *Design for successful marriage* (2d ed.). Englewood Cliffs, N.J.: Prentice-Hall, 1964. Various aspects of dating, marriage, sex, and parenthood in a presentation consistent with the Catholic point of view.

Havighurst, R. J. *Developmental tasks and education* (Rev. ed., Chap. 6). New York: Longmans, 1972. A summary of the developmental tasks of early adulthood.

Specific References

Bee, L. S. *Marriage and family relations.* New York: Harper & Row, 1959.

Bowman, H. A. *Marriage for moderns* (7th ed.). New York: McGraw-Hill, 1974.

Burgess, E. W., et al. *Family: From traditional to companionship* (4th ed.). New York: Van Nostrand, Reinhold, 1971.

Pressey, S. L. Two basic neglected psychoeducational problems. *Amer. Psychologist,* 1965, **20**,, 391–395.

U.S. Bureau of the Census. *Current population reports.* Series P-20, No. 242. 1972.

Wernick, R. *The family.* New York: Time-Life Books, 1974.

Winch, R. F. *Modern family* (3d ed.). New York: Holt, 1971.

Chapter 18

Middle Adulthood

HIGHLIGHTS

Middle adulthood encompasses the years of greatest achievement in occupational roles. These are generally the most productive and satisfying years of a person's life economically, socially, and in many other aspects. Self-confidence and the sense of competence reach a peak.

As the metabolic rate decelerates, control of weight becomes a problem for the majority of Americans, whose diets are hypercaloric. Emotional strain and tension begin to take their toll from persons who have not learned emotional and temperamental moderation. Many people must make realignments in order to avoid immature or perverse behavior.

This period is the most stable in terms of personality development, with few major modifications. But significant alterations are necessary in the self-concept to allow a person to adjust to the realities of his or her life and the inevitability of aging.

The middle stage of life begins when a person attains his peak in performing most of his vocational obligations and many recreational activities. It combines completion of the upward development with an increased integration among motivational tendencies, abilities, and skills. Most types of education and training are finished before one enters this stage. However, the rapidity of cultural change calls for a redistribution of educational efforts throughout the life span (Baltes & Labouvie, 1973, p. 206). Vocational, marital, and other related social experiences are accumulated, and the pattern of life is largely set. It is estimated that women enter this phase of life at about thirty years of age, men at about thirty-five. The middle period of life encompasses approximately fifteen of the most productive years of life. It gradually shades into the phase of late adulthood when declines are not fully offset by recoveries and there is no further growth of diverse human potentialities.

During the middle adult years, most people progress in vocational, marital, civic, and socioeconomic areas. Consolidation of previous gains also occurs at this stage. Intensity of experiencing declines as compared with the adolescent and early adult stages of development. Since one's children are becoming adolescents and adults, marrying, and moving away, responsibilities associated with child and adolescent guidance and education reach a peak and then begin to decline. With the increase of life expectancy, some extension of the middle-age span is observed. This extension and shorter working hours offer increased opportunity for chosen activities and for personality growth.

SOCIOECONOMIC CONSOLIDATION

Adult persons in our culture tend to rank themselves and others in terms of economic success. The outward evidence of this success often takes the form of more expensive homes, cars, and boats or prestige schools for their children. Too often, an adult's sense of worth as a person—husband, wife, parent, or worker—is greatly affected by the degree of economic success enjoyed by the family. Very often, adults at this stage of life find that they have too many projects and not enough time to do them. Setting priorities is important if one's activities are to be effective, and to reduce tension.

By the time one reaches the middle adult years, the personality has gained in stability and there is little room for change. Nevertheless, some traits that were barely noticeable earlier now come into prominence. For example, the man or woman of forty-five who suddenly volunteers as a scout leader may long have had a great deal of civic interest but had to delay its practical expression until responsibilities at home and at work lessened. Learning and experience contribute greatly toward the consolidation of previous gains in the personality structure. Some persons go beyond mere consolidation by exerting considerable personal effort toward optimal development of their talents and potentialities. Throughout these years notable progress in personality and self-integration is often achieved.

Unlike the personality structure, social traits are likely to fluctuate considerably, largely because the growth of social traits is closely related to gratification coming from interaction with other individuals and groups. The young adult recently out of school, beginning the busy whirl of his family life, belongs to few clubs and groups and may barely make his social presence known outside his neighborhood. As the family grows to school age, many adults join groups related to their children's activities, such as scouting organizations and parent-teacher associations. Membership in formal clubs or groups reaches its peak at the end of the middle adult years and slowly tapers off with the advancing years of late adulthood. Social confidence and poise usually rise with the increase of friendships and the development of leadership traits. On the other hand, suspicion and hostility tend to rise when one lacks support in his personal goals or has to compete extensively with others. Interest and participation in civic and political affairs also rise constantly to a peak at about the age of fifty: unlike social traits, however, civic and political activity is maintained at a high level until very old age.

With most persons, economic status improves throughout the middle adult years, owing to progress in vocational standing, seniority rights, and various fringe benefits, on the one hand, and to decreasing capital expenses, such as buying and paying for a house or furnishings and providing infant and child care, on the other. Such expenditures usually burden the years of early maturity but decline at this age. The improving financial situation affords opportunities to acquire articles that promote the comforts of living. Some long-desired luxuries are now procured, often bringing both personal satisfaction and a rise in social status.

Progress in occupational standing usually levels off at about forty, if not before; consequently, economic advance also tends to level off. Excluding people in certain fields of business, in certain professions, and in some governmental positions, the economic plateau and lack of progress in vocational performance is felt by a large majority of the working population; this is often a key factor in a self-reappraisal toward the end of these years, which is elaborated on in a later part of this chapter.

A minority of adults—those who are poorly endowed and emotionally or socially unstable—continue to have employment difficulties for several reasons. Lacking a trade or technical knowledge, they make up part of the unskilled or semiskilled labor force, readily engaging in seasonal and dead-end jobs and changing occupations frequently. They are plagued by intermittent layoffs or unemployment as automation processes and seasonal or general recessions eliminate many job opportunities for the unskilled. The ranks of this category generally decrease with advancing age. Inability to maintain a job at this age is often reflected in decline of ascribed status. This undermines self-reliance and gives rise to hostile tendencies toward what are seen as sources of injustice.

HEALTH AND ACTIVITY

The prime of life does not end abruptly upon entrance into the middle adult years. The body with all its organs and systems continues to function near its optimal level throughout this phase, marked only by very gradual impairment, which often has its origins in the earlier years. Our most vital senses, sight and hearing, illustrate this. From childhood the lens of the eye begins to lose its capacity for accommodation, but visual acuity remains much the same until the age of about forty, when sharpness of vision quite suddenly declines. Many begin wearing glasses for reading and other fine work. While the majority of people never lose the ability to hear low-pitched tones, there may be progressive loss of hearing for high-pitched tones, clearly noticeable after forty. For this reason the enjoyment of music diminishes slightly.

The human body is a finely adjusted complex of systems, organs, and servomechanisms that maintain physiological homeostasis. With age, this balance is more easily upset, and recovery from an illness or disturbed condition becomes more difficult. Because of a slowing of metabolism in the early forties, many people experience difficulty in maintaining their normal weight. Since in many cases physical exercise declines, the tendency to become overweight increases. Other undesirable metabolic tendencies begin to show up. Signs of diabetes may appear, the amount of uric acid in the blood may rise above the average range, and early lesions and tumors indicative of neoplasm or cancer may also occur. Gallbladder and kidney stones form more easily than before. These are some of the factors that suggest a need for thorough physical examinations in this stage of life, even if one feels well. The question then is "Physically, how well am I?" In addition to blood tests, many systems of the body should be examined, including the gastrointestinal, genitourinary, gynecological (female genitalia), and cardiovascular systems. From about forty years of age it is reasonable to consider periodic physical examinations, which from time to time will reveal a need for x-rays or other expansions of the basic medical checkup. It is important for anyone this age to realize the detrimental effects of any strain, whether physical, emotional, or social.

Full physiological vigor and soundness are frequently manifested throughout the middle adult years but gradually decline in the later years. Morbidity and death rates remain low until the age of about fifty, when a progressive acceleration begins. It is interesting to note that the main causes of death change with age. Accidents are the major cause until middle age; from that time on, degenerative conditions such as heart disease and cancer become increasingly prominent.

Activities in hobbies and other interests often decline after marriage from their previous high level. Near the end of the middle adult years, one finds leisure time lengthening considerably. During the years that follow, those who cultivated hobbies in their youth readily take them up again, often in a modified form to reduce physical exertion. Apparently there is an urgent need for education in the use of leisure time, for while the need for activity and exercise of one's powers persists, many people do not take up sports and hobbies to gratify this need.

Interests, or activities one has a "liking for" rather than a wish to participate in, show definite age-related trends. Sedentary and noncompetitive diversions, such as listening to music and visiting historical places and museums, become increasingly popular, but active competitive activities show a decline.

Mass media of communication—radio, television, newspapers, magazines—claim the largest portion of leisure time. The effect of mass media on the personality is great, although declining. There is much doubt, however, that vicarious experience in the ready-made fantasy world they provide is a wholesome replacement for personal efforts that gratify basic human needs and maintain physical vigor.

In a world of timesaving devices, automation, and shortening working hours, leisure-time activities supply the satisfactions of acquired human interests and concerns. Time-killing pastimes should be omitted in favor of more creative, constructive, and noncompetitive activities. This constitutes one of the major developmental tasks of this stage, a preparation for the late adult years and for old age.

PARENTAL ASPECTS

During the middle adult years most parents find that their offspring are no longer children but adolescents and young adults. Parental guidance and protection are vastly altered, because teenage children find less of their time and activities associated with home. As the children carry out higher educational responsibilities and join their friends in various group activities, they gain in motivational strength leading to greater self-reliance and increasing independence from their parents. Some parents are delighted to see this development in their children and are pleased that they are less and less needed. Others may try to reverse the trend by getting deeply involved in teenagers' activities. Their concerns and upsets evoke powerful emotional currents; ego involvement deepens, and many parents, mothers in particular, live their son's or daughter's life more fully than their own. Some parents encounter difficulties with teenagers by requesting their assistance with chores, such as caring for younger siblings, or by expecting them to return at a prescribed hour.

Despite the problems of parents with their adolescent children, the presence of children at home is challenging and reassuring. Many kinds of enjoyment increase at the time the adolescent is approaching adult status. Mutual growth in understanding and appreciating each other's interests and activities is they key to successful cooperative efforts in pastimes, projects, and the resolving of situations that previously provoked conflict. Parents often regret that this phase is rather short-lived. When they begin to feel really gratified by their offspring and hope this equilibrium will continue, their children, having completed their education and job training, are ready to move away, to marry, and to establish families of their own.

Late in the middle adult years and the years immediately following, the stage of the "emptying nest" begins with the marriage of adolescent or postadolescent children, and moments of bleakness appear and multiply. Signs of the "dull residue of existence" may or may not show up, depending on the dominant attitudes and goals set earlier for self-fulfillment in this period of life. Changing one's attitude and outlook must begin while the children are still in their teens. For example, the mother who has the ingrained attitude that her children desperately need her assistance enters this stage with much anxiety as she realizes she is no longer needed. Again, the father who has forced his son to depend on him for all his material needs is likely to enter this stage with regrets that his son fails to live up to his expectations.

Frequently the development of new or modified attitudes is not sufficient for middle-aged adults who are trying to adjust to an empty home. By increasing their activity in organizations and clubs and by greater participation in civic and religious responsibilities, they can help themselves to gain added maturity and equanimity in the face of the problems arising at this stage. New satisfactions in expanding the comfort and attractiveness of the home, in closer relationship with the spouse and other relatives, in enjoying freedom for social visits, hobbies, and travel—all these can contribute to adjustment and to the enjoyment of this phase of life.

REEVALUATING THE SELF-CONCEPT

When a person begins to notice difficulty in making progress at work or a hobby or notices that his attempts to learn something new are less effective than in the past, and especially when declines in these areas become obvious, he will understandably become concerned. In addition to these changes, he may experience less satisfaction in usual recreational pursuits and recognize the law of diminishing returns in energy expenditures. Such experiences readily lead to the conclusion that he is growing old or that his health is declining.

The already anxious person magnifies incipient signs of old age and even invents some. Time and again a disease or ailment of middle age and the following convalescence offer some free time for self-reappraisal. Anxiety is intensified when the idea finds subjective support that he is losing some of his most appreciated qualities and abilities, such as sexual capacity and appeal or memory and ability to learn. Emotional ambivalence and oscillation, reminiscent of the turmoils of early adolescence, may appear and often lead to the second major crisis in life. Deep philosophical and religious questions pertaining to the meaning of life and to the value of the goals pursued during adulthood arise and press for answers; failure to resolve these problems can boost anxiety to disturbing levels. If routine activities lose their motivational strength because they seem purposeless, neurotic defenses such as withdrawal or depression may develop, or there may be an attempt to engage in experiences that had much personal meaning earlier.

Throughout childhood, adolescence, and early adulthood, the person has felt that he was in the process of becoming what he wanted or ought to be. Usually through the thirties he continued to advance his status or improve his self-concept, in some respects at least. He labors and expects a "break" that will enable him to realize his personal aspirations. At about forty, however, many realize that they are merely holding their own or actually losing ground, long before their ambitions have been achieved.

The reevaluative step at this age often leans to self-justification and placing the blame on closely related persons and circumstances. A husband blames his wife for unsatisfactory support in his strivings toward goals, while the wife sees many inadequacies in her husband in her attempts to safeguard her integrity. Statistics show that in the period 1950 to 1970, one in every ten marriages broke up after the twentieth year of marriage (*Vital Statistics,* 1974). As seen in Figure 18-1, living arrangements also change with age. In March 1970, in the United States, most young people lived in households as members of a family (94.5 and 93.5 percent for the age groups fourteen to twenty-four and twenty-five to forty-four). In later adulthood the percent declined to 88.2 for people aged forty-five to sixty-four and to 67.4 for the older population (sixty-five and over). The percent of people living alone was 8.6 for the age group forty-five to sixty-four, and it increased to 25.5 for the oldest group.

At this age, heightened reflection furthers the "executive process" of personality, leading to the use of an expanded array of cognitive and social strategies, including more selectivity, manipulation, mastery, and competence. The stocktaking and restructuring of experience in the light of what one has already learned become salient attitudes in middle age (Neugarten, 1968, p. 98). For the majority of adults, the early forties are the years of retest and

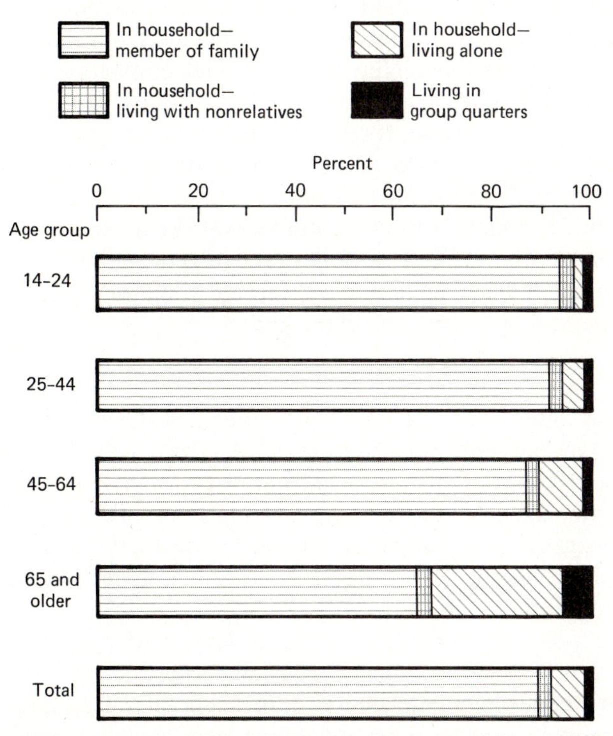

Figure 18-1 Population distribution by age and type of living arrangement: United States, 1970. [*Source:* U.S. Bureau of the Census, 1971.]

realignment for belated achievement of at least some selected life goals. New attempts at vocational promotion or improvement and at marital reintegration usually encounter obstacles. After two or three years of trial, the middle-aged person feels forced to make a final estimation of his level of success. Not many come to positive statements and see meaning in their renewed effort. Many recognize that they continue to fall short of their still-optimistic aspirations. This is a phase when a person can profit considerably from further adult or college education. Because of the rapid increase of knowledge, many people, including professionals, need refresher courses and seminars.

Some people show inability to reconcile themselves to the lack of appreciable gains during this phase of life. The tendency to blame themselves or others often reaches neurotic or, in a minority of cases, psychotic dimensions. Some react by escaping into alcoholism, psychosomatic illness, or paranoiac hostility. Attempts to destroy obstacles and force success at times lead to adult delinquency. Not many are able to change goals in terms of their ability or to acknowledge failure stoically. Frequently compensation is sought through association with the more fortunate or by pressing one's children to raise the family status.

As the middle and late adult years merge, there are fewer and fewer opportunities for self-assertion. This is a key psychological factor, prompting feelings of inadequacy and anxiety. If one cannot prepare himself to meet this problem, one will react by demanding exceptional performance from others, including those under one's supervision. New interpersonal conflicts arise, which increase tension and often make the situation unbearable. If complications of previously ignored problems arise, further entanglements ensue. As a result, symptomatic relief is unknowingly sought and a delayed neurotic pattern takes hold.

For many persons, the period of reevaluation of self is a temporary upheaval. The futility of such outbursts and turmoil is recognized, and renewed efforts to regain control over themselves and over external situations lead to stabilization and readjustment long before the stage ends.

Recapturing Youth

Some drives and urges that have been satisfactorily controlled during adulthood now reappear with new strength and frequency. Analysis will show a relationship to pubertal and adolescent drives and impulses, many of them sexual in nature. The earlier sexual and aggressive tendencies, inhibited or repressed into subconsciousness, now reappear because they remained unsublimated. In addition, their vitality has accumulated through formation of complexes involving related urges.

For many adults, an Apollonian, or intellectually ordered and tempered, way of life gives way to Dionysian expressiveness and disregard of the moderate and conventional. Older men entertaining adolescent girls, frequenting X-rated pictures, or reading sex magazines are examples of such behavior. With age creeping up on them, some feel an urge to engage in sexual and related gratifications before it is too late. Oral gratifications are excessively sought after, and returns to masturbation also occur. This is more likely to happen to those whose maturity and personality integration were never completed. Deeply entrenched traits of mature behavior, adequate character formation, and an explicit philosophy of life are powerful deterrents to such attempts to recapture youth by returning to immature or perverse modes of behavior.

When rational and irrational fears about appearing old accumulate, devices for disguising the signs of age—such as toupees (wigs to cover baldness), hair dye, and face lifting (tightening of sagging tissues and muscles by plastic surgery)—become helpful. Such devices may allow one to appear ten or fifteen years younger than he or she is. For many people, the realistic attitudes characteristic of middle age come too late for them to accept fully their changing appearance, especially in the latter part of middle adulthood.

Consistent progress in maturity enables a person to control such drives and urges, while a proper *Weltanschauung* assists in directing the energies toward constructive and socially acceptable channels of activity and self-expression. Activities motivated by integrated character traits tend to bring greater ego satisfaction than mere behavioral discharges of drives and impulses. For this reason, gratifications gained from altruistic, religious, and related activities are deeply satisfying to human nature and contribute much to the preservation of general welfare and, particularly, of mental health.

Among helpful sources for an understanding of the conflicts and adjustments of this middle phase of adult life are E. Bergler's *Revolt of the Middle-aged Man* (1967), Roy A. Burkhart's *Freedom to Become Yourself* (1956), Eda Le Shan's *Wonderful Crisis of Middle Age* (1973), and Bernice L. Neugarten's *Personality in Middle and Late Life* (1964). This is a period of life that has received little attention from social scientists.

COMPENSATING FOR DECLINE

A program of objective self-examination will typically show many beginnings of decline, slow and barely noticeable in the early part of the middle-adult years but increasing as the stage advances. There is a moderate decrease of psychomotor speed and strength. The biochemical equilibrium is in some cases disturbed by diseases of middle age, such as kidney stones, gallstones, and respiratory or circulatory difficulties. If the adult years have been spent in menial or factory work or mining, signs of old age may now begin to appear prematurely. Poorly used cognitive functions deteriorate rapidly. General organismic decline may become conspicuous toward the end of this stage. Wechsler's evaluation (1958, p. 206) points to a gradual but significant decline in brain weight and lower intelligence-test scores as age advances. This is partially confirmed by a major reassessment study (Matarazzo, 1972, pp. 78, 105–120).

Cognitive decline is usually more gradual than physical decline, while personality deterioration is rather exceptional. Occasionally considerable gains are observed, especially among professional people. Wherever vocational

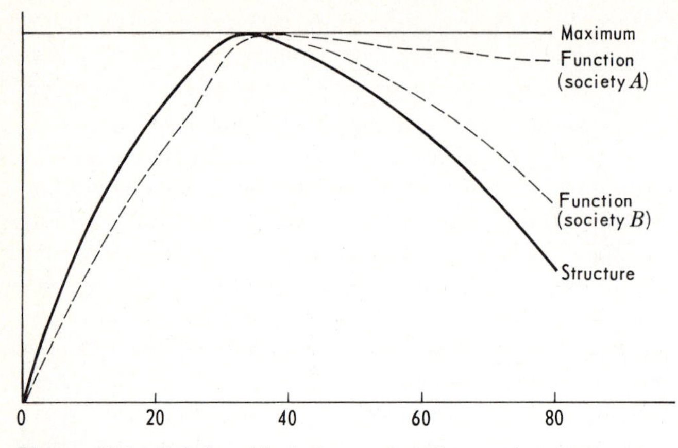

Figure 18-2 Relationship between function and structure through the human life cycle. Society A gives older people high status and the opportunity to continue their social functioning under favorable circumstances. Society B is "youth-oriented" and devaluates middle and old age. [*Source:* R. J. Havighurst. The social function of middle-aged people. *Genet. Psychol. Monogr.*, 1957, **56**, 297–375 (fig. on p. 345). By permission.]

specialization and progress are possible, morale is easily maintained; a high level of self-realization is then made an actuality. Indirect forms of compensation may help to preserve mental health and alertness among workers and the poor. Through diversified interests and hobbies, the range of activities and achievements may be considerably expanded and gratifications other than vocational ones may be secured.

Decline in the social dimensions of the individual presents a vivid contrast to the various other dimensions considered. In a survey of the community life of American adults, Havighurst (1957), interviewing 234 persons in greater Kansas City (Kansas and Missouri) with respect to their functions and competence in their communities, found that the period from forty to seventy is a plateau period with a slight decline toward the later years. The structural aspects of social life decline sharply during this period, especially after sixty years of age. Figure 18-2 presents Havighurst's hypothetical curve of biological and social functioning in two forms of society.

DEVELOPMENTAL TASKS*

Among the most effective ways of compensating for early decline are maintaining awareness of the developmental tasks of this stage of life and engaging in age-related activities to supplement and supersede the more energy-consuming

*This section draws heavily on an excellent presentation of developmental tasks for fathers and mothers by Duvall in *Family Development* (1971, pp. 414–424).

activities of the earlier stages of life. Seven tasks relate to the middle adult developmental level:

1 *Helping growing and grown-up children to become responsible and socially mature adults:* freeing their time for social and recreational opportunities by taking care of their small children (a frequently enjoyable duty); encouraging grown children to participate in civic activities and supporting them with practical and moral aid when they need it.

2 *Developing new satisfactions with one's spouse:* exploring new hobbies, club activities, and community projects; expressing appreciation for the spouse's efforts and successes; sharing the spouse's feelings and thoughts, aspirations and disappointments.

3 *Creating a pleasant, comfortable, and attractive home:* acquiring household facilities for comfort and ease of upkeep; remodeling and decorating in terms of the family's interests and values; assuming responsibilities related to entertainment of members of the extended family and old and new friends.

4 *Increasing social and civic activities:* keeping informed on civic affairs and national and international events; taking an active part in church and civic organizations.

5 *Finding new occupational and other satisfactions:* coming to terms with one's degree of success and working with less tension and increased experience and ease; contributing to the success of others by timely advice and assistance; letting younger persons take over some areas of responsibility without feeling a threat to self-respect or status; planning constructively for eventual retirement.

6 *Making satisfying and creative use of increased leisure time:* enjoying the chance to engage in activities for which time was unavailable before; sharing leisure-time activities with one's spouse and friends; balancing recreation in terms of activity and passivity, group participation and privacy, self-indulgence and service.

7 *Accepting and adjusting to the physical and cognitive changes of the middle years:* getting regular medical and dental examinations; using eyeglasses and hearing or other aids when prescribed; maintaining a program of physical exercise appropriate to one's age and stamina; observing adequate diet and maintaining appropriate appearance; restricting consumption of tobacco and alcohol; expanding and systematizing records to counterbalance the decline of memory; maintaining a variety of interests, with emphasis on the intellectual, artistic, and religious phases of life; making use of modern counseling and psychotherapy before problems, worries, or depression have detrimental effects on personality integration; reaffirming moral and religious values and engaging in related practices that have real and transcendent meaning.

GROWTH OF PERSONALITY AND IDENTITY

It is possible to gain additional roles in middle adulthood, and in many fields people can continue to specialize and to rise in status; but for most, personality and character are the only vital areas of growth at this stage of life. Because of the great influx of societal and cultural changes in the modern world, everyone is pressed to refashion his or her personality at each phase of adult life.

Some changes in attitude result from self-reappraisal when a man or woman admits the impossibility of regaining youth with its intense gratifications. If the developmental tasks of early adulthood have been sincerely undertaken, a person usually makes significant advances in orienting himself or herself to the constructive channels that remain open. Broader perspective, a capacity for detached appraisal, and calm evaluation of each step contribute to success in many social and business enterprises.

While indifference and apathy erode the salience of useful traits, a certain degree of emotional detachment is helpful because it affords a more objective approach to life, with less ego involvement and fewer deep conflicts and worries, than an emotional attitude can command. With some detachment, past experience and new information serve as guides toward success in various undertakings. Routine activities also retain significance if they are attuned to the hierarchy of lasting values. Gaining internal peace and stability in identity are probably the most appreciated achievements of a lifetime. Significant steps can be made in this direction, since boundaries for identity formation are usually set by the person himself.

It is commonly expected that everyone, in late adulthood if not before, will gain in moderation and balance and move toward assertion of a scale of humanistic values. The natural tendency toward lowered emotionality as years increase assists in the development of a preponderant reliance on rational judgment. A higher consistency of personality traits in middle adulthood is thus a frequent outcome of earlier oscillations and a search for lasting values and ideals.

The mature person in the forties should conclude that life will not continue to be a supercharged carrousel; he or she will not be able to recover so effectively from mistakes caused by spontaneity and impetuosity. One ought to examine motives and behavior more carefully, becoming more steadfast in cherished principles and ideals. Before a person enters the late adult years, a certain degree of inflexibility often develops, so that a general reliance on the habitual begins to expand into various behavior areas. This tends to promote the order and consistency of behavior and conduct by which the late adult years are frequently marked.

QUESTIONS FOR REVIEW

1 At what age does middle adulthood begin? Describe the leading characteristics of this period.
2 Explain the factors that contribute most toward the socioeconomic standing of persons in middle adulthood.
3 How do sight and hearing change in the early forites? Why?
4 Why do weight problems often start during the middle adult years?

5 How do fathers and mothers react to having their adolescent and postadolescent children leave home?
6 Describe the reevaluation of the self-concept by middle-aged persons. What has this reevaluation to do with their aspirations?
7 What factors frustrate attempts to recapture youth in the late middle adult years?
8 What means can be taken to reduce physical and cognitive decline during middle years?
9 List several developmental tasks for middle-aged people, and analyze two or three of them.
10 Describe changes in several personality traits during the middle adult years.

REFERENCES

Selected Reading

Duvall, E. M. Family development (4th ed.). Philadelphia: Lippincott, 1971. A comprehensive and well-written work on the expanding and contracting families of the present; stresses dynamic interactions among family members and family tasks at various stages of family life.

Neugarten, B. L., et al. (Eds.) *Personality in middle and late life: Empirical studies.* New York: Atherton, 1964. A comprehensive source on later adulthood by a number of authors. Includes personality patterns, ego functions, and sex roles.

Specific References

Baltes, P. B., & Labouvie, G. Adult development of intellectual performance: Description, explanation, modification. In C. Eisdorfer and M. P. Lawton (Eds.), *The psychology of adult development and aging.* Washington, D.C.: American Psychological Association, 1973.

Bergler, E. *The revolt of the middle-age man* (2d ed.). New York: Wyn, 1967.

Burkhart, R. A. *The freedom to become yourself.* Englewood Cliffs, N.J.: Prentice-Hall, 1956.

Duvall, E. M. *Family development* (4th ed.). Philadelphia: Lippincott, 1971.

Havighurst, R. J. Social function of middle-aged people. *Genet. Psychol. Monogr.,* 1957, **56,** 297–375.

Le Shan, E. *The wonderful crisis of middle age.* New York: McKay, 1973.

Matarazzo, J. D. *Wechsler's Measurement and appraisal of adult intelligence* (5th ed.) Baltimore: Williams & Wilkins, 1972.

Neugarten, B. The awareness of middle age. In B. L. Neugarten (Ed.), *Middle age and aging.* Chicago: University of Chicago Press, 1968.

Neugarten, B. L., et al. (Eds.) *Personality in middle and late life: Empirical studies.* New York: Atherton, 1964.

Wechsler, D. *The measurement and appraisal of adult intelligence* (4th ed.). Baltimore: Williams & Wilkins, 1958.

Vital statistics of the United States: 1970 (Vol. 3). Washington, D.C.: 1975.

Chapter 19

Late Adult Years

HIGHLIGHTS

The period of late adulthood begins when declines in sensory acuity, health, and achievement are not fully recovered. To an ever-increasing degree, losses become irreversible. For women, the menopause, occurring in the late forties, signals the entrance into the last phase of adulthood.

The family situation usually changes drastically as the last children reach maturity and leave home. As many have always done at this time, too, a number of women enter the labor force. Men attain their highest occupational status, which often adds to their strain and anxiety as they now face and compete with much younger people.

Many persons change their life-styles by becoming more motivated from within. They often consider reduction of physical and social activities but usually continue them until an illness strikes. Since there is no attractive alternative to generativity, a source of internal conflict often develops. A new countdown to a vaguely set zero begins: How much time is left to accomplish lifelong aspirations and goals?

There is less transfer and more interference in learning new ways of doing things at home and at work, and the increasing rigidity and reliance on the past yield poor dividends. Self-repair and self-improvement do not come as easily as in early adulthood. A decrease in memory lessens overall resourcefulness. Preparation for retirement is a new concern at this time.

People enter the phase of late adulthood when declines in health and performance are not sufficiently recovered—when large losses are compensated for by lesser gains. For many this happens in their late forties, when some losses become irreversible. The chief task of this period is the maintenance of the achieved level of performance and participation. Because of accumulated experience, most people are able to function at their highest vocational level. This often involves considerable responsibility and extensive self-application, so frequent strains are not unusual at this stage of life, especially among executive and managerial personnel. Progress at this stage consists chiefly in economic gains, expanded civic participation, and stabilization of a life-style. Some progress in fashioning the personality also occurs as many traits become more salient and are more effectively expressed.

The intense responsibilities related to the guidance and education of teenage children have usually terminated by the onset of late maturity, if not before. Following the departure of the last children, women may gain additional leisure time. The menopause usually occurs as women enter this phase of life. Men move into a less noticeable climacteric at the age of about fifty-five, when their sexual capacity declines. At this age prostate enlargement afflicts many men. Physical deterioration accelerates to a moderate rate, and health problems usually become more frequent. Lessened ability to retain new information, coupled with some losses in memory, foreshadows mild cognitive decline in the late part of this period.

MOTIVATIONAL CHANGES

A general decrease in drive is a major factor in changing the total motivational structure at this level of life. The variety of interests of earlier adulthood tends to decrease moderately as some interests and concerns wane, while additional interests are not easily acquired. The functional decline varies with the function measured. Speed of movement begins to fall off slightly at thirty to thirty-five, while coordination remains fairly stable until about sixty. Health often declines in the early forties, and strength at about forty-five. Individual variation in general biological aging and in nodal points must always be taken into consideration. Because of genetic factors as well as diseases and accidents, metabolic changes and various declines may start much earlier or much later for some persons. Frequently, moderate deterioration in sight and hearing makes it necessary to adjust many activities at this time. When decrements in performance occur, it is often difficult to distinguish which losses are due to decreased ability to perfom a task and which are due to loss of motivation (Anderson, 1964; Jakubczak, 1973).

Sexual Decline

For males, a slight decline in sexual prowess during the late adult years accelerates at about fifty-five and produces the sexual involution. Changes in the male gonads, including gradual decreases in spermatogenesis and the secretion of androgens, show little effect on bodily functioning or the personality. Men simply find their desire for regular sexual outlets decreasing as sexual activity becomes less gratifying than in the earlier years of adulthood. A single failure in maintaining penile erection often has a disproportionate effect on the male over forty or fifty. As a result, he readily withdraws from sexual opportunities rather than face repetition of the "ego-shattering," experience of incipient sexual impotency. Many aging males are able to engage in some form of sexual activity into their seventies, even when full erection is not attained until ejaculation occurs (Bischof, 1969, pp. 102–105). Sexual impotence of the mid-fifties sets in after excessive alcohol or marijuana consumption. The impact of illness often contributes to the failure to maintain erection long enough to gratify the partner. Lack of sexual excitement over a long period of time adds to the decline in male sexual responsiveness.

In females, the menopause lasts for a period of about two years, when natural cessation of the menstrual cycle takes place. Women traverse the "change of life" in the mid-forties or—more often—late forties with little or no diminution of erotic excitability. For many of them, however, the climacteric is quite upsetting, especially if it occurs abruptly—a sudden decrease in estrogen supply disturbs the physiological equilibrium. Failures in biochemical controls elicit powerful emotional reactions and moods that make adaptation impossible for some time. During this phase many women are subject to marked excitability, hot flashes, sweating, dizziness, sensitivity to heat and cold, and other symptoms of a confusing nature. Usually they become more irritable, restless, or depressed. There is a noticeable decrease in self-control and the feeling of adequacy. Adolescent maladjustments to self and others are often reactivated. These trying days may continue for several months or for a few years. When menopause changes are gradual, there is time for a woman to get used to a new picture of herself. Tablets or injections of estrogen reduce most symptoms. In addition to hormonal medication, forbearance and understanding from the woman's husband and older children are most helpful in assisting her through this critical phase of life.

Preceding the menopause a considerable number of women experience increased sexual preoccupation, often accompanied by a strong desire to have another child. Many women explore new cross-associations, and some enter into sex-related activities in order to "make up for lost time." Erotic excitability and decreased self-control permit vivid symptomatic behavior. Fears about the loss of femininity and about growing old compound emotional behavior. Following the menopause, many women lose allergies that had bothered them for years. Behavioral stability and self-reliance also rise and are maintained for about a decade. Except in the phases of stress, neither sexual desire nor orgasm are lost, although difficulty often arises from the husband's slackening ability to maintain erection. Most women enjoy sex more after the menopause than before. Affectional and sexual desires last for life, but for the latter there is a moderate decline in old age (Bischof, 1969, pp. 99–116).

Occupational concerns

Motivational changes are also greatly affected by family changes when the last child leaves home and becomes independent of the parents. This is especially true for the mother. It is difficult for most parents to accept the fact that their children do not need them and that they want to move away. The parental home is now too large, yet the older couple often hesitate to sell it and buy a house appropriate to their personal needs. Memories are cherished, and the large home, they maintain, will serve well for the visits of grandchildren.

This "empty nest" situation presents a new opportunity for self-evaluation. The parent can no longer use children to excuse a failure to develop the intellect and personality through reading and adult education. Too many adults are out of touch with their former friends and relatives and with civic affairs. If they have not followed the developments of science, they know less than present-day adolescents. Vocationally, some are embarrassed at finding themselves in competition with young people whose life experience is often less than half theirs but who are effective contributors in their specialized fields. Expertise is typically expected in late adulthood; and one must measure up to this expectation, if only to maintain ego adequacy. Even at this late phase of life the decision to grow or to vegetate is a very real one. It is not easy to write off one's high aspirations for occupational superiority. A dearth of pertinent knowledge at this age is a frequent product of early neglect of additional growth and specialization. Late adulthood is not the time to turn completely to the utilization of past experience, though it may be difficult to retain an orientation toward the future throughout this phase of life.

The vocational situation is often much worse for women who were homemakers for many years but now desire employment. If a woman completed her academic or professional studies about two decades earlier, her vocational efficiency, with the skills acquired so long ago, is usually inadequate; too many subjects, methods, and techniques now used were unknown when she went to school or entered employment. Current skills and techniques must be acquired. For professional reestablishment she often needs refresher courses to pass state board examinations. Additional course work and training are indispensable for most professions and many trades. Throughout adulthood the desire to learn new subjects declines, and an ever-increasing number of persons reject additional training opportunities and take what jobs are available to them. The developmental task of *generativity* (Erikson, 1963, pp. 266–268) presses for new vocational achievements, since its polarized alternative (stagnation) is repulsive to most adults at this age. Mature men and women want to be needed for projects of significance to them. After the completion of childbearing, employment is the major source of productivity and gratification. The achievement curve often moves up until the early forties and down in later adulthood. Constriction occurs as a middle-aged person has less energy to invest, less new experience to relish, and less reason to exert himself or herself (Kuhlen, 1968).

At this stage it is good for men and women to further mutually supportive relationships with their grown children's families and with aging parents. In cooperation with other individuals and organized groups, they can help promote neighborhood civic, religious, and cultural activities. Engagement in such activities is personally gratifying and offers reassurance of being needed. Late adulthood is a time to take stock and to begin monitoring physical vulnerability. Life must be restructured in terms of "time left to live" rather than of "time since birth." There is "only so much time left," as realization that death is very real adds anxiety even when one has many strategies for dealing with life. The sense of competence and expertise combine to give a sense of superior competence (Neugarten, 1968, pp. 94–98).

DEVELOPMENTAL TASKS AND CONCERNS

Adjusting to family changes and preparing oneself for old age are the major tasks of this phase of life. Health, family, and one's role in life are probably the chief concerns marking this level of adulthood. The developmental tasks are (1) helping grown-up children, (2) gaining in social influence and leadership, and (3) adapting oneself to the accelerating process of aging. One can help grown-up children perform their responsibilities by developing mutually supportive relationships with them, as well as with their close friends and relatives. It is possible and desirable to gain new friends through services to and occasional entertainment of relatives and new acquaintances. Since gains in leadership are feasible at this age, an older person, by cooperating with neighbors and by joining congenial organized groups, may do much in promoting worthy civic objectives and selected political and cultural values.

Keeping up with scientific, political, and cultural changes provides multidimensional interests for people of this age. They may investigate the arts, sciences, philosophy, religion, law, and politics. Consider scientific and technological developments: their proliferation is accelerating at a spectacular rate. As Robert Oppenheimer stressed, knowledge, which used to double in millenia, then in centuries, now doubles in a decade. One either reads many articles disseminating scientific achievements or loses touch with modern science. With the current rate of discoveries and technological advances, not wisdom but ignorance accumulates unless one keeps in step with scientific developments by reading daily papers and scientific magazines. With regard to the world and our relationship to it, while geography became much more complex with the rise of many new independent nations, the world became much smaller through greatly improved transportation facilities. A young person experiences little difficulty in accepting scientific achievements and the fast pace of the technological age, but an older person whose world image was formed during his teens and twenties is baffled by the developments that confront him through television or direct experience. To avoid this an adult must keep pace with changes in society and science.

Only by applying oneself to intellectual and cultural concerns is one able to maintain a comparatively high level of cognitive functioning. Owens's follow-up study (1966) points to an absence of significant decline of mental powers during the sixth decade of life. He tested a sample of 127 college freshmen in 1919. From this sample 96 subjects were retested in 1950 and again in 1961, using the Army Alpha (a group intelligence test) each time. The findings showed that the decade from age fifty through age sixty was one of relative constancy in mental ability, since the apparent downward trend was statistically not significant. The author draws the conclusion that cognitive decline, like cognitive development, is conditioned to a large extent by the nature and intensity of environmental stimulation.

SELF-CONCEPT AND IDENTITY

During the years of late adulthood, most persons consider themselves middle-aged and try to convince others that they are not old and are still capable of doing the same things they did when they were young. Gradually or suddenly, however, they begin to feel different as physical speed, vigor, concentration, and alertness decline. Trying to stay young and to keep as active as they were earlier is difficult. Hence many persons, after some unsuccessful trials, reverse their attitude, deliberately slow down, and seek additional comforts, often in accordance with a physician's advice after an illness. They begin studying what is good for them and what is not good for them and try to conform to the former and avoid the latter. Aging occurs in spite of desire and without consent, and it is irreversible. The direction of the total development cannot be changed significantly.

Despite their accumulation of experience and dexterity, many older persons begin to have some difficulty in applying themselves. This difficulty in part springs from the accelerated rate of technological changes, requiring acquisition of new abilities and skills. As mentioned earlier, readiness to learn new subjects subsides in later years. Then too, significant declines in observational accuracy and speed add their share to the problems of the older person. At this age, "to learn the new they often have to unlearn the old and that is twice as hard as learning without unlearning" (Lehman, 1953). Generally there is less transfer and more interference as age advances. While learning is always desirable, it is much easier to learn at fifteen or at thirty than at fifty-five or sixty. Pressures to learn are dreaded by many middle-aged persons. They prefer to rely on experience.

Creativity and inventiveness subside greatly. Notable and superior contributions to most fields of human creative endeavor are usually made by young adults. It has been found that as a rule the important contributions to the field of chemistry were made by scientists who were not more than twenty-six to thirty years of age. For practical inventions, physics, botany, and electronics, the ages were thirty to thirty-four; for surgical techniques, genetics, and psychology, thirty to thirty-nine; for medical discoveries, bacteriology, physiology, and pathology, thirty-five to thirty-nine (Lehman, 1953, p. 342). Both past and present generations of scientists who produced more than their proportionate share of high-quality research did so at no more than thirty to thirty-nine years of age. The curves for contributors of both past and present generations ascend and descend at almost identical rates (Lehman, 1968).

The frequent tendency to assert oneself through consistency and rigidity is not very helpful in a culture of rapid successions and changes. General reliance on the past and an inclination to apply the habitual earn fewer and fewer dividends. For ambitious persons, the decreasing amount of success stirs up emotional currents that breed intensified problems of adjustment. Without some success, tranquility cannot be preserved for any long period of time. Inability to gain additional competence is always difficult to accept. A significant number of poorly adjusted persons start using projection as a defense mechanism. It is easier to blame somebody else or to claim a "degeneration" of society than to accept one's inability to learn and profit from the new changes so eagerly embraced by the young.

Exposure to the many influences and vicissitudes of life has produced deeply engraved traits and attitudes in the older person. Many traits carry a personal significance; there is ego involvement. The oscillations of adolescence and of the early forties produced some pervasive revisions of the early personality structure. Now, in late adulthood, the final stabilization is taking place. Identity is fully defined, and change is more often resisted. Step by step the movement toward the rigidity of old age gains in power. The person is less willing to change ways of acting and believing. Reliance on the past and the habitual is now more frequent, and only rarely does he or she compromise with new fads, fashions, and renewals. The older person often rejects opportunities to acquire skills related to modern technological advancement. "Personality patterns are firmly established long before middle age and they tend to continue throughout adult life" (Peck & Berkowitz, 1974). In relating to the environment, the middle-aged person usually changes from an active-assimilatory to a passive-accommodatory style with more orientation to the inner world (Birren, 1969; Neugarten, 1966).

The general future orientation of adolescence and early adulthood declines during this stage. Now recollections of the past begin to play an increasing role and occupy more of the time. Hopes for future success are weakened by the disappointments of the recent past, and the need for continual expansion declines. Ideas of one's exceptionality, giftedness, and superior worth are toned down as reality imposes itself on the thoughts and fantasies. Except for the most gifted and successful people, disruptions and uncertainties dictate maintenance rather than raising of one's occupation, status, or reputation.

Social-class differences in adjustment have been attested to by investigators in this field. People of lower social class often have poorly developed and unresourceful personalities that interfere with their adaptability from middle age through senescence. A majority of middle-class people have fairly differentiated and adaptable motivational systems—an asset in the progressive modern society. Upper-class people have additional advantages springing from accumulated knowledge, civic leadership, and financial resources. Yet for people of all classes there is a slowly rising difficulty in adjustment with advancing age as ego integration slowly or moderately declines. David L. Gutmann's projective (TAT) study (1974) of 287 white urban men and women aged forty and seventy shows that with advancing age both men and women have increased difficulty in the management of their inner life. Active mastery decreases with age, and passive and magical (not realistic) mastery increases. Older men and women are more frequently prone to illogical thought and to motivated misperception of stimuli. Reduced self-confidence and reduced satisfaction with life were also observed.

A study of empirical stress and health (Gray, Baker, Kesler, & Newman, 1965) of two elderly samples (*N*=301 and 258) disclosed that persons in their fifties and sixties who had experienced "stressful situations during the previous ten years tended to have poorer physical health than did comparable persons who had experienced less stress during this period." This was true for both a sample of severely disabled persons and a sample of older persons who were not disabled and who lived in the community. Apparently a person cannot manage much stress effectively without detrimental effects to health, ego strength, or identity. This is consistent with the assumption of "wear and tear."

The late years of adulthood are marked by a lessening capacity for self-repair. This applies to bodily functioning and health as well as to social relationships and cognitive processes. People can lose and lose much, but gains are rare and do not come as easily as in early adulthood. One is pressed for conservation of health, material goods, and personality gains. Attempts at self-improvement bring barely observable increments. At this age a person must prepare for even bigger slides downhill. This will be shown in Chapters 20 and 21, on senescence.

QUESTIONS FOR REVIEW

1. Discuss changes in the family situation during the late years of adulthood.
2. Explain the basic motivational changes occurring in late adulthood.
3. Describe the effects of the menopause and compare them with the climacteric as men experience it.
4. What are the major effects on parents when all their children have left the home?
5. Describe the vocational problems of men and women in late adulthood. How does past experience affect their reemployment?
6. List the developmental tasks of late adulthood and explain one of them.
7. Give reasons for the difficulty in keeping up with scientific and cultural changes at this age.
8. Explain why learning of new subjects is difficult at this level of life.
9. Give some reasons for character stabilization and for the tendency toward conservation.
10. Explain the role of the past in personal adjustment during late adulthood.

REFERENCES

Selected Reading

Bischof, L. J. *Adult psychology.* New York: Harper & Row, 1969. This study of maturity includes self-image, marital life, and vocational concerns.

Kimmel, D. *Adulthood and aging.* New York: Wiley, 1974. A comprehensive source on various phases of adulthood, including old age.

Neugarten, B. L. (Ed.) *Middle age and aging: A reader in social psychology.* Chicago: University of Chicago Press, 1968. A leading source on aging written by a large number of psychologists and social scientists. Contents include age and status, family relationships, theories of aging, leisure and retirement, health, illness, and death.

Specific References

Anderson, J. E. Psychological research on changes and transformations during development and aging. In J. E. Birren (Ed.), *Relations of development and aging* (Chap. 2, pp. 11–28). Springfield, Ill.: Charles C Thomas, 1964.

Birren, J. E. Age and decision strategies. In A. T. Welford & J. E. Birren (Eds.), *Decision making and age.* Basel: Karger, 1969.

Bischoff, L. J. *Adult psychology.* New York: Harper & Row, 1969.

Erikson, E. H. *Childhood and society* (2d ed.). New York: Wiley, 1963.

Gray, R. M., Baker, J. M., Kesler, J. P., & Newman, W. R. E. Stress and health in later maturity. *J. Gerontol.,* 1965, **20,** 65–68.

Gutmann, D. L. An exploration of ego configurations in middle and later life. In B. L. Neugarten (Ed.), *Personality in middle and late life: Empirical studies* (Chap. 6, pp. 114–118). New York: Atherton, 1964.

Jakubczak, L. F. Age and animal behavior. In C. Eisdorfer and M. P. Lawton (Eds.), *The psychology of adult development and aging,* pp. 98–111. Washington, D.C.: American Psychological Association, 1973.

Kuhlen, R. B. Developmental changes in motivation during the adult years. In B. L. Neugarten (Ed.), *Middle age and aging: A reader in social psychology,* pp. 115–136. Chicago: University of Chicago Press, 1968.

Lehman, H. C. *Age and achievement.* Princeton, N.J.: Princeton, 1953.

Lehman, H. C. The creative production rates of present versus past generations of scientists. In B. L. Neugarten (Ed.), *Middle age and aging: A reader in social psychology,* pp. 99–105. Chicago: University of Chicago Press, 1968.

Neugarten, B. L. Adult personality: A developmental view. *Human Developm.,* 1966, **9,** 61–73.

Neugarten, B. L. (Ed.) *Middle age and aging: A reader in social psychology.* Chicago: University of Chicago Press, 1968.

Owens, W. A. Age and mental abilities: A second adult follow-up. *J. educ. Psychol.,* 1966, **57,** 311–325.

Peck, R. E., & Berkowitz, H. Personality and adjustment in middle age. In B. L. Neugarten (Ed.), *Personality in middle and late life: Empirical studies* (Chap. 2, pp. 15–43). New York: Atherton, 1964.

Part Eight

Late Phases of Life

The late period of life is characterized by deterioration, constriction, and decline. There is little learning and even less development of any kind. Recent medical progress and economic improvements have been accompanied by increasing life expectancy. Specialized medical aid, social security benefits, and spreading pension plans permit the majority of old people a more pleasant way to complete their lives. Chapters 20 and 21 will consider the losses in various abilities and the contingent limitation of activities that mark senescence.

The commonsense theory of *activity* stresses the need to maintain vocational interests and to augment recreational activities in order to occupy the time fully and make these later years gratifying and productive. As the difficulties of adjustment multiply for the aging person and he or she becomes increasingly dependent on others, ingenuity, humor, and wisdom will often be called into play to make the final stage of life bearable. In contrast, the "wear and tear" assumption calls for continued reduction of activities. Formulated as the theory of *disengagement* (Cumming & Henry, 1961; Henry, 1965),* it holds that aging is accompanied by a mutual withdrawal on the part of both the

*For references on this and the following page, see Specific References at the end of Chapter 20.

senescent and others in the social system to which he or she belongs. This "disengagement," initiated either by the individual or by others, expands social distance and decreases the intrapsychic family of the senescent. Maddox (1964) argues that for many persons beyond fifty, disengagement adequately reflects the modal tendency in the life process. Havighurst (1968) and Henry (1965) observe that social withdrawal is accompanied by increased preoccupation with the self and decreased emotional involvement with persons and objects in the environment and that it is a natural rather than an imposed process. As Neugarten (1970, 1972) explains, we need to understand the social clock that is superimposed over the biological clock, the two together establishing the rhythms of development and decline.

Chapter 20

Biological and Cognitive Changes

HIGHLIGHTS

Senescence is the last period of life, during which the process of aging accelerates. Marked deterioration of some organic systems and poorer self-application characterize old age.

Many old persons lose much of their perceptive power, become disoriented, and regress to mere biological and emotional need gratification. Senility *is the general term applied to this condition when psychological reactions include hypochondriasis and delusional states.*

Biological decline is unavoidable, as practically all bodily systems deteriorate in both structural and functional aspects. Accident-proneness increases as physical coordination declines. Biological errors, noncycling cells, and metabolic disturbances—all add to and accelerate aging.

Cognitive decline is most apparent in losses of memory, often compensated for by imagination. Faulty performance in many activities elicits powerful emotional reactions, and rumination over past performance often leads to increased restlessness and moments of helplessness and despair.

The centrality of self helps the old person to preserve ego integrity and to find satisfaction in his or her self-fulfillment in areas of past endeavor. A reviewing of life that offers this satisfaction adds much to graceful aging.

As the later years of life approach, there is a marked decline in the physical abilities. The onset and rate of deterioration vary, however, from one organic system to another and from one person to another. The sequence depends on many factors, including genetic endowment and specific past experiences and influences, such as diet and exercise, illnesses and injuries. The kind of life lived determines the pattern of aging. The process of aging is also related to personal and social adjustment during the years of adulthood. Satisfactory adjustment in adulthood promotes the integration of feelings of self-fulfillment that aid in maintaining higher activity and achievement levels during middle age. If identity has been clarified and self-esteem magnified, there are high dividends for many years.

The exact time of the onset of old age is difficult to specify. Some people show noticeable changes in traits as early as forty; others still appear "young" at seventy. Just as no single criterion can be employed to delineate adolescence and adulthood, so no decisive criterion can be given for the onset of old age, and individual variation is great. Generally the mentally deficient, the retarded, and those with physical handicaps deteriorate early, and often at a rapid rate. At the age of thirty or even earlier, the defectives begin to exhibit signs of old age. Their life expectancy is very short. On the other hand, many well-endowed persons seem to be capable of resisting accelerated decline until seventy or eighty. Although the spry senior may be in better physical condition than persons who are ten or fifteen years younger, he is not as strong or vigorous as he was earlier, though he may manage to maintain a youthful attitude. The average group takes the middle ground. They begin to deteriorate earlier than the well-endowed, but their rate of decline is moderate. Statistics show that women's average life expectancy is significantly longer than men's and is continuing to rise on a worldwide basis. Ultimately each person has his own rate and pattern of aging, similar to many others yet always distinct in some traits and features. Empirical studies (Britton & Britton, 1972; Kastenbaum & Burkee, 1964) suggest that approximately half those between sixty and seventy identify themselves as middle-aged rather than old. Identity losses involved in acknowledging that one has become old are apparently so great that the fact is not readily accepted.

DISTINGUISHING SENESCENCE AND SENILITY

Senescence should be clearly distinguished from senility. *Senescence* is a period of life somewhat arbitrarily identified by chronological age. The age of sixty-five or seventy is now considered the point introducing this last stage of life. Retirement with its new adjustment problems may usher it in. An accelerated rate of deterioration is a more definite criterion. This does not imply marked deterioration of organic systems, cognitive powers, or acquired skills, since many adult qualities and traits are often preserved; but there is usually lessened activity and poorer self-application.

Senility, on the other hand, although closely allied with senescence, implies a substantial loss of organic and functional integration. The retention of earlier adult powers and characteristics is low. Senility is closely associated with a considerable loss of physical and cognitive functioning, whether in old age or prematurely. Impairment of brain tissues and of motor coordination, high irritability, and considerable loss of memory, orientation, and self-control are typical signs of senility rather than old age. A regression to satisfaction of merely biological and emotional needs is frequent. Senility is fairly common in old age. Severe cerebral arteriosclerosis will result in symptoms of senility, but senility is frequently due to psychological rather than organic factors. Delusional states are frequent. Many senile persons become faceless, like old coins whose stamping has been worn away, with little identity and even less character.

Generally the period of old age is one of widespread and sometimes drastic change. Possibly only the years of early adolescence offer a comparable challenge. The late years of life, like adolescence, are characterized by physical, social, and emotional upheavals. But as in the early years, adequate preparation for such changes can prevent them from being too stressful and too disruptive. Indeed, the late years can be a period of considerable equanimity and happiness.

Creaking joints, increasing weakness of sense organs, decrease in one's usual energy and speed, rapid change in quantity and color of hair and teeth, wrinkled skin, and distinct decline in sexual potency and pleasure—all combine in varying degrees to make a person very aware of his own aging. He begins to think about the role of the elderly and often makes some adjustments to it. Social expectation and cultural pressure act together, often forcing a new mode of life upon him before a personal need exists or declining powers demand it.

THEORIES OF AGING

G. Stanley Hall (1922, pp. 403–411) felt that the human being remains essentially juvenile even during old age. Senescent helplessness is more affected than real. Old age calls on us "to construct a new self just as we had to do at adolescence, a self that both adds to and subtracts much from the old personality of our prime." At this age there are "no less tendencies to polymorphic perversity than before these were constellated into the normal sex life of maturity" (p. 394). Hall affirms that "it is the very nature of love to grow sublimated as years pass" and that "one man's norm would be another man's disaster" (pp. 393, 397).

Erik Erikson (1963, pp. 268–269) offers a *bidirectional* interpretation of old age. If the core crises during adult stages of life have been resolved in favor of intimacy and generativity, high *ego integrity* is thus made possible for old age. The possessor of this integrity can readily defend the dignity of his or her own life-style against any threats. However, if the adult life was marked by a predominance of isolation and stagnation, old age will be filled with despair. Substantial ego damage and labile emotions are likely to produce the symptoms of hypochondriasis, delusions, or other moderate to severe disturbances marking senility.

Bidirectionality is also a feature of Charlotte Bühler's theory. Bühler (1968, pp. 400–403), in her self-fulfillment theory of old age, stresses critical self-evaluation in terms of lifelong accomplishments or lack of them. Failures and losses are unavoidable, but many lives attain partial fulfillment—satisfaction is found in one or two basic areas of life. When the amount of frustrating experience has been large, resignation often marks old age.

The theory of *disengagement,* originated by Elaine Cumming and her associates (1960), is apparently the first fairly well elaborated psychosocial interpretation of the basic constituents of aging. Originally this theory was based on three hypotheses: (1) rate of interaction and variety of interaction lessen with age; (2) changes in amount and variety of interaction are accompanied by concomitant changes in perception of the size of the life space; and (3) a change in the quality of interaction accompanies decrease in the social life space, from absorption in others to absorption in self and from evaluative to carefree. The empirical data, based on interviews with 211 fifty- to ninety-year-old subjects, suggest that disengagement starts during the sixth decade with a shift in self-perception accompanied by a reduction in the variety of interactions with others (1960, pp. 25–34).

W. E. Henry (1965) claims that the central issue of this theory is the "resurgence of focus upon *interiority*" accompanied by a decline in the importance formerly attributed to other persons and to events. Indeed, the disengagement of the aged is "an intrinsic process and it is inevitable," since it occurs in all life-styles. An elaborate statement of the disengagement theory may be found in Cumming and Henry's *Growing Old* (1961, pp. 210–218). Both disengagement and activity theories are too simple to account for a great variety of adjustment patterns in late life. A comprehensive theory of aging is yet to be formulated.

Generally, people age in ways that are consistent with their life histories. There are many distinct psychosocial patterns of aging. The Kansas City Studies of Adult Life (Neugarten, 1964; Neugarten, Havighurst, & Tobin, 1968) distinguished eight patterns of aging in one group of seventy- to seventy-nine-year-olds, classified as reorganizers, focused, disengaged, holding-on, constricted, succorance-seeking, apathetic, and disorganized. In adjusting to both biological decline and social disengagement, the "aging person continues to draw upon that which he has been, as well as that which he is" (Neugarten, 1973).

BIOLOGICAL AGING

Like early development, the aging process is genetically programmed—limitations to preserve an ordered state of the organism increase with age. Fundamentally, biological aging is marked by a lower metabolic rate, which slows down energy exchange within the organism; hence its resources for behavioral self-expression are gradually curtailed. Energy, when overused, is not fully recovered; structures and powers are impaired by overexertion or prolonged activity. Functional deceleration results from increased cell age owing to a lessened capacity for cell division. Finally the proliferative capacity of cells ends—their clocks stop. Tissue aging is thus a result of transition from cycling to noncycling cells. Cycling cells are actively moving through the cell cycle $G_1 \rightarrow S \rightarrow G_2 \rightarrow M$ (G_1 and G_2 are gaps; S is the period of nuclear DNA synthesis; and M is the period of mitosis); noncycling cells are blocked at G_1 or G_2. An increase in the number of noncycling cells in any tissue accelerates its aging (Gelfant & Smith, 1972). For example, there is a moderately accelerating decrease in brain weight, and brain and heart muscles show an increased amount of insoluble pigment. The rise of insoluble compounds interferes with vital cell functioning. Age impairs the synchrony of various brain centers and leads to a decline in their overall functional efficiency (Still, 1969).

The outstanding geneticist Zhores A. Medvedev formulated the *biological error* theory of aging. At the molecular level, aging is furthered by the accumulation of errors in the synthesis of proteins and nucleic acids. Random errors occurring in replication, incorporation of various analogs, heat damage, and radiation—all favor subsequent regression. Altered DNA molecules, for example, retain acquired changes in their progeny and pass their errors over into RNA. Structures damaged during synthesis are retained and accumulate. Some breakdown in the interactions between DNA and RNA produces RNA synthesis no longer controlled by DNA, thereby reducing the overall genetic regulation (Medvedev, 1962, pp. 258–261). Finally the double-stranded DNA helix breaks in its process of replication (Cutler, 1972).

The general health situation often becomes precarious in the older age group, and chronic diseases are more frequent and severe. Heart, lung, kidney, gallbladder, and genitourinary disturbances and other ailments are more frequent than in earlier years. Injuries and wounds heal at a much slower rate, and sense receptivity becomes much less efficient. Difficulties with vision such as presbyopia (far sightedness due to rigidity of the crystalline lens) and hearing (presbyacusia, for instance) are frequent and severe (Jarvik, 1975).

The decline of cellular functioning depends on the extent of damage to the nucleic acids of the chromosomes. With injuries to the DNA, defective RNA messenger molecules are produced that are unable to synthesize the enzymes for maintaining certain basic cell functions, especially at the time of mitosis (Birren, 1964, p. 52). Because of carcinogenic substances, the chemical oncogens (initiators) form simplified or neoplastic cells, including neoplasms, or benign and malignant tumors (Rous, 1967), among them various forms of cancer. Cancer is a malignant tumor of proliferating new cells marked by infiltration into adjoining tissues and dissemination to remote sites, resulting in secondary growths. The progressive increase in cancer cases reaches a peak in the sixth decade and thereafter declines. Cardiovascular diseases, on the other hand, continue to increase progressively until very old age (Timiras, 1973, pp. 471–472). If diagnosed early, many forms of cancer respond well to chemotherapy, radiotherapy, and surgery, but long-term cures are expected in only about 40 percent (NCI, 1973) of all cases of cancer. Any accumulation of deleterious mutations accelerates aging (Curtis, 1963). Other causes of aging include somatic mutation, chromosomal aberration, and accumulation of errors in the synthesis of proteins (Medvedev, 1962; Sinex, 1966).

During the late adult years, biological aging is a gradual debilitating process. It is also a process that cannot be stopped or reversed. Practically all bodily systems deteriorate in both their structural and functional efficiency. Functional abilities chiefly depend on the circulatory system, which supplies the total organism with oxygen, fluids, and nutrition. The walls of blood vessels—arteries, veins, and capillaries alike—harden and narrow as age advances. This in turn interferes with optimal circulation of the blood. The hardening of the capillaries disturbs the supply of nutrients to various bodily systems and organs, including the central nervous system, and gradual muscle and tissue atrophy begins, decreasing the strength, weight, and immunity to infection of such vital organs as the brain, lungs, and heart. When the heart loses weight, the blood pressure mounts. At a certain point in this process a physiological insufficiency of the heart results—the cardiac reserve approaches zero.

Physical work or exercise easily strains the circulatory system; climbing stairs increases the heartbeat and oxygen demand considerably, and continuation of such activity will begin to disturb the heart and general organic equilibrium. While the amount of oxygen used is an index of bodily strength, a person in late adulthood soon demands more frequent periods of rest that decrease oxygen exchange. There is lessened overall utilization of oxygen with advancing years. The lung capacity is only about half what it was from about twenty-five to thirty years of age.

A smaller caloric intake is another sign of organismic aging and points to a loss in work capacity. Clinical experience indicates that biological aging can be slowed down by well-controlled athletic activities. Moderate and regular physical exercise throughout the years of adulthood slows down the degenerative process and helps to preserve organic structures and physical welfare for several years longer than normal and also raises life expectancy appreciably. Leisurely cycling or walking, if practiced during late adult years, may be continued into old age. The exercise improves circulation to all parts of the body (Leaf, 1973).

Sensorimotor coordination gradually becomes poorer. Reaction time increases, and some movements become awkward; speedy movement, when needed, is difficult or impossible, and accident-proneness increases. Loss of previous poise and attractiveness is striking, especially among women; graceful aging is nearly impossible for many of them.

Every physical impairment or limitation produces notable changes in the personality. Some of these trait modifications are a direct result of physiological functioning—for example, the memory losses following certain arterial disturbances. Other psychological alterations, however, represent somewhat remote aftereffects of physical malfunctioning. Electrophysiological and cognitive functions, for example, are normally relatively independent of each other but become more interdependent in pathological states (Obrist, 1970).

Accumulation of metabolic waste products over time ultimately reaches a critical phase at which various cell functions become disturbed. Hence, increases in debris and waste accelerate biological aging. Moreover, deleterious mutations from irradiation and other sources add to metabolic disturbance. There is a relative augmentation of immunological activity and an increase in serum gamma globulin, and heightened antibody production with age, but the total status of immunology is not improved significantly because much of overproductivity is dysfunctional (Walford, 1964). Longevity of the human being may be reasonably estimated by the multiple correlation of life span with brain and body weight, the coefficient of which is .84. This is not a perfect predictability but it points clearly to the fact that accelerated decline of brain weight makes continuation of human life impossible. Generally, the weight of a mammalian brain is an estimator of its cybernetic capability, on which the longevity of a species depends (Sacher, 1965, pp. 102–103).

It is noteworthy that a person functions, not in terms of the strongest systems of his organism, but in terms of the weakest. Usually one vital organ or system "wears out" early in comparison with other physiological systems, and illness or even death results from such an impairment. Transplanting vital organs can prolong the life of a few selected persons but is not yet a general answer for most people. The forces maintaining life are only as strong as their weakest vital component. Whenever a vital link "breaks," the resulting stress leads to death. Old people engaging in strenuous exercise of any kind must take this principle into consideration (Selye, 1956, pp. 274–276 and 299–301).

One major change resulting from physical impairment is gradual restriction of a person's environment. In infancy and childhood, a major contribution to psychological development was the increasing ability to go beyond the immediate home surroundings. Late in life, the trend is typically reversed. Decline in vision reduces the degree to which a person can depend on the written word for knowledge of the outside world. Driving at night becomes a hazard. Auditory loss likewise reduces the effectiveness of verbal communication, especially in a group situation, where masking effects are usually present. Losses of motor strength and coordination similarly reduce the ability to travel from place to place, even with an automobile, which so greatly facilitates contact with distant persons and places. Visual and orientation deficiencies, increased reaction time, reduced coordination, and liability to backaches and intense fatigue all contribute to a loss of mobility and its serious consequences for social adjustment.

With old age many physiological deteriorations occur. Accelerated loss of hair and change of its color to white, facial wrinkling, "old-age spread," and the "dowager's hump" are some of the easily observed senescent features. Accumulation of fat, especially in the abdominal region, and a lowered position of the head are also related to aging. Arthritic and related processes speed up, changing appearance beyond the biological schedule. This also depends on the earlier patterns of living and adjustment to stressful situations and events during adolescence and adulthood (Bell, Rose, & Damon, 1972; Tournier, 1972).

DECLINE OF COGNITIVE ABILITIES

Retention of information is one of the major cognitive aspects of intelligence. When memory declines at an accelerated rate, it is often replaced at least in part by imagination, leading to confabulation, especially in attempts to report recent happenings. Memory for events of early life is retained fairly well, so that the old person refers more frequently to remote than to recent or current experiences. Failing memory and a shrinking of perspective are two key factors influencing the general orientation and perception of time and space. Time appears to pass much faster than before, and the old person has difficulties in adjusting to consequent inevitable changes. As he forgets the names of streets and buildings and their appearance and loses track of his whereabouts, he sometimes experiences estrangement even in familiar surroundings.

Despite the accumulation of experience, an old person scores lower on intelligence tests than before, indicating a decline of higher functions and performance. In order to preserve IQ constancy, computation is usually statistically adjusted to the normal rate of decline of mental abilities in the later years of life. With his narrowing alertness, it is rare for an old person to accept readily any new ideas or ventures. Making realistic choices becomes difficult. Creativity, if it has developed and been utilized, also declines, probably at about the same rate as cognitive abilities. However, Riegel and Riegel (1972) examined intelligence scores of German adults fifty-five to seventy-five years old by going backward in age, with the time of death as the 0 point. They found that change with age was marked by a sudden drop in performance which occurred within one to five years of the subject's death. Throughout adulthood, they reported, performances of long-term survivals were basically unchanged.

In vocational activity an older person usually experiences a loss in efficiency. By the early sixties, many begin to show some inadequacies in performing their accustomed work, and they become tired more quickly than before. Setting retirement arbitrarily at a particular age, for example sixty-five, is unfair to some who are still capable of performing their jobs well. A lack of proper recreational facilities for the retired person, in addition to the feeling that he or she is not needed, elicits feelings of inadequacy and depression. If the retiree has no outlet for exercising his powers and abilities, the resulting feeling of uselessness is detrimental to his security and status. The psychological effects of "empty time" are damaging to many, and especially to those who lack a variety of interests and hobbies to substitute for employment.

Although the amplitude of emotional experience and the control over feelings and emotions decrease to a great extent, emotional sensitivity does not. As a result, affective irritability rises somewhat and emotionally toned discontent is frequent. There is a tendency to rationalize and to blame others by projection.

Decreased engagement in social activities is often due to decreased satisfaction from such interaction. Difficulty in attending to a conversation and gaps in information concerning current events are two factors contributing to the decline in interpersonal communication. Frequently, pessimistic interpretation of human behavior and events adds to the problem.

For a person who has established early and practiced for many years a wholesome pattern of living and adjustment, the aging process proceeds smoothly, with little distress and anxiety; such a person is often ready to accept aging and to make the best possible adjustment to it. If, on the other hand, emotional upsets have been frequent and defense mechanisms, such as projection, rationalization, and fault-finding, have been commonly resorted to, difficulties usually become magnified in the later years of senescence. Attitudes of selfishness and superiority that may have been disguised earlier now tend to become more marked. The desire to be honored by others is a form of self-aggrandizement that now appears with considerable vividness, and a tendency to boast about past accomplishments is common.

Persons who do not develop healthy control and sublimation in the earlier years of adulthood are likely to crave oral gratification with age. Since affectional needs are less often satisfied at this age, compensation by excessive eating and drinking is frequent. Constant complaining about one's health and finding fault with others, such as younger relatives, are frequent means of compensating. These problems make this stage of life another period of crisis, comparable to puberty. In a significant number of cases, this turning point in life is accompanied by psychosomatic disturbances and, to a lesser extent, by senile psychotic outbreaks.

The common tendency to emphasize minor injuries and symptoms seems to serve several purposes: it provides an excuse to avoid unpleasant obligations, it justifies egocentric demands, and it obtains the concern and attention of others. The tendency to hold on to life somewhat alleviates this despairing attitude. Reactionary and conservative attitudes come into prominence as psychological flexibility and a readiness to experiment disappear. Unfavorable experiences of the past make many older persons timid and cautious. Anxiety, worry, and sensitivity to danger greatly inhibit an old person and promote withdrawal from challenging activities (Neugarten, 1973).

With old age there is also a substantial increase in leisure time. When children marry and leave parents and, especially, when retirement comes, the remaining energy has to be directed toward previously neglected hobbies and new activities. Neglected potential may now be used. Old age is an appropriate time for developing artistic and intellectual interests. Writing, drawing, painting, and a variety of crafts are good ways to engage energy and find enjoyment. Church activities, charity drives, and civic projects are usually gratifying occupations at this age. Active participation in some individual and group activities is of crucial psychological importance for maintaining self-esteem and a sense of belonging. It is advantageous for the old person to have opportunities for serving or assisting others, even for a very limited time (Duvall, 1971, pp. 475–479).

Lifelong emotional reaction patterns, attitudes, and sentiments related to values and various spheres of living influence the kind of emotional disturbance to which the elderly are prone. Adjustment difficulties at the adult level tend to intensify significantly (Kastenbaum & Burkee, 1964). Analyzing the findings of the Kansas City Studies of Adult Life, Havighurst (1968) speaks of the "angry" older person, who was hostile toward the world and who tended to blame others when anything went wrong. A small cluster of "self-haters" openly blamed themselves for their failures.

Most fundamental needs are felt more intensely now than earlier. Recognition and respect, affection and achievement, security and self-esteem are all strongly sought by old persons. Most experience difficulties in gaining gratification of these needs. Unduly high demands on the part of old people not infrequently cause friction in social relations.

Duvall (1971, pp. 453–480) lists the following developmental tasks of the aging: (1) finding a satisfactory home for the late adult years, (2) adjusting to retirement income, (3) establishing comfortable household routines, (4) nurturing the husband or wife, (5) facing bereavement and widowhood, (6) maintaining contact with children and grandchildren, (7) caring for elderly relatives, (8) keeping an interest in people outside the family, and (9) finding meanings in life.

CHANGES IN PERSONALITY

Changes in personality structure and organization encompass practically all dimensions in senescence. The usual decrease in motivational strength is linked with a narrowing range of interests and activities. Lesser gratifications result from poorer performance in most fields of endeavor. As powers decline, some interests, habits, and attitudes disintegrate. The general decrease in flexibility and ability to learn is directly proportional to an increase in rigidity and constriction. Many self-expressive activities, including speech and conversational skills, begin to decline. Ruminating over earlier and more satisfactory experiences preoccupies the old person and influences his or her conversation with others. Repetitiousness and habituation to routine activities increase, and adjustments to expressed preferences of others decline.

Success in preserving integration of personality and its operative traits shows widespread individual differences related to former personality development. Those who acquired an attitude marked by a desire to learn whenever opportunities existed now earn high dividends. Likewise, those who faced reality in all its dimensions throughout the adult stages of development and acquired the needed reservoir of abilities, interests, and skills are able to cope with emerging problems and novel situations. Their functional level of self-expression is consistent with their endowments and brings many gratifying experiences of self-actualization. Such ego-enhancing experiences facilitate adjustment to old age. Such persons preserve personality integration without serious conflicts and disillusionment during their advancing years.

Many others, because of unfavorable parental and social influences or sheer lack of personal effort, acquired little knowledge and practical skill for the complex art of modern living. They failed to develop their endowments and, internally, remained in either an acute or a dormant conflict situation. In such cases, personality disintegration takes hold early and often leads to pervasive losses. Early senility is a frequent result, appearing in the later years of adulthood or the early years of old age. Many such persons finish their lives in mental or similar institutions.

Mental health cannot be substantially improved during the late stages of life. Experts in psychotherapy find poor response to counseling and psychotherapy after approximately forty-five years of age. Emotional disorders, hypochondriasis, and general senile dementia are frequent. Apparently many persons failed to deal successfully with moderate and severe deprivations, conflicts, and problems during their twenties and thirties and became predisposed to psychosomatic and mental disorders. Many sixty-year-olds, for example, see their environment as complex and dangerous, and there follows a movement from outer-world to inner-world orientation and preoccupation with self. A shrinking of the life space and avoidance responses develop. All cathexes are turned inward, and the person moves away from both old and new engagements (Cumming & Henry, 1961, pp. 222–226).

QUESTIONS FOR REVIEW

1. Discuss two contrasting theories of aging.
2. Explain the difference between senescence and senility.
3. Give some reasons for biological aging and indicate the more conspicuous signs of it.
4. Explain the need for regulation of physical exercise as one grows old.
5. Present Selye's theory of the causes of stress and death.
6. Give some reasons for the decline of cognitive functions.
7. Why does the increase in leisure time occur suddenly? Suggest proper activities for utilizing leisure time at sixty-five and later.
8. Why is there much habitual activity and redundancy in speech during old age?
9. On what factors does adjustment during old age depend? Analyze the significance of one of these factors in the light of a particular experiential background.
10. How do elderly persons tend to interpret their environment?

REFERENCES

Selected References

Eisdorfer, C., & Lawton, M. P. (Eds.) *The psychology of adult development and aging.* Washington, D.C.: American Psychological Association, 1973. A volume by members of the APA Task Force on Aging; includes analysis of gerontological, developmental, clinical, and experimental studies on aging. The social environment is also examined.

Palmore, E. *Normal aging.* Durham, N.C.: Duke University Press, 1970. Reports from the Duke longitudinal study, 1955–1969, dealing with attitudes, activities, physical and health problems, marriage, death, and other aspects of senescent behavior.

Timiras, P. S. *Developmental physiology and aging.* Part II, *Physiology of aging,* pp. 408–614. New York: Macmillan, 1972. Aging of cells, tissues, organs, and systems analyzed in detail.

Tournier, P. *Learn to grow old.* New York: Harper & Row, 1972. A comprehensive analysis covering work and leisure, acceptance and condition of the old, suggestions for a more humane society, and theories of death. Includes personal experiences of the French physician-author.

Specific References

Bell, B., Rose, C. L., & Damon, A. The normative aging study: An interdisciplinary and longitudinal study of health and aging. *Aging hum. Developm.* 1972, **3,** 5–17.

Birren, J. E. *The psychology of aging.* Englewood Cliffs, N.J.: Prentice-Hall, 1964.

Britton, J. H., & Britton, J. O. *Personality changes in aging: A longitudinal study of community residents.* New York: Springer, 1972.

Bühler, C. Fulfillment and failure of life. In C. Bühler & F. Massarik (Eds.), *The course of human life* (Chap. 24, pp. 400–403). New York: Springer, 1968.

Cumming, E., Dean, L. R., & Newell, D. S. Disengagement: A tentative theory of aging. *Sociometry,* 1960, **23,** 23–35.

Cumming, E., & Henry, W. E. *Growing old: The process of disengagement.* New York: Basic Books, 1961.

Curtis, H. J. Biological mechanisms underlying the aging process. *Science,* 1963, **141,** 686–694.

Cutler, R. G. Transcription of reiterated DNA sequence classes throughout the life-span of the mouse. In B. L. Strehler (Ed.), *Advances in gerontological research* (Vol. 4), pp. 219–321. New York: Academic Press, 1972.

Duvall, E. M. *Family development* (4th ed.). Philadelphia: Lippincott, 1971.

Erikson, E. *Childhood and society* (2d ed.). New York: Norton, 1963.

Gelfant, S., & Smith, J. G., Jr. Aging: Non-cycling cells an explanation. *Science,* 1972, **178,** 357–361.

Hall, G. S. *Senescence: The last half of life.* New York: Appleton, 1922.

Havighurst, R. J. Personality and patterns of aging. *Gerontologist,* 1968, **8,** 20–23.

Henry, W. E. Engagement and disengagement: Toward a theory of adult development. In R. Kastanbaum (Ed.), *The psychobiology of aging,* pp. 19–35. New York: Springer, 1965.

Jarvik, F. Thoughts on the psychobiology of aging. *Amer. Psychologist,* 1975, **30,** 576–583.

Kastenbaum, R., & Burkee, N. Elderly people view old age. In R. Kastenbaum (Ed.), *New thoughts on old age,* pp. 250–262. New York: Springer, 1964.

Leaf, A. Getting old. *Scientific American,* 1973, **229** (No. 3), 44–52.

Maddox, G. L., Jr. Disengagement theory: A critical evaluation. *Gerontologist,* 1964, **4,** 80–82.

Medvedev, Z. A. Aging at the molecular level. In N. W. Shock (Ed.), *Biological aspects of aging,* pp. 255–266. New York: Columbia, 1962.

NCI. *National Cancer Institute 1973 Fact book* (NIH 73-512). Washington, D.C.: Department of Health, Education and Welfare, 1973.

Neugarten, B. L. Adaptation to the life cycle. *J. geriatric Psychiat.*, 1970, **4,** 71–87.
Neugarten, B. L. Personality and the aging process. *Gerontologist,* 1972, **12**(1), 9–15.
Neugarten, B. L. Personality change in late life: A developmental perspective. In C. Eisdorfer and M. P. Lawton, (Eds.), *The psychology of adult development and aging.* Washington, D.C.: American Psychological Association, 1973.
Neugarten, B. L. (Ed.) *Personality in middle and late life.* New York: Atherton, 1964.
Neugarten, B. L., Havighurst, R. J., & Tobin, S. S. Personality and patterns of aging. In B. L. Neugarten (Ed.), *Middle age and aging: A reader in social psychology.* Chicago: University of Chicago Press, 1968.
Obrist, W. D. Cerebral ischemia and the senescent electroencephalogram. In E. B. Palmore (Ed.), *Normal aging.* Durham, N.C.: Duke, 1970.
Riegel, F., & Riegel, R. M. Development, drop, and death. *Developm. Psychol.*, 1972, **6,** 306–319.
Rous, P. The challenge to man of the neoplastic cell. *Science,* 1967, **157,** 24–28.
Sacher, G. A. On longevity as an organized behavior. In R. Kastenbaum (Ed.), *Contributions to the psychology of aging,* pp. 99–110. New York: Springer, 1965.
Selye, H. *The stress of life.* New York: McGraw-Hill, 1956.
Sinex, F. M. Genetic mechanisms of aging. *J. Gerontol.*, 1966, **21,** 340–346.
Still, J. W. The cybernetic theory of aging. *J. Amer. Geriatr. Soc.*, 1969, **17,** 625–637.
Timiras, P. S. *Developmental physiology and aging.* Part II, *Physiology of aging,* pp. 408–614. New York: Macmillan, 1972.
Tournier, P. *Learn to grow old.* New York: Harper & Row, 1972.
Walford, R. L. Further considerations toward an immunologic theory of aging. *Exper. Geront.*, 1964, **1,** 67–76.

Chapter 21

Self-Concept, Needs, and Problems of the Senescent

HIGHLIGHTS

General biological decline in old age leads to deterioration of many somatic systems. For many persons, sensory defects become incapacitating—loss of sight when accompanied by loss of hearing, for instance, is difficult to compensate for. Blood pressure often rises and lowers the cardiac reserve. Arthritis in its many forms afflicts many senescents.

Decline in health marked by various psychosomatic symptoms presses for modifications of the self-concept. Increasing limitations must be accepted by the old person, but exaggeration of any illness furthers helplessness and invalidism.

After retirement a person is free to do "what he always wanted to do." Some find new sources of enjoyment, but many fail to carry out the necessary adjustments as uncertainty and anxiety increase.

Awareness of bodily impairments and decline in health impels the aging toward various forms of self-preoccupation. "What really gives meaning to the present life?" The long-held answers often become ambiguous, and a search for final answers is initiated. Reading the Bible or philosophical works is helpful for many senescents.

Opportunities to test one's own ideas, attitudes, and beliefs depend on contacts with other people, such as relatives and companions of younger days. Emotional and social stimulation usually declines drastically during old age. The needs for love, recognition, and status can be satisfied through lively intercourse with others.

Without making some contribution to the extended family and community, the senescent cannot maintain his or her feelings of self-worth and adequacy. American society, by segregating old persons, makes this a nearly impossible task for most senescents.

With the increasing physical and cognitive decline characteristic of the late years of life, certain personality modifications seem to be inevitable. The aging person's self-concept usually undergoes some changes. Personal needs and suitable means for satisfying them are modified. In addition, the role in society is greatly altered. With all this, diverse adjustment problems are bound to occur. This is a period when a person is called upon to utilize the resources he or she has developed during the preceding decades. Moreover, it is a period during which the assistance and understanding of relatives and the community is sorely needed.

HEALTH AND ILLNESS

As indicated in the previous chapter, the later years of life are characterized by a general decline of the biological systems of the organism. Despite the wide individual differences in the rate and amount of deterioration, some physical impairment is inevitable. Visual, auditory, and other sensory defects become increasingly prevalent and incapacitating. The decline in mineral reserve weakens the bone structure. Reaction time, strength, and endurance are all adversely affected. Arthritis in its various forms afflicts many persons. The cardiac reserve is lowered. All this can be seen in the increasing number of illnesses and accidents, which, together with lowered recuperative power, show that the organism is approaching terminal illness and death. These symptoms of aging are psychologically significant not only in themselves but also in the effect they have on the personality and behavior. If health is defined as "vigor of body and mind" or "a state of complete physical, mental, and social well-being and not merely the absence of disease or infirmity" (World Health Organization, 1946), senescents are unhealthy, even though many have no symptomatic evidence of illness over varying periods of time.

The dependence of the cognitive and emotional functions on the integrity of the neurological and chemical systems of the body is well established. Any gross change in these systems as a result of disease or injury is usually reflected in behavior deterioration. Equally important, however, are the personality changes that reflect the old person's reaction to the physical condition. Such *somatopsychological* changes, as Barker and his collaborators (1953) have termed them, are often just as significant as those biologically induced. Moreover, these changes sometimes occur even in the absence of serious or disabling conditions.

With the rise of physical and psychosomatic symptoms, the person's concept of himself or herself undergoes some restructuring. Just as the bodily changes occurring with the onset of puberty forced the adolescent to revise his view of himself, so too the old person alters the picture he has of himself to a certain extent. While his self-image may remain relatively intact, his abilities do not. Sooner or later he must accept the fact that he is no longer the robust, healthy individual he was in earlier years. No longer is he capable of many activities that were previously part of his daily living. Increasingly he must protect his general well-being. Even if he is still in relatively good health, a reassessment of declining abilities and rising limitations, as well as awareness of potential dangers, seems unavoidable as he sees his friends and peers beset by physical ailments, in many cases resulting in death. He must adjust his exercise, diet, entertainment, reading, and general participation as his physical and cognitive powers decline. After retirement he is freer than others to do as he wishes with his life and resources, but he is usually not eager enough for any basic change to carry it out; anxiety and uncertainty increase, and the status quo is often preserved. For those who can make the change, transfer from city to country, for example, brings a change in the rhythm of life that is very suitable to old age, especially for those who came from rural areas to city life. A return to places close to nature can revivify an aging person and open new roads to acquiescence and contemplation (Tournier, 1972, pp. 108–109).

Basic reorganization of the self-concept is both normal and desirable. Since human behavior is largely a reflection of how people perceive themselves in relation to their surroundings and other people, it is imperative to have a realistic view. The person who refuses to accept the fact of new limitations is obviously rejecting reality. To the extent that his self-concept fails to correspond to the actual condition, the person will be inadequate. Similar problems arise, of course, in the case of the person who exaggerates the physical changes he perceives, who regards himself as completely limited, inadequate, and dependent on others. Old age, invalidism, and illness are often exaggerated and used by some to control the behavior of those who are close to them. When others fail to respond, nursing homes and old age colonies remain the only genuine alternatives.

In addition to the rather direct impact of physiological changes on the personality of a person, whether young or old, other effects also are observed. One subtle effect is the gradual decrease in exciting and pleasant experiences and intellectual stimulation. In childhood, the acquisition of motility, spoken language, and reading skills mean enlargement of the child's psychological world. More and more the child is able to reach beyond his immediate environment and experience many new things. As new restrictions are imposed on the aging person by physical limitations, the perspective of personal experience shrinks. Visual difficulties frequently limit the acquiring of new ideas through the written word. Such deficiencies also tend to reduce freedom to leave the immediate environment. Hearing defects, especially if severe or uncorrected, cut older people off from many personal contacts and from the information and stimulation they ordinarily provide. Motor disabilities, even merely lack of sustained endurance, reduce the opportunities for experiences and social interaction outside the immediate home environment. Any loss or restriction is difficult to accept at any age, and old age is no exception. Emotional upheavals come like waves at sea. Detachment is painful at any age.

MAINTAINING A VARIETY OF INTERESTS

With the general restriction of activities and the consequent limitation of intellectual, emotional, and social stimulation, it becomes increasingly essential that the old person maintain a wide variety of wholesome interests. There is, of course, no specific approach or pattern that can be considered the best, but some genuine sources of activity and amusement are a necessity. Personal adjustment at this time is dependent on past attitudes and habits, as revealed in the interests of the older person. Activities that have long held the attention typically tend to be continued, especially when they do not conflict with specific physical or social limitations. Because of the increasing limitation of activities in the later years and the consequent conflict between interests and abilities, it is imperative that interests be extensive in range. In line with the theory that activity is beneficial, Fromm (1967) urged the older person to become more responsive to the world around him. He should learn to recreate through his genuine interest in the world. Fromm predicted, too, that perhaps fifty years from now, old age will be any age above forty, because nobody will have to work after that age.

Although many elderly persons acquire new and rewarding interests, the task of doing so becomes increasingly difficult with the passage of time. For one thing, the opportunity for adequately testing new areas is restricted. Entirely new, truly satisfying interests characteristically require a considerable time for development. Consequently, occasional or sporadic contact with areas of potential interest is generally of little value. Moreover, because of increasing difficulty in coping with totally new and novel situations, untried areas of interest are not so likely to be sampled. One is again faced with the conclusion that the years before old age are the time during which genuine interests and areas of satisfaction should be developed. Even more essential is the formation of healthy attitudes regarding those interests. All is not lost if a certain activity is restricted; related or new areas can provide sources of personal satisfaction. With such a background, a person is prepared to compensate for whatever the future takes away. Should certain sources of personal reward later be denied him, he is able to face the loss and turn to other areas. He is prepared to enrich and enjoy life rather than lapse into a state of self-pity or constant reminiscence, either of which is unrewarding and leads to stagnation and deterioration of the entire adjustment. Many older persons become interested in personal memoirs but, finding few listeners, resort to writing them. This is a challenging form of self-preoccupation. The intensity of reminiscence often reaches a point at which some people begin to live the past. When this happens, aberrant behavior may occur: an old person may, for example, one day prepare to visit a parent who has been dead for two decades.

INTELLECTUAL AND RELIGIOUS CONCERNS

The pattern of decline during senescence occurs with intellectual activities as well. Without the intellectual stimulation provided by extensive and varied communication with the outside world, the old person is forced more and more to rely upon what had been learned previously. This limiting of experience and the decreasing sharpness of memory for new and novel concepts account for much of the constriction and rigidity of intellectual activities so frequently associated with old age. Tendencies to reminisce and relive the past likewise become understandable. The decline of cognitive capacity reported in the later years may well reflect in part the dearth of stimulation. It is clear that lack of exercise of any capacity or system, physical or mental, ultimately leads to its deterioration.

Because of the lack of varied social stimulation, the excessive amount of leisure time, and the awareness of bodily changes heralding the eventual approach of death, the old person typically is impelled toward further self-examination. Was it all really worthwhile? Why were there so many mistakes, if not blunders? What gives meaning to life? By such questions he is led to a reappraisal of his philosophy of life. Long-held answers often become ambiguous. This situation incites new efforts to search for more acceptable, if not final, answers. Coming to terms with what life means to the individual is an ongoing developmental task from the early adolescent years. Adolescent and early adult experiences are often on the mind of the senescent engaged in reviewing his life.

In keeping with this concern for a philosophy of life and a workable hierarchy of values is the concern with religion. In seeking a permanent system of values and resolution of the fears and vicissitudes of life, it is only natural to examine one's ideological and religious commitment. This growing interest in religion is vividly illustrated by the findings of Ruth Cavan and her colleagues (1949). Although only 71 percent of the men in their sixties who were questioned were certain of an afterlife, 81 percent of those in their late eighties were certain. Interestingly, 83 percent of the younger women and 90 percent of the older women were certain of an afterlife. Moreover, 100 percent of both men and women ninety years old and over had this conviction. O'Reilly (1957) studied a Chicago working-class population of persons over sixty-five. His sample represented 6.5 percent of the 4,511 total. His findings showed an increase in religious activity among older people. However, those who were lonely or unhappy did not turn to religion more than others. A major survey (Moberg, 1965) of religious studies concluded that interest in and concern about man's relationship with God increases even into extreme old age, and that religious feelings and beliefs intensify during senescence, although participation in religious services outside the home tends to diminish.

SOCIAL NEEDS

Throughout the entire life span, human beings live in a social environment. They depend on their fellows not so much for physical support as for psychological support, affection, and cognitive stimulation. This dependence does not diminish during the later years of life; in fact, in many respects it tends to increase. With old age, however, the total interaction of senescents with their relatives and community tends to decrease significantly. They are less able to maintain achieved roles and positions. There are practically no significant roles for them to fill at this age. They usually live apart and away from children and grandchildren, and they are often removed from companions of younger days. As a rule, social isolation is greater than at any other period of life. Whether they are living in their own homes, in nursing homes, or in "senior villages," the social and emotional needs of the elderly are neglected by younger adults and by governmental agencies (Townsend, 1971, pp. 133–138).

Yet, by means of social intercourse, the old person may be provided with a wealth of experience and stimulation. Just as the child's boundaries of experience were vastly extended by meeting many youngsters in the school situation, so the elderly person's boundaries are partially determined by the scope of his personal contacts. New ideas, interests, and attitudes are needed, and only to a limited extent can these stimuli be compensated for by less personal substitutes such as reading, radio, and television. The opportunity to express and test one's own ideas, beliefs, and attitudes is not provided by such media. Nothing less than direct personal contact will suffice—feeling that one is listened to with interest and some admiration. It is essential to communicate ideas and attitudes, to test them in the light of others' reactions. Personal contact is necessary also for the stimulation of feelings and emotions. In the absence of normal social interaction, either gradual impoverishment of affectivity or inappropriateness of reactions may be expected.

The safeguarding of cognitive and affective processes is by no means the only function of social communication. The fundamental needs of recognition, love, belonging, and status depend on interaction with others for their satisfaction. Frustration of these basic needs naturally leads to unhappiness and is accompanied by use of defense mechanisms—for example, aggression, withdrawal, or regression—whenever achievement of goals is thwarted. Just as the adolescent, viewing the gross changes in his own bodily and cognitive structure, needed the reassurance of his peers and a feeling of dialogue with them, so does the senescent who witnesses in himself the widespread losses of advancing age. Contact with others his own age also helps the aging person accept gracefully his decline in physical and cognitive powers. Contact with younger persons has a high stimulus value that enlivens the senescent personality.

SOCIETY AND THE SENESCENT

In addition to the adjustments required of the senescent as a more or less direct result of the aging process, certain problems arise from his or her changing role in society. Some of these are related to family and immediate friends, some to the community and society in general. Whereas earlier in life the person was the head of a household, helping to shape the lives of his or her children and other younger people, in the declining years this role is lost. Instead, frequently the roles of parents and children are reversed to such a point that the parents are dependent on the children to a considerable extent. Such a reversal cannot help altering the older person's self-concept and present new problems of adjustment. Should old people regard themselves as a nuisance or a burden? How much personal freedom should they relinquish? A multitude of questions naturally arise for the elderly. Their life satisfaction and that of those about them depend on the answers they receive to these questions.

Intimately related to this entire problem area is the attitude of the younger members of the family. Is the senescent to be regarded as a liability or, at best, an ever-available baby-sitter? Or is he someone who needs constant care and protection even to the point of being treated much like a child? Should he be sent to a home for the aged or perhaps even to a mental institution? Any behavior that lessens the dignity and self-assurance of the senescent naturally produces serious adjustment problems for him. Old people want to be close to their families and yet maintain their independence from them.

The general attitude of American society toward the senescent is improving. The feeling that old people are of less value than other members of society finds less support today than in past decades. The growing concern over the older population has brought about considerable legislation for their welfare. The change is marked by provision for proper housing of the elderly in many communities, by rising social security benefits, by the pension plans of many companies, and by improvements in the provision of medical care. By furnishing many necessities, such plans enable senescents to be independent of their children. There is still need to utilize the services and contributions of senior citizens for the benefit of the community and society at large. Pressey and Pressey's "inside" study (1966) of four modes of life in old age suggests many improvements to the present situation, especially through increased participation in groups of the elderly and service to other old people by competent senescents capable of such acitvity. Despite their increasing rigidity, old people are aware of their social environment and remain willing to contribute even at advanced ages.

Contributing to the Community

Elderly persons, in their younger days, were full participants in the civic and economic life of the community. They bore full responsibility for their own welfare as well as that of others and along with such responsibility enjoyed corresponding privileges and status. With retirement and increasing physical limitations, however, their role changes, even to the point of financial dependence. Such a reversal of role lowers their status and self-esteem. At the same time, the scope of personal privileges becomes somewhat restricted, creating additional adjustment problems. The transition crisis of retirement is serious for all who have not prepared for it.

Recently, many private and company pension plans have increased benefits to employees who retire early. This opportunity to retire before sixty-five, whether it is used or not, encourages earlier planning for retirement and reduces the stress of mandatory retirement at a specific age. When both husband and wife are working, the early retirement of one of them because of poor health, dissatisfaction with the job, or other reasons is an opportunity for leisurely homemaking, as well as for spending more time together. With pension and social security benefits a couple are usually capable of financial self-support, at least until their savings or investments dwindle. The current rate of inflation makes it difficult to plan financial security for life.

The aging person, during the years of adulthood, was physically, emotionally, and intellectually capable of helping chart the course of his community, economically, politically, and culturally. Because of his age and experience, he has had a greater voice than young adults. With advancing age, however, this position too has changed. Evidence of this is the fact that no one younger than forty-two years of age or older than sixty-eight years has as yet been elected President of the United States (Murray, 1956). For a variety of reasons, including increasing physical limitations, the maturity of the children who formerly looked to him for help and guidance, his partial or complete retirement from gainful employment, or perhaps his inability to modify long-held views to meet changing circumstances, the old person typically forfeits most of the control and direction he previously exerted.

In relinquishing his role in and contribution to the community, the senescent is faced with still another break with society. Without such participation he tends to lose contact with the wide circle of peers and younger persons so necessary for intellectual and emotional stimulation. The result is naturally unfavorable to his personal well-being. Loss of such contact leaves him less prepared to deal with the problems of society and to contribute to the welfare of the community.

RELATING TO THE INCREASING LIFE-SPAN

The social and personal problems related to old age continue to increase. Because of rapid advances in the medical sciences and the improved general conditions in which young and old live, the human life-span has been steadily growing. Figure 21-1 shows that the median age of the population of the United States has risen from less than twenty-one years in 1880 to more than thirty years in 1947. In the 1950s and 1960s a slight decline occurred, but projections for the 1970s and 1980s show an increase.

The older age groups continue to increase remarkably. While the total number of persons in the United States approximately doubled between 1900 and 1950, the number of persons sixty-five years old and over increased fourfold. In 1950, the older group constituted only 4.1 percent of the population; in 1970, almost 10 percent. Moreover, census projections show that this age group will steadily increase from 20,156,000 in 1970 to 23,703,000 in 1980 and to 28,839,000 in the year 2000, a nearly 40 percent increase over 1970 (U.S.

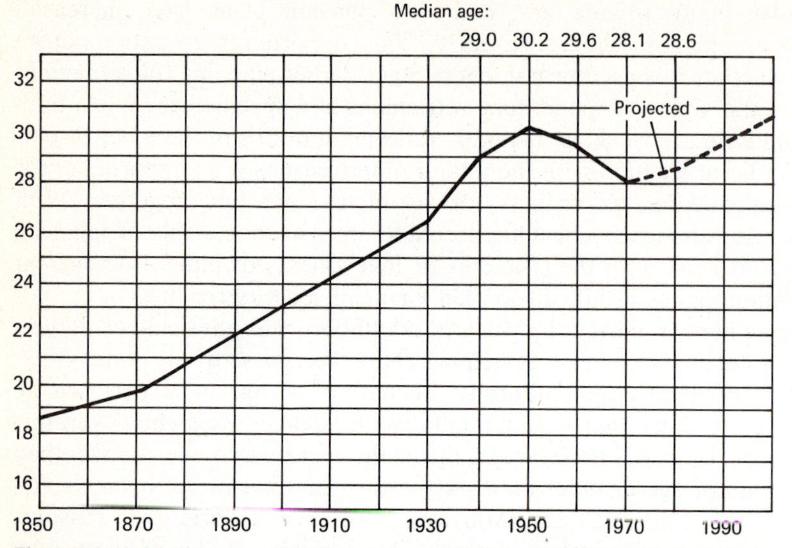

Figure 21-1 Median age of the American population, 1880-1980. [*Source*: U.S. Bureau of the Census. *Statistical abstract of the United States: 1973* (94th ed.); and *Current population reports* (Series P-25, Nos. 448 and 490), 1970 and 1974.]

Bureau of the Census, 1974, p. 6). Figure 21-2 illustrates the past and projected future changes in the size of the senior population.

With the increasing number of persons who live to old age and the increasing number of years people live beyond retirement, new problems beset society. The early retirement trend, along with increasing longevity, strains public and private pension plans as the ratio of workers to nonworkers

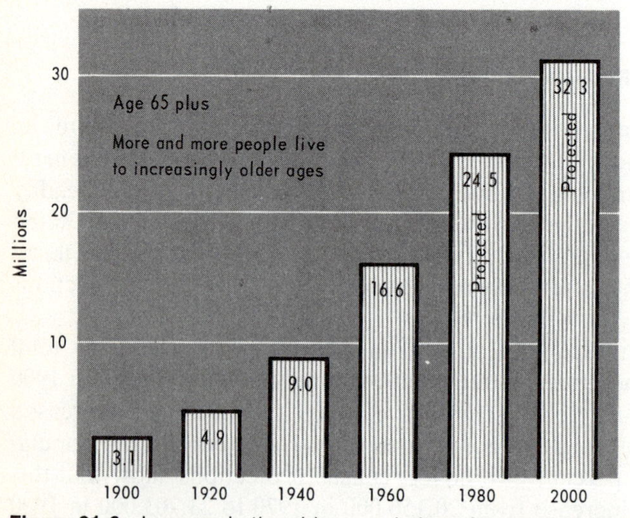

Figure 21-2 Increase in the older population of the United States, 1900-2000. [*Source*: U.S. Bureau of the Census. *Current population reports* (Series P-25, No. 490), 1974.]

continues to shrink. Greater time, effort, and resources are required for researching the medical, psychological, and social problems of the aged. Additional facilities are needed for the treatment and care of the elderly population. In addition, the general well-being of the elderly depends on opportunities for leisure-time activities and for productive endeavor: opportunities to do, to achieve, to feel success, and to make a real contribution to society. Such opportunities demand the cooperation of society as a whole, not merely the aged population. The distinguished historian Toynbee declared: "A society's quality and capability can best be measured by the respect and care given to its elderly citizens." After quoting Toynbee, President John F. Kennedy, in a speech to the House of Representatives in 1963, noted:

> Our senior citizens present this nation with increasing opportunity to draw upon their skill and sagacity and the opportunity to provide the respect and recognition they have earned. It is not enough for a great nation merely to have added new years to life—our objective must also be to add new life to those years.

The problems of increasing age are, of course, not merely social. Each adult must prepare for a greater life-span. Biologists predict that within the next generation the human race can look forward to a life-span of more than 100 years. Will that prolongation of life be a curse or a blessing? Much depends on society's attitude toward the senescent. There will be more time for finishing one's lifework, for autobiographical retrospect, and for contemplation or meditation. Hence every person must anticipate more and longer-term adjustments to old age than were common in past generations. However, with the sympathetic cooperation of society and the acquisition of wholesome attitudes, interests, and activities, one can look ahead not in despair but rather with hope for continued constructive work and accomplishment of one kind or another. Much will depend on one's skills and on efforts to find meaningful activities.

QUESTIONS FOR REVIEW

1. Describe the physical condition of old people and enumerate illnesses frequent at this age.
2. What changes in the self-concept may be expected to result from the gradual decline of the bodily systems?
3. What is the significance of psychosomatic interdependence in regard to behavior?
4. What important effects on personality and behavior does the restriction of the senescent's environment have?
5. Explain the religious concerns of old people.
6. Why is a diversity of interests necessary for the elderly?
7. In what ways are social contacts essential for the well-being of the elderly person?
8. What changes typically occur in the status of the elderly within the community?
9. What are some basic provisions society makes for older persons in the United States?
10. Indicate how the proportions of different age groups in the American population are shifting.

REFERENCES

Selected Reading

Birren, J. E., Butler, R. N., Greenhouse, S. W., Sokoloff, L., & Yarrow, M. (Eds.) *Human aging: A biological and behavioral study.* Bethesda, Md.: National Institute of Mental Health, 1963. This Public Health Service publication, No. 986, includes articles—among them several medical studies—by twenty-two contributors.

Britton, J. H., & Britton, J. O. *Personality changes in aging: A longitudinal study of community residents.* New York: Springer, 1972. A sociological study of over 200 older persons with a mean age of seventy-one at the start and eighty at the close of the assessment.

Kastenbaum, R. (Ed.) *New thoughts on old age.* New York: Springer, 1964. A presentation of theoretical perspectives, clinical explorations, and assessment of characteristics of older people.

Riley, M. W., & Foner, A. (Eds.) *Aging and society.* Vol. 1, *An inventory of research findings.* Vol. 2, *Aging and professions.* Vol. 3, *Sociology of age stratification.* New York: Russell Sage, 1968, 1970, 1971. Sociological view of man and gerontology is expounded in this compendium. Volume 1 contains an overwhelming amount of data well integrated into four sections: sociocultural context, the organism, personality, and social roles. Volume 2 organizes material around education, nursing, religion, and other professions. Volume 3 examines the relationship of the aging process to the society in which it occurs. It includes chapters on age stratification, the concept of population, research on age, the work force, higher education, the life of individuals, and change of generations.

Wolff, K. *The biological, sociological, and psychological aspects of aging.* Springfield, Ill.: Charles C Thomas, 1959. A study of the various aspects of aging in terms of their effects, as well as suggestions for the welfare of the elderly population.

Specific References

Barker, R. G., et al. *Adjustment to physical handicap and illness: A survey of the social psychology of physique and disability* (Rev. ed.). New York: Social Science Research Council, 1953.

Cavan, R. S., Burgess, E. W., Havighurst, R. W., & Goldhammer, H. *Personal adjustment in old age.* Chicago: Science Research Associates, 1949.

Fromm, E. Psychological problems of aging. *Child Fam.,* 1967, **6,** 78–88.

Moberg, D. D. Religiosity in old age. *Gerontologist,* 1965, **5**(2), 78–87, 111–112.

Murray, A. *U.S.A. at a glance.* Boston: Houghton Mifflin, 1956.

O'Reilly, C. T. Religious practice and personal adjustment of older people. *Sociol. soc. Res.,* 1957, **42,** 119–121.

Pressey, S. L., & Pressey, A. D. Two insiders' searchings for best life in old age. *Gerontologist,* 1966, **6,** 14–17.

Tournier, P. *Learn to grow old.* New York: Harper & Row, 1972.

Townsend, C. (Proj. dir.) *Old age: The last segregation.* New York: Grossman, 1971.

U.S. Bureau of the Census. *Statistical abstract of the United States: 1974.*

World Health Organization. Constitution of the World Health Organization. *Public Health Reports,* 1946, **61,** 1268–1277.

Part Nine

Recapitulation and Conclusions

It is helpful to reexamine the entire life-span and to note the crucial developmental factors and processes of various phases of life in relation to our technology and culture. As has been pointed out, each stage of human development has needs, tasks, and hazards of its own; each offers opportunities for growth and adjustment but also for disturbance and regression.

Through parental influence and personal endeavor one learns the societal expectations and cultural norms and adjusts oneself to them, and by personally motivated pursuit of selected goals one greatly influences one's conscious and unconscious search for one's own identity, for status in the community, and for the meaning of one's labors and sufferings, if not one's total existence. The intricacies of our present society, civilization, and technology—with the special opportunities and dangers of the space age—intensify challenges to personal growth and to the assumption and maintenance of adult responsibility.

Chapter 22

Synopsis of Human Development throughout Life

HIGHLIGHTS

All human beings are members of the same species, Homo sapiens, *whose genotypes depend on the intricate arrangement of DNA molecules passed on from past generations to the present.*

As a phenotype, a person is molded by the home environment and other surrounding influences. Psychic birth—creation of the self—results from the interaction of genetic and environmental factors within the setting of the home.

The prenatal period determines the organismic foundation for life; the early years of life outline the personality structure for childhood, adolescence, and adulthood, even though pervasive changes occur during puberty, often before an adult pattern of life is achieved, and even later in life.

Adulthood is the prime of life if the person is prepared for its developmental tasks. Remaining single or selecting a life mate and starting a new family, gaining economic independence by vocational activity and investment, contributing to the community, managing a home, and adjusting to changes in the family—all challenge a person's resources for healthy adjustment.

Old age, like adolescence, is marked by hazards and adjustment problems. Retirement and the decline of many abilities press the person toward readjusting his or her life-style and goals. The danger of narcissism increases and health problems multiply for the majority of senescents, and ideas of death often have a depressing power. Death is a natural phenomenon, yet it contains something mysterious.

The process of human development as a whole is a differential growth and cyclic structuring producing a personality and identity that are greatly influenced by individual endowments and environmental contingencies. Like a spiral, it presses upward for more than two decades, then maintains the level that has been achieved for about two decades; then reaches a long period of slow decline; then deteriorates more rapidly until a point is reached when a major organ or system fails and death occurs.

The life-cycle approach, pioneered from different perspectives by Quetelet, Hall, Bühler, and Erikson, among many others, finds its culmination in Lidz's metaphoric depiction (1968) of major phases of life. For a person of our century, going through life is "not like climbing up a hill and down the other side"; it is "more akin to a Himalayan expedition during which camps must be made at various altitudes, guides found, the terrain explored, skills acquired, rests taken before moving to the next level, and the descent also made in stages." During this expedition "childhood longs to die into youth and youth into maturity and so the latter in its turn should long to pass away into age" (Hall, 1922, p. 437).

From a biological point of view all human beings are members of a single species, *Homo sapiens.* People of the whole world share the same gene pool of the species. It is true that geographic, racial, national, linguistic, economic, religious, and other factors subdivide the breeding population into many segments, but they do not separate them completely (Dobzhansky, 1973, pp. 24–26). The human being is a highly distinct animal. Among the more than two million species that have inhabited the earth, *Homo sapiens,* according to Dobzhansky (1967, p. 108), is

> the only one who experiences the ultimate concern. Man needs a faith, a hope, and a purpose to live by and to give meaning and dignity to his existence. . . . Above all, he yearns for love and relatedness to other people; he wants to gain and hold his self-respect, and if possible the respect and admiration of others.

As the human being grows and matures, he displays many qualities far removed from his animal origins and has enormous plasticity in adapting himself to his environment and culture. He is "a highly adaptive animal and it is largely that property of his which we call 'intelligence' that makes possible the wide range of his adaptations" (Birren, 1973, p. 150). For the purpose of self-protection, he usually becomes a master of defense mechanisms and of maneuvering others to suit himself. He is a creature both "made" by others and making others in the image of himself (Berelson, 1964, pp. 663–666). Perhaps there is a natural desire in people "to develop fully the potential that is born in them; to experience all the emotions that the range of human existence has to offer; and always, above all else, to be permitted to become the individual that *their* hearts and minds tell them they should be" (Good, 1974, p. 171).

Human beings are highly complex organisms and even more complex personalities. Developmental psychology is a very young branch of psychological science. In the presentation of human development and adjustment, therefore, it must include many hints, leads, and theories if it wants to deal with human existence in the real world. Insight into the unconscious and the irrational is provided by psychoanalytic theory—which is supported by many evolutionary principles. The behavioristic orientation of Watson and Skinner, on the other hand, deals with external aspects of behavior and adds much information about human responses in variously structured situations. The behavioristic approach includes various techniques to modify behavior by manipulation of contingencies. Bruner, McClelland, Kagan, Piaget, Shane, and White show what is needed for various competencies and skills to be acquired, what is needed to adjust and relate to the modern world of today. Among many others, Maslow and Rogers include humanistic topics and concerns in their theories, such as love, peace, creativity, self-fulfillment, and higher values. These contributions are necessary to comprehend the fully human person as he or she copes with conflicts and problems. Ultimately, all these views must be integrated into a comprehensive perspective—a challenging task for future years of psychological research (Lugo & Hershey, 1974, pp. 199–208).

KEY FACTORS IN HUMAN DEVELOPMENT

Parents play the major role at the outset and during the early periods of development. The "psychic birth" of most persons occurs within the setting and atmosphere of the family (Jung, 1954). Peers and companions are potent modifiers during the later stages of life. Early patterns of feeling and emotion toward various dimensions of reality largely determine later attempts at self-direction and search for identity and status in life.

In this child-conscious century one would suppose that children in the United States would receive a more than satisfactory start in life. Some facts, however, seem to contradict this assumption. A large number of children are reared without one or both parents; in fact, almost 15 percent of children live with a stepmother or a stepfather. Figure 22-1 illustrates familial conditions as they existed in 1970. An undetermined percentage of families, although still physically functioning as family units, are barred by frequent dissension and psychological isolation from a meaningful sharing of attitudes, interests, and activities. Beginning and continuing life without both parents often results in serious deprivation of proper human models for self-identification. After reviewing nearly a thousand studies on trait and personality development from birth to adulthood, Benjamin S. Bloom (1964) concludes that the most important period for personality formation is that between birth and the start of schooling. Individual tendencies in motivation become more and more striking with increasing age. Early dispositions are unfolded in accordance with provisions and contingencies that stimulate their development. A striking illustration is provided by Skeels's empirical study (1966), which demonstrated reversal of a defective developmental trend by intervention and provision of enriched social stimulation for an experimental group of young children who gained competence as adults; the control group fared poorly in institutions.

Psychological bisexuality and other human qualities are developed within

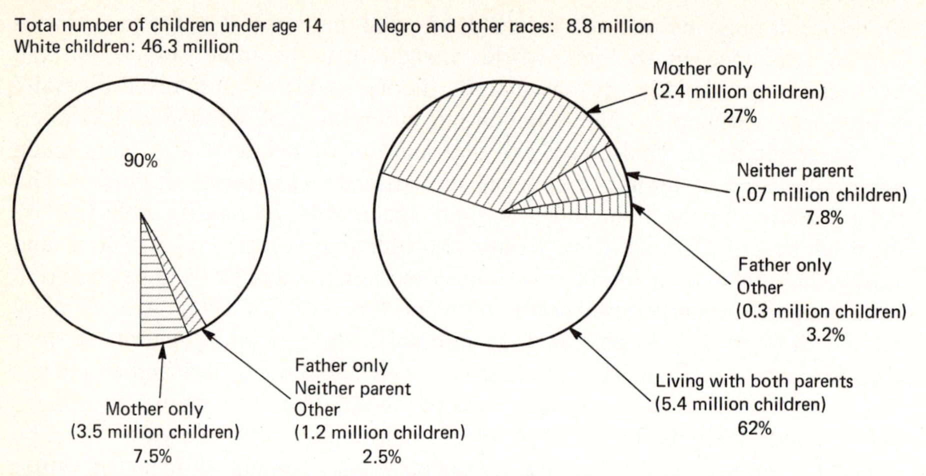

Figure 22-1 Living with and without parents: United States, 1970. [*Source*: *Profiles of Children*, 1970, pp. 23, 142.]

the confines of the total environment. The way one perceives the surroundings defines one's way of behaving in that environment (Insel & Moss, 1974). When the child is highly influenced by both the father and the mother, he or she usually develops the qualities and traits pertinent to both sexes. The boy or girl learns to identify with both male and female figures. The deepest needs of the child (affection, acceptance, and security) are gratified through dynamic interaction with the parents. Other influences are largely modifying factors, but a strong relationship with someone outside the family can substitute for either father or mother and in some cases for both. Ultimately each adult is a mosaic of many influences bound by a concept of self.

As presented in Chapters 11 and 16, the present trend toward lengthened education helps in some ways to prepare for the complexity of modern life. Children and young people today need a longer and better educational preparation for efficient adult living. Indeed, education is becoming a lifelong process. From the primary grades, it should include courses in family life as well as in values conducive to dependable citizenship. Learning from experience is even more important. Therefore, not only children and adolescents but also adults must have the freedom to explore and to make mistakes so that they can learn from them. The range of experience ought to be broad enough to create genuine opportunities for testing the total spectrum of potentialities. Figure 22-2 outlines the fundamental influences on most individuals; these affect us for much of our lives, but in varying degrees. Young people function within parameters defined by parents, siblings, and peers; and within a matrix of less personal contingencies generated by community, school, and society.

FOUNDATIONS FOR LIFE

The prenatal period, infancy, childhood, and adolescence form the foundation for an adult pattern of life. Each makes significant contributions to adult traits

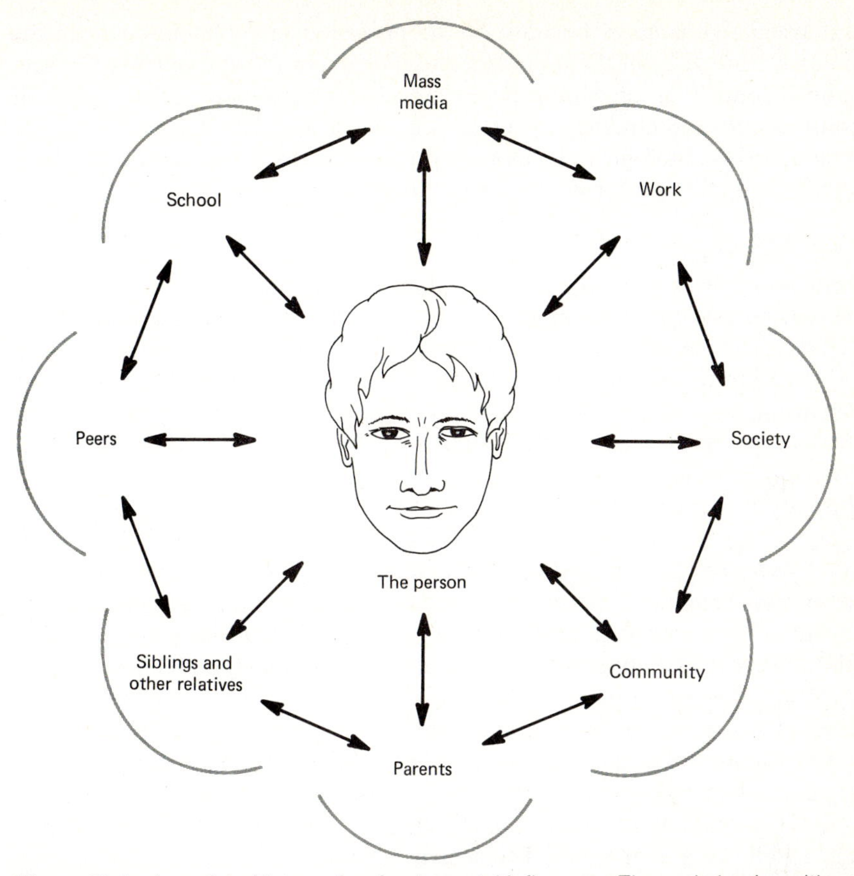

Figure 22-2 A model of interacting fundamental influences. The cycle begins with parents, who are affected by their past and all fundamental influences. The young person is affected by and adjusts to all major influences present in his or her milieu.

and characteristics by the influence it exerts on subsequent periods. Thus prenatal growth lays a foundation for development during infancy; what happens in infancy affects life and adjustment during childhood; and developments during puberty and adolescence can also be traced to childhood and, to a degree, to infancy. There is considerable support for the psychoanalytic theory that the basic personality pattern is established during the first five years of life; yet many extensions and modifications occur later. The section that follows retraces the most influential developments contributing to the total personality.

Prenatal Period

The prenatal period is a stage of extreme dependence during which physiological structures, motility, and sensitivity to stimuli have their beginnings; it normally continues until full term, when the infant is ready to function outside the mother's uterus. Although the developments occurring during the prenatal stage primarily represent physiological growth, this stage is also of great psychological significance, because of the profound relationship between the physiological integrity of the organism and its behavioral functioning. Without autonomic control of physiological processes, for example, other forms of integrative control, including conscious regulation of behavior, are impossible (Kitamura, 1974). Biological development provides the foundation for behavioral and personality patterns and characteristics.

Birth and Infancy

The process of birth gives the newborn infant the first exposure to a personal and increasingly autonomous existence. At this traumatic point the newcomer's needs have to be met by others, who may or may not have satisfactory attitudes and information to safeguard his welfare. The infant may be welcomed by his parents, or he may enter into a discordant group of persons barely managing to live under the same roof. In the latter case, the infant's adaptation to his environment is a difficult task.

Infancy is a preparational phase of life; all the major developments that mark human life appear before this stage gives way to childhood. The infant needs a variety of physical and social stimuli before he can discover what they mean for him. Scientists interested in human development and adjustment do not fail to acknowledge the crucial role of the first two to three years of life. Certain universal essentials must be supplied if continuing development is to occur. Some of these "raw materials" are tangible, like food and shelter, and some are intangible, like acceptance and love. Both are indispensable in promoting feelings of trust, security, and individuality. When general sensitivity to stimuli begins to rise, from about the age of three months, the infant needs a large amount of sensory stimulation for seeing, hearing, touching, patting, cuddling, and moving. The infant needs a multitide of stimuli long before he can discover them on his own initiative.

During infancy the pattern of living and adjusting gains much in structure and complexity. If the parents' personality traits and the atmosphere of the home do not produce any significant damages to the early structural organization, the physical and emotional well-being of the infant is enhanced by a sound foundation and a growing strength to meet challenging situations during the later periods of development. Opportunities for experiencing sufficient stimulation by human beings and objects, as well as exercise in initiative and exploration, are of major assistance in psychological development.

Childhood

Childhood depends on the development and experiences of infancy. The timing of progress in developmental tasks serves as a basis for predicting further developments and overall adjustment. Figure 22-3 illustrates the average time for the beginning and completion of several developmental tasks. It may be noted that some tasks are comparatively simple, while others are complex. An infant, for example, readily takes solid foods and chews them. To learn to eat solid foods the infant merely needs opportunities at the right time. The task of

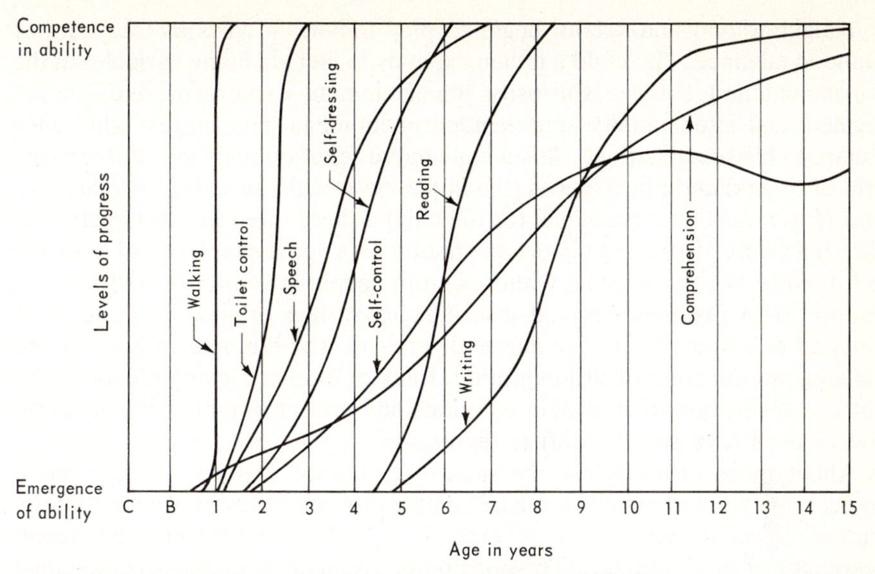

Figure 22-3 Hypothetical curves of various developmental tasks.

self-control begins early but extends well into the twenties. A great deal of help and encouragement is needed to direct drives and impulses successfully into acceptable channels of expression. The child ought to be provided with value stimulation and clarification in order to further the socialization process.

Outstanding qualities of the child are self-awareness and the beginning of self-regulation. While organismic regulation is established long before birth and is strengthened during the early weeks of postnatal life, desirable forms of behavior regulation take years of effort. As the child is exposed to many environmental influences, his selectivity increases, and, more important, his self-concept appears as a third major force directing his activities and adjustment (genetic makeup and environmental stimulation being the others). The child's growing self-awareness is the searchlight that enables him to further explore what is desirable for the structure of his own identity. Thus the self-concept becomes an important influence on the formation of other traits and attitudes. "From childhood on, a person's own choices determine to a considerable extent which possible courses of development are to be followed, which to be closed out" (Tyler, 1965, p. 506).

During the preschool years, the child constructs within himself *his* type of world for himself to live in. Internally and externally, the created *Eigenwelt* ("own world") becomes the arena where he plays out his own dramas and struggles for self-adjustment. Many events have great importance for the child—for example, recognizing the importance of speech as a means of communication, leaving the confines of home for the first vacation trip, and beginning to attend school. In middle childhood, the consequential event is finding a companion in whom to confide. Each event of this order reorganizes the child's *Eigenwelt* significantly and affects the self-concept.

Neighborhood and school begin to play important roles as the years of childhood advance. The child's milieu expands to include many variables in the environment and culture. Curiosity impels him to explore most observed, imagined, and intellectually apprehended phenomena. His suggestibility adds its share to his socialization. Thus a child moves continuously into *Mitwelt* (the world of others) and into *Umwelt* (the objective world), as well as into his own world *(Eigenwelt)*. It is beneficial for the child if these three worlds largely fuse rather than differ from and distort one another. In some cases a child lives too much the life of others, or he withdraws too deeply into the world of his own creation. Most disturbances and disorders have their origins in the years of childhood or earlier. Excessive parental demands, for example, may strain the child's ego to the point of disintegration. Lack of mothering and affection may stunt the absorption of desirable emotions and attitudes, inhibit the desire to grow, or even reverse it by diffuse regression.

About three years before the span of childhood expires, children move into peer society and establish a basic pattern for identifications with contemporaries. Success with peers is crucial for later adolescent adjustment. Acceptance of one's duties and responsibilities is an important lesson that must be learned within the childhood years. Personal competence in working toward perceived goals and ideals increases during the preadolescent years. An effective pattern of living and adjusting to situations encountered makes the child eager and ready to anticipate higher levels of maturation during the pubertal years.

Adolescence

This period includes the pubertal growth spurt and later developments leading up to an adult pattern of life, with its salient and well-embedded traits and features. Rapid physiological developments, while producing a temporary loss of organismic equilibrium and behavioral self-control, also indicate acquisition of adult powers and capabilities. Intense self-observation, heightened emotional experiences, and moral-religious stirrings usually occur in the period of adolescence. It is a period of intensified conflict and ambivalence, or regression and progression.

Self-consciousness and emotionality reach new heights with turbulent rates of physical growth and sexual maturation. Heterosexual associations contribute to the further rise of social awareness and self-consciousness. Bold explorations of sexuality and drugs are frequent. The voice of conscience and perfection often becomes vivid at this stage of life and presses the adolescent toward additional self-improvement. Feelings of uncertainty, guilt, and remorse run high as the adolescent ventures into experiences conflicting with his or her high standards and values or those of important people in the environment. Most teenagers wonder what life is all about and what kind of people they would like to become.

Maturation of intellectual power is followed by much doubt and rumination, with attempts to discover answers to emergent and seemingly conflicting values. A highly critical attitude toward parents and authority is often a marked feature of the adolescent. Tendencies toward perfectionism and toward social discrimination are also heightened. The search for identity for the male is marked by strivings for achievement and independence from parents; the female seeks to establish a network of intimate interpersonal affiliations and loves (Douvan & Adelson, 1966, pp. 343–347). Female roles are being sharply challenged and restructured by the recent women's liberation movement. Both sexes look for and usually find adults on whom to model their personalities. Both male and female adolescents wonder what kind of goals and objectives in life they should embrace.

The need to encourage capable adolescents to remain in school and develop their potentialities as fully as possible has been recognized in many quarters. School opens the eyes of many young persons to various cultural opportunities and chances for higher levels of self-application and socioeconomic advance.

Sexual maturation and interest in members of the opposite sex is frequently accompanied by ups and downs, with both encouraging and discouraging feelings about personal adequacy in heterosexual relationships. For many adolescents emotional ambivalence and irregularities in sexual function and drive confuse the overall picture of maturation. Frequent attempts to conform fully to the standards and actions of their peers contribute to the formation of cliques and crowds which, in turn, provide opportunities for deep interpersonal identifications.

Toward the end of adolescence, the adolescent's self-concept is clearer and more complete. The person is then capable of visualizing his present roles and of projecting himself into future roles and goals. Now he seems to realize what he wants to become, but later changes can be expected. Many life-directing decisions are made before entering into an adult style of life. The direction of development is often significantly changed by a confidant, a spouse, or the decision to disregard certain aspects of experience. Apparently, the adolescent chooses one of many possible life-styles as he accepts and strengthens his adult identity.

ADULT PHASES OF LIFE

Barely perceptibly the individual is becoming an adult as he tackles the tasks of vocational choice and marriage. Competent regulation of drives and emotions and efficient application of powers and abilities are expected from the young adult. Most deliver, some fail. Experience and adjustment on an adult level depend greatly on strengths and weaknesses in earlier life, especially the solutions to the conflicts and frustrations of adolescence. Finding satisfactory outlets for emotional and sexual drives is essential for a healthy adult pattern of living. Constructive self-regulation of energies and drives usually involves considerable use of compensation and sublimation. Adulthood is the longest period of life, when aging progresses slowly up to sixty-five or seventy years of age.

During adult life, marital discord and vocational disappointments are two frequent sources of maladjustment. Lack of genuine insight into one's abilities, assets, weaknesses, and liabilities is one of the principal etiologic conditions that produce and intensify neurotic patterns of adjustment. Difficulties in attaining and controlling mature genital sexuality and the capacity for intimacy play a large role in general difficulty in coping with life. The achievement of vocational stability facilitates adaptation to one's total environment. Long-term adjustment is best maintained by wholesome self-application and achievement in important areas of life.

During the adult stage of life most abilities peculiar to the individual constellation of endowments are developed to a high level. Their integration and application in terms of vocational opportunities represent major developmental tasks. One has to find his niche in life and settle down before his abilities and energy begin to decline noticeably during the middle phase of adult life. During the years of early adulthood, there are good possibilities of correcting personality weaknesses by one's own efforts or with the assistance of professional counselors.

If a person makes efforts to find opportunities to apply his abilities and skills, and if he develops civic and cultural interests, cognitive decline is often retarded. Adherence to physical and mental hygiene is a necessary adjunct to proficient living during the adult years of life. Moreover, a religiously oriented philosophy of life is generally helpful in promoting meaningfulness in activities and an adaptation to reality in all its aspects.

DECLINING PHASES

Aging is a lifelong process, but its disadvantages are concentrated in old age. During the years of later adulthood a gradual decline begins, and illnesses strike many persons at this age. Decline accelerates as late adult years merge into senescence. Though unevenly, decline or deterioration affects most structural and many functional powers. Structural deterioration usually precedes functional decline. Some organs and systems deteriorate at a faster rate than others. For example, while there is an otherwise high level of vitality, failing kidneys may lead to death. A heart attack may also occur at an early stage of decline. Amplitude of memory narrows extensively, while reasoning power may show no significant decline until the late seventies. Any deterioration of organs or powers, however, has significant repercussions on the total organism and personality.

The majority of elderly persons tend to cling to the image of themselves as they were in early or middle adulthood and continue to set long-range goals and to propose additional self-realization for the future. Approaching these goals is often difficult because of decline of ability and the occurrence of illness. The older person, then forced to deal with his actual self, often encounters an existential vacuum or identity crisis, especially when age or illness is considered a misfortune rather than an integral part of life.

Difficulties in maintaining former social status and in making adjustments multiply. Withdrawal from physical and social activities should not create an abrupt change in the pattern of living. In order to avoid major financial setbacks, many older persons continue to work as long as employment is available to them. Both unmodified continuation and sudden dropping of work tend to be unhealthy. Not fully realizing the extent of their declining abilities, some people strain their hearts by engaging in strenuous physical work such as shoveling snow or working in the garden. A survey of studies of elderly people shows some polarization: many embrace the "activity model" while others tend toward disengagement; many gain in stability while others show changeability (Kuypers, 1972).

Final Decline and Death

To the young adult, death is a strange phenomenon, difficult to comprehend. His or her thoughts about it are accidental and short in duration. A funeral is often attended as an external necessity rather than as a moving experience. The situation changes in later years. On an occasion such as a funeral, a senescent is likely to be emotionally involved, and ideas of identification or self-reference become disquieting.

During the senescent years of life, ideas of death begin to enter consciousness more often. For many persons the idea of death is depressing, something they avoid facing and suppress. When the terminal decline begins, usually about one year before death, there is a noticeable departure from the relative stability of earlier years. Psychomotor skills, cognitive functions, and self-control all fail greatly, and incoherence and gloominess often set in. Even untrained observers notice the difference.

The idea of being near death produces anxiety and stimulates preparation for it. To many people the depressing idea of bodily extermination gives way to a resigned or an anticipative outlook toward death and the hereafter. When Socrates was sentenced to death and given a glass of poison to drink, his philosophical attitude moved him to utter a still immortal message: "Only my body will die; my soul will eternally exist and be judged by a Supreme Being in accordance with the good and the bad deeds of my life." This insight of a pre-Christian philosopher reveals much about the senescent's belief concerning death as a gate into a transcendent existence.

Death is a part of the total human condition. For Martin Heidegger (1962), it is "the mortal mode" man lives through from birth onward, since "as soon as a man lives, he is old enough to die. . . . Death teaches us that life is a value, but an incomplete value." Abortions, cases of sudden infant death syndrome or crib death, and teenage accidents and suicides attest to Heidegger's idea of the omnipresence of death in human life. Death is a natural phenomenon, yet it is also a mystery for people who are alive, for there exists something other than this visible world, which can be explored only with fantasy, excitement, and wonder (Tournier, 1972, pp. 233–236).

Reveries that survey life are often vivid in old age, especially just before death when discomforts and pain subside. Reviewing past events is beneficial if they have been in keeping with the philosophy of life and its key values. Duvall (1971, p. 430) puts the matter in these words: "Nothing can bring greater

Table 22-1 Death Rates per 100,000 Population

Rank	Cause of death	Death rate		Percent of total deaths	
		1970	1972	1970	1972
	All causes	940.4	942.2	100	100
1	Diseases of heart	360.3	361.3	38.3	38.3
2	Malignant neoplasms, including neoplasms of lymphatic and hematopoietic tissues	162.0	166.6	17.2	17.7
3	Cerebrovascular diseases	101.7	100.9	10.8	10.7
4	Accidents	54.2	54.6	5.8	5.8
5	Influenza and pneumonia	30.5	29.4	3.2	3.1
6	Certain causes of mortality in early infancy	20.9	16.4	2.2	1.7
7	Diabetes mellitus	18.5	18.8	2.0	2.0
8	Arteriosclerosis	15.9	15.8	1.7	1.7
9	Cirrhosis of liver	15.8	15.7	1.7	1.7
10	Bronchitis, emphysema, and asthma	14.9	13.8	1.6	1.5
	All other causes	145.7	148.9	15.5	15.8

Source: Monthly vital statistics report, 1971, **19** (13); and 1973, **21** (13).

satisfaction than finding that life all adds up, and that together the two (husband and wife) know who they are and where they are headed in the business of living." Religiously oriented people envision death as the entrance into a life with God. Atheists tend to see death as the end of their personal existence.

In 1961, for the first time, life expectancy for the United States population exceeded seventy years. In 1970, the life expectancy for the newborn was 74.6 years for females and 67.1 for males. In the early 1970s it continued rising at a moderate rate. Many biological researchers expect major breakthroughs that will significantly decelerate the process of aging.

Table 22-1 identifies the leading causes of death in the United States in 1970 and 1972. It will be noted that heart diseases account for more than one-third of all deaths. From 1954 to 1964 the death rate for kidney infections had increased by about 93 percent. As in the 1960s, most of the yearly variations in death rate in 1970 and 1972 were slight rises or decreases; tuberculosis mortality continued to decline, while the death rate from malignant neoplasms (cancer) increased fairly steadily. In 1972, the figures for cardiovascular (heart) and cerebrovascular diseases reached 462.2 and malignant neoplasms 166.6 per 100,000 population.

REVIEW OF HUMAN DEVELOPMENT

Table 22-2 lists many occurrences essential to each stage of life, including developmental tasks, hazards, and important characteristics of motivation, personality, and the self. This schematic presentation permits an overall view

Table 22-2 Review of Key Human Developments*

Level of development and approximate age	Physiological growth and psychomotility	Dynamics and motivation	Developmental tasks	Major hazards	Personality and the self-concept
Prenatal:					
Zygote, 0 to 2 weeks	Conception		Implantation	Subchromosomal aberrations; defective heredity	
Embryo, 2 weeks to 2 months	Differentiation of tissues and bodily systems		Total organismic growth as a foundation for postnatal developments	Penetration of drugs and other toxins	
Fetus, 2 to 9 months	Emergence of reflexive motility; approaches postnatal functioning power	Maintenance of organismic equilibrium; reflex movement	Biochemical controls; rise of viability	Endocrine and circulatory malfunctions; certain maternal diseases; birth complications	Transition from tranquility to motility and sensitivity
Neonatal (early infancy)	Increase of sensitivity and beginnings of sensorimotor coordination; integration of gross and refined central nervous system functions	Satisfaction of bodily needs; affective excitement; OR—orienting toward stimuli	Preservation of life: adjustments to new external and internal conditions, e.g., maternal approach, temperature, food, etc.	Physiological disequilibrium and infections	Adjustability vs. excitability; overactivity vs. quietness

Continued

*The roles various factors play at a particular phase of life usually do not begin or end abruptly as the individual moves into another stage, also some of them reappear at later stages but play different roles, e.g., maternal intake of drugs is a hazard to the embryo's growth, but adolescent drug misuse is a hazard created by the young person himself. In a schematic review of human development, overgeneralization or overspecification of various factors must be recognized.

Table 22-2 (Continued)

Level of development and approximate age	Physiological growth and psychomotility	Dynamics and motivation	Developmental tasks	Major hazards	Personality and the self-concept
Middle infancy, 2 to 15 months	Rapid growth in size and weight; refinements in neuromuscular coordination; further integration of lower and higher central nervous system mechanisms	Greater interest in environment; recognition of mother and familiar objects; intense need for mothering; rapid emotional differentiation; strong drive for activity	Gaining of control over neuromuscular and vocal systems; acquisition of new attention-getting techniques; establishment of emotional trust	Lack of physiological stability; perceptual deprivation; lack of mothering and social-emotional stimulation	Adaptability to parents; awareness of own individuality
Late infancy (toddler stage), 15 to 30 months	Advance and completion of phylogenetic motor patterns; control over fine muscles	Greater initiative in exploration of surroundings; emergence of childhood motivation; increasing resistance to parental demands	Progress in initiative; acquisition of speech facility; establishment of toilet controls	Difficulties in relating emotionally to parents and siblings; distrust and tears; maternal deprivation	Narcissism; rising awareness of self; strong attitudes toward self and others; acquisition of strong likes and dislikes
Early childhood, 2½ to 6 years	Acquisition of ontogenetic motor patterns; increase in gracefulness; decline in rate of physiological growth	Interest in distant environmental and social relationships; fantasy preoccupation; make-believe; "why" questioning	Increased use of verbal communication and social play activities; distinguishing right from wrong	Insecurity and childhood diseases; withdrawal from social stimulation	Great increase in social response; growth in self-consciousness and attitudes toward self; rising identification with parents

Middle childhood, 6 to 9 or 10 years	Greater control over fine muscle groups, e.g., dressing self, ball games; decrease of physical growth	Growth of realism in attitude and adaptability; recognition of role relationships; interest in friendships	Control of negative emotions; development of a scale of values; cooperative attitude; sense of sex identity	Lack of achievement and self-acceptance; attitudes of inferiority and defeatism	Extroverted and enthusiastic; appearance of surface traits; growth in personal responsibility
Late childhood (preadolescence); girls, 9 to 11½, boys, 10 to 12½ years	Rate of physiological growth ebbs, then increases; ready acquisition of various motor skills	Sex identity strengthened; adventure and novelty sought; scientific questioning arises	Adaptation to peer society; experience of group security	Poor peer relationships; lack of industriousness	Greater preoccupation with self; loosening of emotional identification with parents; wondering about years ahead
Puberty (early adolescence): girls, 11½ to 14, boys, 12½ to 15½ years	Turbulent growth of many organs and systems; approach to adult size and proportion; biochemial balances disturbed; external awkwardness increased	Strivings for independence; negativism; emotional vacillation, ambivalence, and moods; emergence of powerful sexual drives; erotic fantasy; strivings for intimacy with peers	Self-reorganization; gains in independence by emancipating self from family; control over sexual impulses and base emotions	Isolation and excessive daydreaming; lack of self-assertiveness; extreme rebellion; peer rejection	Increase of introversion; indecision; search for human models and oneself
Mid-adolescence: girls, 14 to 16, boys, 15½ to 18 years	Reduced rate of physiological growth; large gains in fine motor control and strength	Powerful drive for social companionship, including members of opposite sex; expansion of intellectual quests and reasoning	Acceptance of masculine or feminine role; identification with peers	Peer rejection; perfectionist aspirations; moodiness; use of dangerous drugs	Lack of integration; ambivalence and antagonistic strivings; magnified social awareness; search for standards

Continued

Table 22-2 (Continued)

Level of development and approximate age	Physiological growth and psychomotility	Dynamics and motivation	Developmental tasks	Major hazards	Personality and the self-concept
Late adolescence: girls, 16 to 20; boys, 18 to 22 years	Appearance of adult characteristics; adult level of performance; biochemical equilibrium	Approach of heterosexual adjustment; striving for maturity and popularity	Selection of occupation; improved self-control; formation of *Weltanschauung*	Rejection of self and neurotic or delinquent solutions of conflicts	Crystallization of self-concept in terms of social and moral norms; concern over future
Early adulthood: women, 20 to 30; men, 22 to 35 years	Optimum level of physiological development and psychomotor controls	Attainment of relatively persistent hierarchy of motives; active social and civic participation	Achieving economic independence; selecting mate and starting family; performing paternal or maternal role	Fixation of pubertal and adolescent attitudes and modes of adjustment; lack of flexibility	Integration of behavior-organizing factors into personally acceptable pattern; increase in extroversion
Middle adulthood: women, 30 to 45; men, 35 to 50 years	Moderate decrease in speed and strength; increasing appearance of physical limitations	Interest in children, comforts, and stability; concern about vocational status and responsibilities	Management of home and care of children; sense of responsibility in community; gains in leadership	Lack of unified philosophy of life; inability to maintain economic or social standards of living; lack of readiness to release children; family dissension	Decrease in flexibility; reliance on the habitual and ideological

Late adulthood: women, 45 to 60; men, 50 to 65 years	Problems in health preservation; decline in physical strength and endurance; sight and hearing difficulties	Decrease of interest and drive; moods and worrying more frequent; leisure-time activities sought; decrease in desire to learn new subjects	Adjustments to family changes as children leave home; preservation of adult personality traits and abilities; preparation for retirement	Excessive reliance on past; acceleration of aging and lessened capacity for self-repair	Increased rigidity and decrease of resourcefulness
Senescence: women, 60 to death; men, 65 to death	Further deterioration of sensory activity and motor skills; lessened capability for even daily routine	Desire to be of use; partial withdrawal from social functions; loss of interests; restriction of activities; life review and expectation of death	Maintaining frequent contact with children and grandchildren; maintenance of health; integration through maintenance of self-esteem	Physical strain; chronic illness; isolation from relatives; skepticism and depression; excessive preoccupation with self; loss of meaningfulness of life	Difficulties in relying on past and habitual; self-fulfillment along some lines or apathy and rapid disengagement

of human growth and decline. Of course, the student here may miss certain notches in the life of any person, as when the toddler learns how to open the door, when the child learns how to speak and be understood, when he finds a friend to confide in, when the teenager begins to wonder what it feels like to engage in a sexual act with someone of the opposite sex, or when the idea of living together for life with a loved person is contemplated. There are many exciting events in life that psychology barely touches on. Novels and biographies usually emphasize this kind of moving experience.

QUESTIONS FOR REVIEW

1 At what age and in what ways do parents greatly influence their children?
2 What basic differences are there between children who live with both parents and those who live with one?
3 What are the fundamental essentials for continuing development?
4 What are the qualities that distinguish the child from the infant and the adolescent from the child?
5 How does the child increase control over his behavior? What is the role of parents in this respect?
6 Compare and contrast major adolescent and adult traits.
7 How does vocational development contribute to adult adjustment?
8 Identify some significant changes occurring during middle adulthood and relate them to the developmental tasks of that age.
9 What is the life expectancy of children born in 1970, compared with projections for the future, and what is the impact of longer life on developmental stages?
10 Identify and discuss several leading causes of death and their impact on senescents' goals.

REFERENCES

Selected Reading

Berelson, B. & Steiner, G. A. *Human behavior: An inventory of scientific findings.* New York: Harcourt, Brace & World, 1964. A fairly systematic presentation of 1,045 findings in the field of behavioral science.

Coleman, J. C., & Hammen, C. L. *Contemporary psychology and effective behavior.* Glenview, Ill.: Scott, Foresman, 1974. A comprehensive study of human nature, its resources and abilities; behavior and self-identity theories; coping with stress; development and interpersonal relationships; marriage and family; the quest for values; and means of assisting the person in his or her search for adjustment.

Neale, R. E. *The art of dying.* New York: Harper & Row, 1973. Includes fear of life and death; various ways of dying, including suicide; emotions, concepts, and beliefs about death. Selected bibliography is annotated.

Specific References

Berelson, B., & Steiner, G. A. *Human behavior: An inventory of scientific findings.* New York: Harcourt, Brace & World, 1964.

Birren, J. E. A summary: Prospects and problems of research on the longitudinal development of man's intellectual capacities throughout life. In L. F. Jarvik, C. Eisdorfer, and J. E. Blum (Eds.) *Intellectual functioning in adults: Psychological and biological influences,* pp. 149–154. New York: Springer, 1973.

Bloom, B. S. *Stability and change in human characteristics.* New York: Wiley, 1964.

Dobzhansky, T. *Genetic diversity and human equality.* New York: Basic Books, 1973.

Dobzhansky, T. *The biology of ultimate concern.* New York: New American Library, 1967.

Douvan, E., & Adelson, J. *The adolescent experience.* New York: Wiley, 1966.

Duvall, E. M. *Family development* (4th ed.). Philadelphia: Lippincott, 1971.

Good, P. *The individual.* New York: Time-Life Books, 1974.

Hall, G. S. *Senescence: The last half of life.* New York: Appleton, 1922.

Heidegger, M. *Being and time.* New York: Harper & Row, 1962.

Insel, P. M., & Moss, R. H. Psychological environments expanding the scope of human ecology. *Amer. Psychologist,* 1974, **29,** 179–188.

Jung, G. G. *The development of personality* (F. F. C. Hull, Trans.). New York: Bollingen Foundation, 1954.

Kitamura, S. Integrative regulations of psychological functions. *Tohoku Psychologica Folia,* 1974, **33** (1–4), (1–12).

Kuypers, J. A. Changeability of life-style and personality in old age. *Gerontologist,* 1972, **12,** 336–342.

Lidz, T. *The person: His development throughout the life cycle.* New York: Basic Books, 1968.

Lugo, J. O., & Hershey, G. L. *Human development: A multidisciplinary approach to the psychology of individual growth.* New York: Macmillan, 1974.

Skeels, H. M. Adult status of children with contrasting early life experiences. *Monogr. Soc. Res. Child Developm.,* 1966, **31,** (3, Whole No. 105).

Tournier, P. *Learn to grow old.* New York: Harper & Row, 1972.

Tyler, L. E. *The psychology of human differences* (3d ed.). New York: Appleton-Century-Crofts, 1965.

Glossary

Achievement quotient (AQ) The ratio between a person's scores in scholastic performance and the standard.

ACTH Adrenocorticotrophic hormone produced by the pituitary gland to stimulate corticoid production in stress situations.

Adjustment Processes and behaviors that satisfy a person's internal needs and enable the person to cope effectively with environmental, social, and cultural demands.

Adjustment, emotional A state of emotional maturity proper to the age of a person and marked by a relatively stable and moderate emotional reactivity to affect- and mood-eliciting stimuli.

Adjustment, social Reaction patterns toward others conducive to harmonious relationships within family and other reference groups.

Adolescence The developmental period from the onset of major pubertal changes to adult maturity.

Adrenals A pair of ductless or internal-secretion glands attached to the kidneys and secreting adrenalin and cortin, important in emergency and stress situations.

Adult A postadolescent person whose growth is completed in most aspects of development and who is capable of satisfactory reality testing and adjustment to self and environment.

Affect A vital feeling, mood, or emotion characterized by specific physiological (psychophysical) changes and states.

Age, mental (MA) The level of intellectual efficiency as determined by a test of intelligence; the age at which a computed score on an intelligence test occurs.

Age norm The average for a given age as revealed by sample group performances at this age.

Aging The continuous developmental process beginning with conception and ending with death during which organic structures and functions of an immature organism first grow and mature, then decline and deteriorate.

Alienation A process of separation or an emotional state in which familiar persons and relationships appear strange or unacceptable and meet with reactions of criticism and withdrawal.

Allergy Heightened sensitivity to pollen or any other foreign substance, causing respiratory, skin, or gastrointestinal irritation, including swelling. Hay fever and hives are frequent allergic conditions.

Altruism Deep unselfish concern for others, often expressed in helping them or in charitable activities.

Ambivalence Internal tendency to be pulled (usually psychologically) in opposite directions, e.g., acceptance-rejection, love-hate, participation-withdrawal.

Amnesia Defensive forgetting caused by a strong conflict or inability to face a certain event or experience, with subsequent repression.

Amniocentesis Withdrawal of embryonic cells from the amniotic fluid; the cells may then be cultured for such purposes as assessment of genetic defects.

Anesthesia Lack of psychophysical response to sensory stimuli; unawareness of pain.

Anlage An original basis for or a disposition toward a specific developmental trend or factor.

Anoxia Deficiency in the supply of oxygen to tissues, especially the brain, causing damage to their structural integrity.

Anxiety neurotic Distress and helplessness due to ego damage or weakness, accompanied by an expectation of danger or misfortune.

Aptitude A recognizable context of capacity or potentiality for specific achievements if the person is given proper training.

Arthritis Inflammation or deformation of one or several joints, accompanied by pain, stiffness, and swelling, often chronic.

Aspiration, level of The intensity of striving for achievement, or the standard by which a person judges his or her activity in reference to expected end results.

Atrophy Progressive decline of a part, or decrease in size, or degeneration.

Attitude An acquired persistent tendency to feel, think, or act in a fixed manner toward a given class of stimuli.

Autistic Self-centered; with perception, feeling, and thinking unduly controlled by personal needs, desires, and preferences at the expense of sensitivity to others or to situational demands.

Autogenous Self-originated, as distinguished from what is initiated by outside stimuli and learning.

Behavior Any kind of reactivity or self-generated activity, including complex patterns of feeling, perceiving, thinking, and willing, in response to internal or external, tangible or intangible stimuli.

Birth injury Temporary or permanent injury occurring during the birth process. Many disabilities are attributed to brain damage resulting from birth injury.

Carcinogenic Capable of eliciting cancer.

Cathexis Attachment of affects and drives to their goal objects; direction of psychic energy into a particular outlet.

Cephalocaudal The direction of growth from head to extremities (or tail).

Character The acquired ability to act and conduct oneself in accordance with a personal code of principles which is based on a scale of values, and facility in doing so.

Child A person between infancy and puberty.

Childhood The period of development between infancy and puberty (or adolescence).

Chromosome A minute threadlike body in the cell nucleus that carries many molecules of DNA, RNA, and protein and small amounts of other substances.

Cirrhosis Replacement of regular tissue, especially of the liver, by fibrous tissue—a frequent liver disease of alcoholics.

Compeer An age-mate (*see* Peer).

Conception The merging of the spermatozoon and ovum in human fertilization.

Conditioning As used in the present work, a mode of training in which reinforcement (reward or punishment) is used to elicit desired (rewarded) responses.

Conduct That part of a person's behavior, including insufficiencies and reverses, which is guided by ethical, moral, or ideological standards.

Confabulation An attempt to fill in the gaps of memory by unwitting falsification.

Conflict An intrapsychic state of tension or indecision due to contrary desires, ungratified needs, or incompatible plans of action; such tension may also exist between conscious and unconscious choices.

Congenital Referring to characteristics and defects acquired during the period of gestation and persisting after birth.

Constitution The organization of organic, functional, and psychosocial elements within the developing person that largely determines his or her condition.

Conversion As used in the present work, transformation of anxiety and energies elicited by a conflict into somatic symptoms.

Culture A country's manner of living, characterized chiefly by intellectual and societal aspects of a given civilization; its methods of child rearing and education, customs and mores, traditional civic and religious practices.

Daydreaming A form of withdrawal from unpleasant or frustrating reality into the realm of fantasy and reverie, frequently of a pleasant, wish-gratifying type.

DDT Acutely poisonous pesticide. If ingested by man it produces heightened excitability, muscular tremors, and motor seizures; deadly in larger amounts.

Defense dynamism or mechanism Any habitual response pattern that is spontaneously used to protect oneself from threats, conflicts, anxiety, and other conditions that cannot be tolerated or coped with directly.

Development, level of A period in a person's life marked by specific clusters of traits, interests, and attitudes and by a similarity in interests and concerns to other persons in that period of life.

Developmental-level approach In psychology, the approach in which the total personality of the person is considered at each phase of life.

Developmental psychology A division of psychology that investigates the growth, maturation, and aging processes of the human organism and personality, as well as cognitive, social, and other functions, throughout the span of life.

Developmental task An increase in the ability to produce more complex behavior patterns in any dimension of growth specific to one of the successive levels of human development, adequate performance and application of which promote adaptation to reality and an attitude of personal adequacy.

Differentiation The process by means of which structure or function becomes more complex or specialized; the change from homogeneity to heterogeneity.

Dimension A coherent group of processes having a particular denominator—e.g., intelligence, emotion, and language dimensions of personality.

Dimensional approach In psychology, the approach in which a specific aspect or area of personality is considered throughout various phases of life.

DNA Deoxyribonucleic acid molecule, containing the genetic code—"the molecule of life." Each cell in its nucleus contains DNAs arranged in the form of a double helix.

Drive The tension and arousal produced by an ungratified need and directed toward a chosen object or end.

Dyadic As used in the present work, pertaining to active relationships between two persons, e.g., mother and child, father and son.

Dynamic Pertaining to forces and potent influences that are capable of producing changes within the organism or personality.

Dysfunction Disturbance or impairment of the functional capacity of an organ or system, physical or mental.

Ectoderm The outermost cell layer in the embryo, from which structures of the nervous system and skin develop.

Ego The core of personality, which exercises control and directs drives and impulses in accordance with the demands of reality.

Embryo As used in the present work, a human organism in the early phases of prenatal development, from about two weeks to two months after conception.

Emotion A conscious state of experience characterized by feeling or excitment and accompanied (and frequently preceded) by specific physiological changes and frequently by excitation of the organism to action.

Encoding The transformation of external stimuli into internal signals that stimulate behavior appropriate to them.

Endocrine glands The ductless glands of internal secretion, such as the pituitary, thyroid, and adrenals.

Endoderm The innermost of the three cell layers of the embryo, from which most of the visceral organs and the digestive tract develop.

Endowment Capacity for development, physical or mental, conditioned by heredity and constitution.

Envy A distressful feeling aroused by the observation that another person possesses what one desires to have.

Epigenesis Appearance of new phenomena not present at previous stages in an organism's development from fertilized egg to adult maturity.

Etiology Investigation of the origins, causes, and contributing factors of a trait, attitude, or disease.

Euphoria An intense, subjective sensation of vigor, well-being, and happiness, which may exist despite some problem or disability.

Extrovert A type of personality in which thoughts, feelings, and interests are directed chiefly toward persons, social affairs, and other external phenomena.

Fantasy A function of imagination marked by engagement in vicarious experiences and hallucinatory actions; reveries, daydreaming.

Fetus As used in the present work, the human organism in advanced stages of prenatal development, from two months after conception to birth.

Fixation The persistence of infantile, childish, pubertal, or adolescent response patterns, habits, and modes of adjustment in successive phases of development.

Frustration The experience of distress and morbidity induced by failures and by thwarting of attempts to gratify one's needs or ambitions.

Gene A complex protein molecule constituting the smallest unit of inheritance of a single trait or characteristic in the chromosomes of reproductive cells.

Genetic psychology The branch of psychology that studies the human organism and its functions in terms of their origin and early course of development.

Genome All the genes found in a haploid set of chromosomes.

Genotype Genetic composition of an individual or group.

Group, reference The group a person belongs to or is interested in belonging to, e.g., peer groups, usually with a molding influence on the person.

Growth Increment to an organism or its structures; structural or functional change toward a more differentiated state.

Guidance Refers to a variety of methods, such as advising, counseling, testing, and use of special instruction and corrective teaching, by means of which a person may be helped to find and engage in activities that will yield satisfaction and further adjustment.

Habit An acquired or learned pattern of behavior, relatively simple and regularly used with facility, that leads to a tendency to use such acts rather than other behavior.

Habituation Decreased awareness due to the process of becoming accustomed to a particular stimulus or set of circumstances.

Hedonism As used in the present work, a psychological system of motivation explaining all behavior and conduct in terms of seeking pleasure and avoiding pain.

Heredity The totality of physiological influences biologically transmitted from parents (and ancestors) to the offspring at conception.

Heterogeneous Showing marked differences in reference to some significant criterion or standard.

Heterosexual Emotionally and sexually centered on the opposite sex; seeking and finding erotic gratification with a person of the other sex.

Homeostatis Cannon's term for the relative constancy—e.g., in temperature, blood pressure, and pulse rate—that the body must maintain to function properly.

Hominids Members of several extinct species of the primate order from which man is descended.

Homogeneous Showing marked similarity or low variability in the qualities or traits considered.

Homosexual Centered on the same sex; marked by a tendency to find sexual and erotic gratification with a person of the same sex.

Hormone A chemical substance, produced by an endocrine gland, that effects certain somatic and functional changes within the organism.

Hypothesis A tentative interpretation of a complex set of phenomena or data on the basis of some supportive facts to be verified by research.

Id In psychoanalysis, the instinctive and impulsive drives that seek immediate gratification according to the pleasure principle by which they operate.

Ideal A standard approaching some level of perfection, usually unattainable in practice.

Identification An unconscious effort to gratify certain deep-seated needs through affiliation with and imitation of another person, group, or ideal.

Identity Sense of sameness despite growth, aging, and environmental change.

Imprinting In many animals, a species-specific innate disposition accompanied by a strong drive to follow (imitate) the parent or its surrogate in a very early phase of life. In human beings, basic emotional and social patterns are followed but to a lesser degree and with less precision.

Incubation A period in assimilation and the problem-solving process during which certain presented ideas gain in motivational strength and begin to condition a part of behavior, especially during childhood.

Individuation Differentiation of behavior into more distinct and less dependent parts or features.

Infancy The first two to three years of human life, during which all major human abilities originate, marked by almost total dependence on others.

Infantile Pertaining to the lowest level of postnatal maturity; mode of behavior or adjustment resembling the infant level.

Inferiority attitude or complex An emotionally conditioned and frequently unconscious attitude with reference to one's organism, self, or personality, characterized by serious lack of self-reliance and notions of inadequacy in many situations.

Inhibition Prevention of the starting of a process or behavior by inner control despite the presence of the eliciting stimulus.

Innate Existing before birth and accounting for a particular trait or characteristic.

Intelligence As used in the present work, the practical application of sensorimotor and cognitive functions as shown by standardized performances that are measurable.

Intelligence quotient (IQ) The index of mental capacity obtained by testing. Originally a numerical ratio between mental age and chronological age, it now more commonly refers to the statistical concept based on standard scores and on normal distribution and deviation from the mean.

Introjection A basal (crude) form of identification in which a person assimilates simple behavior patterns of other persons.

IQ *See* Intelligence quotient.

Juvenile Pertaining to an older child or adolescent.

Kinship Blood relationship between two or more persons; usually includes marriage and adoption ties.

Latency period In psychoanalysis, the period, from approximately four to eleven or twelve years of age, during which interest in sex is not apparent.

Libido In psychoanalysis, the total undifferentiated life energy (Jung), sexual in nature (Freud).

Life cycle The total time from birth to death, divided into a number of stages and phases and emphasizing recurrence of certain important events.

Malfunction *See* Dysfunction.

Marijuana (also spelled **marihuana**) A product derived from *Cannabis sativa* containing tetrahydrocannabinol (THC), which has sedative-hypnotic effects and induces the experience of a "high" and some release of tension.

Matrix A framework or enclosure that gives form, meaning, or perspective to what lies within it.

Maturation Developmental changes primarily due to heredity and constitution and manifested in organismic functioning; organismic developments leading to further behavioral differentiation.

Maturity The state of maximal function and integration of a single factor or a total person; also applied to age-related adequacy of development and performance.

Median The measure of central tendency above and below which 50 percent of cases fall; the fiftieth percentile.

Mendelian ratio The proportion of dominant to recessive phenotypes.

Mental conflict *See* Conflict.

Menopause The stage in a woman's life when menstruation ceases, usually in the late forties or early fifties.

Mental hygiene The art and science of mental health; application of the principles and measures necessary for its preservation and promotion.

Mesoderm The middle of the three fundamental layers of the embryo, which forms a basis for the development of bone and muscle structure.

Metabolism The physiochemical changes within the body for supplying, repairing, and building up (anabolism) and for breaking down and removing (catabolism).

Method A logical and systematic way of studying a subject.

Morpheme A combination of phonemes that makes up a meaningful unit of a language.

Mother fixation Deep identification with the mother to the virtual exclusion of other females as models or idols.

Motive Any factor that stimulates or contributes to a conscious effort toward a goal.

Mutation Any change of a gene, usually from one allele to another.

Need Any physicochemical imbalance within the organism, due to a lack of particular nutrients, that arouses tension and drives. By analogy, psychological and personality needs are recognized. Primary or genetically determined needs and derived needs (generated by the operation of primary needs) are usually distinguished.

Negativism A primary mode of expressing one's will by persistent refusal to respond to suggestions from parental or authority figures.

Neonate A newborn infant.

Neuromuscular Pertaining to both nerve and muscle, their structure and functions.

Neurotic Mentally and emotionally disturbed; characterized by recurrent symptoms often caused by unconscious conflicts.

Normative Based on averages, standards, or values.

Object permanence The ability to recognize the continuing existence of an object which is no longer visible or audible. It is achieved at about eleven months by the majority of infants.

Ontogenesis As used in the present work, origin and development of an individual organism and its functions throughout life, especially what and how the individual person learns in a specific culture (*cf.* Phylogenesis).

Organismic age The average of all basic measures of a person's development at a particular time, such as carpal development, dental development, height, weight, and lung capacity.

Orienting reflex (OR) The initial response to any novel situation, maximizing its stimulus value.

Orthogenesis Theory that the germ plasm is gradually modified by its own internal conditions and that the organism (and personality) has a sequential and specific, species-related course of development unless blocked.

Ovum The female germ cell or egg cell produced by one of two ovaries.

Parallel play The side-by-side play of two or more children with some independence of action yet heightened interest because of each other's presence.

Peer A person of about one's own level of development and therefore an equal in play or any other mode of association.

Percept A unit of the perceiving response or reaction; immediate knowledge of the object perceived.

Perfectionism The tendency to demand frequently of oneself or others a maximal quality of achievement, without proper consideration of limiting factors.

Personality Generally, acquired consistencies of behaivor. More specifically, the multilevel functioning of those qualities, traits, and characteristics which distinguish a human being and determine his or her interaction with social and cultural factors.

Phenotype The observable features of a person, including genetic traits and characteristics.

Phyletic Pertaining to species.

Phyletic scale A line of descent from the lowest to the highest living species.

Phylogenesis Evolution of traits and features common to a species or race; development of a species from its origin to its present state.

Projection A defensive dynamism by which a person attributes to others one's own qualities and traits, usually undesirable ones, such as hostility or dishonesty.

Proximodistal Pertaining to movements near the body axis that differentiate and specialize earlier than the more distant ones.

Psychosomatic Pertaining to the effects of psychological and emotional stress on health and pathology; used to indicate that a phenomenon is both psychic and bodily.

Psychotherapy The various techniques for the systematic application of psychological principles in the treatment of mental or emotional disturbances or disorder.

Pubertal Pertaining to or related to the developmental period of puberty.

Puberty The period of physical (especially sexual) and cognitive maturation, characterized by rapid somatic growth and the assumption of adult traits or features.

Pubescent Pertaining to the early part of puberty; a person who exhibits significant characteristics of that period of maturation.

Rationalization A defensive dynamism by which a person justifies his or her activities or conduct by giving rational and acceptable, but usually untrue, reasons.

Readiness, principle of Refers to the neurological and psychological disposition to attend to and assimiliate a category of stimuli to which sensitivity and learning responses were previously lacking.

Reference group *See* Group, reference.

Regression Returning to an earlier and less mature level of behavior and personality functioning.

Reinforcement Any facilitating influence or condition for strengthening selected behavior patterns (*cf.* Conditioning).

Resistance As used in the present work, opposition offered by a child or adolescent to the suggestions, orders, or regulations of the parents.

RNA Ribonucleic acid molecule playing the role of a messenger for vital DNA functions; also occurs as transfer RNA and ribosomal RNA.

Role conflict The situation in which a person is expected to play two or more roles that cannot be integrated into the self-system.

Self-concept A person's awareness of and identification with his or her organism, cognitive powers, and modes of conduct and performance, accompanied by specific attitudes toward them.

Self-direction Independent selection of goals and of the proper means and actions to attain them.

Self-realization The lifelong process of unhampered development marked by self-direction and responses in terms of one's capabilities or potentialities.

Senescence The period of old age.

Senility Marked loss of physical and cognitive functions in old age or before.

Sentiment An affective and cognitive disposition to react in a certain way toward a particular value, object, or person.

Sex typing The learning of behavior patterns appropriate to the sex of the person, e.g., acquisition of masculine behavior traits for a boy.

Sibling One of two or more offspring of the same parents; a brother or sister.

Socialization A progressive development in relating and integrating oneself with others, especially parents, peers, and groups.

Somatic Pertaining to the body or organism.

Sperm The male germ cell, or spermatozoon, containing chromosomes, DNA, RNA, protein, and other substances.

Strain The condition within a system or organ exposed to stress, e.g., overactivity or deprivation.

Sublimation A defensive dynamism by which the energies of a basic drive are redirected to a higher and socially more acceptable plane of expression (a mark of normal development).

Superego In psychoanalysis, that part of the personality structure which is built up by early parent-child relationships and which helps the ego to enforce the control of primitive instinctual urges and later functions as a moral force analogous to an early form of conscience.

Temperament The affective disposition and expression of emotional energies in terms of speed of reaction, depth and length of emotional experiences, and relevant behavior.

Tension A state of acute need, deprivation, fear, apprehension, etc., that keeps an organism or certain organs—e.g., adrenal glands—in a state of intensified activity.

Teratogenic Capable of causing organic malformation.

Toddler The child between the ages of about fifteen and thirty months.

Trait A distinctive and enduring characteristic of a person or the person's behavior.

Trauma Any somatic or psychological damage, including stressful and terrifying experiences.

Unconscious The area of motivational structure and thought process of which a person is not directly aware.

Valence In gestalt psychology, a term referring to the subjective appraisal of an object or situation in the life space, by virtue of which the object is sought (positive valence) or avoided (negative valence). The term was introduced by K. Lewin.

Value The worth or excellence found in a qualitative appraisal of an object by reliance on emotional and rational standards of the person or of selected reference groups.

Viability Ability of the organism to survive; e.g., the prematurely born infant's capacity for survival outside the uterus.

Vital capacity age (VA) Relationship between lung capacity and age.

Weltanschauung A configuration of attitudes and views toward all dimensions of reality, both material and metaphysical; key tenets of a philosophy of life.

Zygote As used in the present work, a new individual formed by the union of male and female gametes, and the resultant globule of cells during the first phase of human prenatal development after conception, lasting approximately two weeks.

Selective List of Journals

Below are the titles of selected journals, with their Library of Congress (LC) catalog numbers and formal abbreviations, which publish theoretical and research studies in human development.

Title	LC number	Abbreviation
Adolescence	HQ796.A35	*Adolescence*
Aging	HQ1060.A27	*Aging*
American Academy of Arts and Sciences. Proceedings	Q11.B7	*Proc. Amer. Acad. Arts Sci.*
American Academy of Political and Social Science. Annals	H1.A4	*Ann. Amer. Acad. pol. soc. Sci.*
American Association for Advancement of Science. Summarized Proceedings	Q11.A51PR	*Sum. Proc. Amer. Asso. Advancem. Sci.*
American behavioral scientist	H1.A472	*Amer. beh. Scientist*
American educational research journal	L11.A66	*Amer. educ. Res. J.*

Title	LC number	Abbreviation
American journal of human genetics	QH.A31.A1A5	*Amer. J. hum. Genet.*
American journal of orthopsychiatry	RA790.A1A5	*Amer. J. Orthopsychiat.*
American journal of psychology	BF1.A512	*Amer. J. Psychol.*
American journal of science	Q1.A52	*Amer. J. Sci.*
American psychologist	BF1.A518	*Amer. Psychologist*
American scientist	LJ115.A57A5	*Amer. Scientist*
American sociological review	HM1.A52	*Amer. soc. Rev.*
Annals of human genetics	HQ750.A1A5	*Ann. hum. Genet.*
British journal of psychology	BF1.B7	*Brit. J. Psychol.*
Canadian journal of psychology	BF1.C3	*Can. J. Psychol.*
Character and personality	BF1.J66	*Char. Pers.*
Child and family	HQ769.C4	*Child Fam.*
Child development	HQ750.A1C45	*Child Developm.*
Child development abstracts and bibliography	Ref. HQ750.A1C47	*Child. Developm. Abstr. Bibl.*
Child study, the bulletin of the Institute of Child Study, University of Toronto	LB1101.C4	*Child Study Univ. Toronto*
Children	HV741.C536	*Children*
Developmental psychology	BF723.J	*Developm. Psychol.*
Exceptional children	LC3951.J6	*Except. Children*
Family	HV1.J56	*Family*
Genetic psychology monographs	LB1101.G4	*Genet. Psychol. Monogr.*
Genetical research	HQ431.A1G395	*Genet. Res.*
Genetics	QH431.G43	*Genetics*
Gerontologist	HQ1060.G4	*Gerontologist*
Hereditas	QH431.A2	*Hereditas*
Human biology	GN1.H8	*Hum. Biol.*
Journal of abnormal and social psychology	BF1.J79	*J. abnorm. soc. Psychol.*
Journal of applied behavioral science	H1.J53	*J. appl. beh. Sci.*
Journal of applied psychology	BF1.J55	*J. appl. Psychol.*
Journal of educational psychology	L1051.A2J6	*J. educ. Psychol.*
Journal of educational research	L11.J75	*J. educ. Res.*
Journal of experimental child psychology	BF721.J64	*J. exp. child Psychol.*
Journal of genetic psychology	L11.P4	*J. genet. Psychol.*
Journal of genetics	QH431.J621	*J. Genet.*
Journal of gerontology	HQ1060.J6	*J. Gerontol.*
Journal of heredity	S494.A2J7	*J. Hered.*

Title	LC number	Abbreviation
Journal of personality	BF1.J66	*J. Pers.*
Journal of personality and social psychology	HM251.J56	*J. Pers. soc. Psychol.*
Journal of personality assessment	BF698.J65	*J. Pers. Assessm.*
Journal of school psychology	LB3013.6J.6	*J. School Psychol.*
Journal of social psychology	HM251.A1J6	*J. soc. Psychol.*
Marriage	HQ1.M35	*Marriage*
Marriage and family living	HQ1.J6	*Marriage Fam. Living*
Merrill-Palmer quarterly of human growth	HQ1.M4	*Merrill-Palmer Quart.*
Psychological abstracts	BF1.P65	*Psychol. Abstr.*
Psychological bulletin	BF1.P75	*Psychol. Bull.*
Psychological monographs	BF1.P8	*Psychol. Monogr.*
Psychological record	BF1.P68	*Psychol. Rec.*
Psychological review	BF1.P7	*Psychol. Rev.*
Review of educational research	L11.R35	*Rev. educ. Res.*
School and society	L11.S36	*Sch. Soc.*
Science	Q1.S35	*Science*
Scientific American	T1.S5	*Sci. Amer.*
Society for Research in Child Development. Monographs	LB1103.S6	*Monogr. Soc. Res. Child Developm.*
Zeitschrift für Entwicklungspsychologie und pädagogische Psychologie	L31.Z46	*Z. Entwickl. pädag. Psychol.*

Name Index

Subject Index